Ltd: Tsz-shan Kwok (B). **Shutterstock.com**: Igor Smichkov (A); **SP10f Alamy Images**: Tim Graham (C). **SPL**: Tony McConnell (A); **SP10g Alamy Images**: Arcaid Images (A). **Getty Images**: Picture Post / Hulton Archive (Bl); Peter Macdiarmid (Br); **SP10h Getty Images**: EdnaM (A); Bill Hogan / Chicago Tribune / MCT (B); **SP10i 123RF.com. Alamy Images**: Mode Images (A). **Corbis**: Berliner Verlag / Steinach / dpa (C); **SP11a Getty Images**: Benjamin Cothren / EyeEm (B). **SPL**: Charles D. Winters (D); **SP11b Alamy Images**: Andrew Cribb (A). **Fotolia.com**: Marsel (C). **Shutterstock.com**: NADA GIRL (B); **SP11c Rex Shutterstock**: Rebecca Harley (D). **SPL**: Ted Kinsman (B)

SP12/SP13 NASA: ESA; **SP12a Alamy Images**: sciencephotos (A); charistoone-stock (B); **SP12b Getty Images**: Caialmage (A). **Pearson Education Ltd**: Trevor Clifford (B); **SP12c NASA; SP13b Alamy Images**: Bax Walker (B); **SP13c Alamy Images**: Marek Uszynski (B)

SP14/SP15 Fotolia.com: Jose Gil; **SP14a Alamy Images**: sebastien bonaime (A). **SPL**: Phil Degginger (C); **SP14a CP Alamy Images**: Juice Images (A); **SP14b Getty Images**: Creatas Video (B). **Shutterstock.com**: Belozorova Elena (A); **SP14c Alamy Images**: Zoonar GmbH (A). **FLPA**: Donald M. Jones / Minden Pictures (C); **SP14c CP Alamy Images**: Steve Morgan (A). **Pearson Education Ltd**: Trevor Clifford (B); **SP14d Corbis**: JAXA / Michael Benson / Kinetikon Pictures (A); **SP14e Cochranes of Oxford Ltd**: (D). **Ray Maker - dcrainmaker.com**: (A). **Rex Shutterstock**: KeystoneUSA-ZUMA (C); **SP15a Getty Images**: Johner Images (B); DAGOC / Michael Steele (A); **SP15b 123RF.com**: alex_star (E). **SuperStock**: Corbis (B); **SP15b CP Alamy Images**: John James (A). **Pearson Education Ltd**: Trevor Clifford (B); **SP15c Alamy Images**: Martin Shields (C). **Shutterstock.com**: GROGL (A); **SP15d Nature Picture Library**: Franco Banfi (D)

All other images © Pearson Education

Figures
SP3e graph D from https://www.gov.uk/government/uploads/system/uploads/attachment_data/file/454484/DUKES_2015_Long_Term_Trends.pdf, DECC Contains public sector information licensed under the Open Government Licence v3.0; Figure SP6d. from Data from: https://www.phe-protectionservices.org.uk/radiationandyou/, © Crown copyright. Contains public sector information licensed under the Open Government Licence (OGL) v3.0. Open Government Licence

Text
SP5 Preparing for your exams from question 5b from P11 paper, June 2014, Foundation Tier, Edexcel; SP6 Preparing for your exams from question 6d from P21 paper, June 2013, Higher Tier, Edexcel; SP7 Preparing for your exam from question 6b from P11 paper, March 2012, Higher Tier, Edexcel

Contents

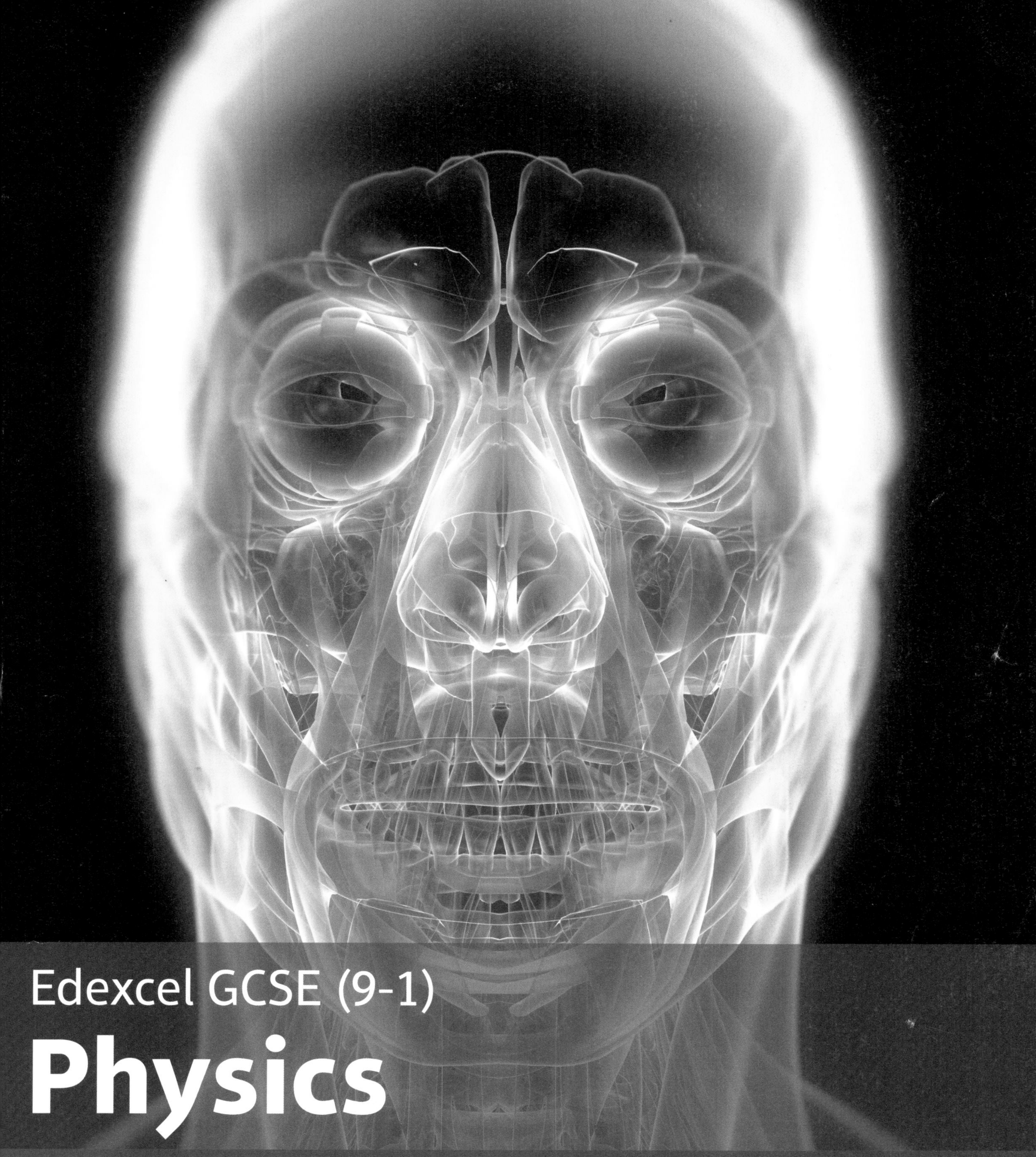

Edexcel GCSE (9-1)

Physics

Mark Levesley Penny Johnson Carol Tear

ALWAYS LEARNING PEARSON

Published by Pearson Education Limited, 80 Strand, London, WC2R 0RL.

www.pearsonschoolsandfecolleges.co.uk

Copies of official specifications for all Edexcel qualifications may be found on the website: www.edexcel.com

Text © Penny Johnson, Carol Tear and Pearson Education Ltd 2016
Series editor: Mark Levesley
Designed by Poppy Marks and James Handlon, Pearson Education Limited
Typeset by Phoenix Photosetting, Chatham, Kent
Original illustrations © Pearson Education Limited 2016
Illustrated by KJA Artists Illustration Agency and Phoenix Photosetting, Chatham, Kent
Cover design by Poppy Marks and Colin Tilley Loughrey
Picture research by Caitlin Swain
Cover photo © Science Photo Library: Science Picture Co

The rights of Penny Johnson and Carol Tear to be identified as authors of this work have been asserted by them in accordance with the Copyright, Designs and Patents Act 1988.

First published 2016

19 18 17
10 9 8 7 6 5 4

British Library Cataloguing in Publication Data
A catalogue record for this book is available from the British Library

ISBN 9781292120225

Printed in Slovakia by Neografia

A note from the publisher
In order to ensure that this resource offers high-quality support for the associated Pearson qualification, it has been through a review process by the awarding body. This process confirms that this resource fully covers the teaching and learning content of the specification or part of a specification at which it is aimed. It also confirms that it demonstrates an appropriate balance between the development of subject skills, knowledge and understanding, in addition to preparation for assessment.

Endorsement does not cover any guidance on assessment activities or processes (e.g. practice questions or advice on how to answer assessment questions), included in the resource nor does it prescribe any particular approach to the teaching or delivery of a related course.

While the publishers have made every attempt to ensure that advice on the qualification and its assessment is accurate, the official specification and associated assessment guidance materials are the only authoritative source of information and should always be referred to for definitive guidance.

Pearson examiners have not contributed to any sections in this resource relevant to examination papers for which they have responsibility.

Examiners will not use endorsed resources as a source of material for any assessment set by Pearson.

Endorsement of a resource does not mean that the resource is required to achieve this Pearson qualification, nor does it mean that it is the only suitable material available to support the qualification, and any resource lists produced by the awarding body shall include this and other appropriate resources.

Acknowledgements
The following authors contributed text to previous Pearson publications and the publishers are grateful for their permission to include elements of their work: James Newall, James de Winter and Miles Hudson. The publishers would like to thank Peter Ellis for his original contributions.

The authors and publisher would like to thank the following individuals and organisations for permission to reproduce photographs, figures and text:

Photographs
(Key: b-bottom; c-centre; l-left; r-right; t-top)

SP1 Nature Picture Library: Fred Oliver; **SP1a Alamy Images**: Michael Wheatley (A); Jim West (D). **Getty Images**: Bryn Lennon (B); **SP1b Getty Images**: John Chappie (A); **SP1c Alamy Images**: Stocktrek Images, Inc (A). **The Defence Picture Library**: (C); **SP1d Getty Images**: Rusty Jarrett (A); **SP1a Alamy Images**: Michael Wheatley (A); Jim West (D). **Getty Images**: Bryn Lennon (B); **SP1b Getty Images**: John Chappie (A); **SP1c Alamy Images**: Stocktrek Images, Inc (A). **The Defence Picture Library**: (C); **SP1d Getty Images**: Rusty Jarrett (A)

SP2 Shutterstock.com: Tatiana Belova; **SP2a Alamy Images**: Julie Edwards (A); Andrey Nekrasov (D). **Rex Shutterstock**: Imaginechina (B). **Shutterstock.com**: Martijn Smeets (C); **SP2b Getty Images**: Henrik Trygg (B). **Rex Shutterstock**: Eddie Boldizsar (A). **Shutterstock.com**: Ivan Garcia (D); cleanfotos (C); **SP2c Science Photo Library Ltd (SPL)**: (A); **SP2d Alamy Images**: David Wall (D). **Getty Images**: AFP / Mohd Easfan (A). **Reuters**: Daniel Munoz (B); **SP2d CP Corbis**: Marc Sanchez / Icon Sportswire (A); **SP2e 123RF.com**: Aleksandar Mijatovic (A). **Reuters**: Antonio Bronic (D). **Shutterstock.com**: Keith Publicover (B); **SP2f Alamy Images**: RosalreneBetancourt 4 (A); **SP2g Alamy Images**: incamerastock (D); epa european pressphoto agency b.v. (A). **Corbis**: Amanda Brown / Star Ledger (B); **SP2h Alamy Images**: Stocktrek Images, Inc (D); epa european pressphoto agency b.v. (A). **Corbis**: Transtock (B); **SP2i Alamy Images**: Cliff Hide News (A). **Getty Images**: Taxi (B). **SPL**: TRL Ltd. (C); NASA (D)

SP3 123RF.com: Quintanilla; **SP3a Corbis**: Transtock (B). **SPL**: Edward Kinsman (A); **SP3b Alamy Images**: Sciencephotos (A). **Robert Harding World Imagery**: Ashley Cooper (B); **SP3c SPL**: NREL / US Department of Energy (B); Alan Sirulnikoff (A); **SP3d Getty Images**: Oli Scarff (A); Auscape (C); **SP3e Alamy Images**: epa european pressphoto agency b.v. (C); Craig Ruttle (B). **NASA**: (A); **SP3f 123RF.com**: (B). **Alamy Images**: epa european pressphoto agency b.v. (C). **SPL**: Hans-Ulrich Osterwalder (A)

SP4 NASA; **SP4a Reuters**: Toby Melville (A); **SP4b 123RF.com**. **Reuters**: Carlos Gutierrez (A); **SP4c Alamy Images**: A&J Visage (B). **SPL**: D. Levesque, Publiphoto Diffusion (C); **SP4c CP NOAA**: NOS / Office of Coast Survey (A); **SP4d Alamy Images**: ohn briscoe (C). **Fotolia.com**: Stephen Coburn (A). **Getty Images**: alacatr (D).; **SPL**: R Degginger (B); **SP4e AfterShokz**: (C); **SP4f Alamy Images**: **Nature Picture Library (A)**. **SPL**: US Navy (C); **SP4g Getty Images**: Bill Greenblatt (B). **National Geophysical Data Center**: (A)

SP5 SPL: Edward Kinsman; **SP5a SPL**: Michael Szoenyi (C); **SP5b Fotolia.com**: nilanewsom (B). **Shutterstock.com**: Chris Howey (C); **SP5c Getty Images**: Jeff Pudlinski / EyeEm (A); **SP5d Getty Images**: Image Source (C). **Martyn F. Chillmaid**: (D). **SPL**: Cordelia Molloy (A); **SP5d CP Alamy Images**: sciencephotos (B). **Nature Picture Library**: ARCO (A); **SP5e SPL**: David Parker (D). **Shutterstock.com**: Pichugin Dmitry (A); **SP5f Getty Images**: Tim Ireland (B). **Pearson Education Asia Ltd**: Joey Chan (A); **SP5g Getty Images**: Roger Ressmeyer / Corbis / VCG (A); **SP5g CP Reuters**: Arnd Wiegmann (C). **Shutterstock.com**: Mark Roger Bailey (A); **SP5h Corbis**: Monty Rakusen / Cultura (C). **Press Association Images**: Khalil Senosi / AP (A); **SP5i Corbis**: Marilyn Angel Wynn / Nativestock Pictures (E). **Getty Images**: Spencer Platt (D). **Shutterstock.com**: thaikrit (A); Suzanne Tucker (C); Jaroslav Moravcik (B)

SP6 Fundamental Photographs: Richard Megna; **SP6a Shutterstock.com**: dade72 (E); **SP6b Reuters**: Pichi Chuang (E); **SP6c Shutterstock.com**: donsimon (A) **SP6d SPL**: Public Health England (A). **Shutterstock.com**: wellphoto (D); **SP6e NASA. Press Association Images**: AP / Phil Coale (D); **SP6f Shutterstock.com**: Unicus (C); **SP6g Wellcome Library, London**: (A); **SP6h SPL**: Cordelia Molloy (A); **SP6i Getty Images**: Toshifumi Kitamura / AFP (D). **Reuters**: Desmond Boylan (C). **SPL**: Sputnik (A); Gustoimages (B); **SP6j SPL**: Dr Robert Friedland (D); Alain Pol, ISM (B); **SP6k Alamy Images**: Paul White - North West England (B). **USN photo courtesy of US Navy Arctic Submarine Laboratory**: (A); **SP6l SPL**: Patrick Landmann (C); **SP6m Getty Images**: Roger Ressmeyer / Corbis / VCG (D). **NASA**: ESA / J. Hester (Arizona State University) (B)

SP7 NASA: JPL-Caltech; **SP7a SPL**: NASA (D). **Shutterstock.com**: orla (C); **SP7b ESA**: (A); **SP7c SPL**: Detlev van Ravenswaay (A); **SP7d Alamy Images**: Mark Sykes (B); **SP7e SPL**: NASA / WMAP Science Team (C)

SP8/SP9 Getty Images: Oliver Furrer; **SP8a Alamy Images**: frans lemmens (A). **Masterfile UK Ltd**: robertharding (D). **Shutterstock.com**: LorraineHudgins (C); **SP9a Getty Images**: Stephen J Krasemann (D); Mike Kemp (B); Ken Welsh (A); **SP9b AP Wide World Photos**: Li jianshu - Imaginechina (C); **SP9c Fotolia.com**: wjarek (D). **Getty Images**: Frank Bienewald (A)

Teaching and learning

The **topic reference** tells you which part of the course you are in. 'SP5g' means, 'Separate Science, Physics, unit 5, topic g'.

The **specification reference** allows you to cross reference against the specification criteria so you know which parts you are covering. References that end in P, e.g. P7.2P, are in Physics only, the rest are also in the Combined Science specification criteria.

If you see an **H** icon that means that content will be assessed on the Higher Tier paper only.

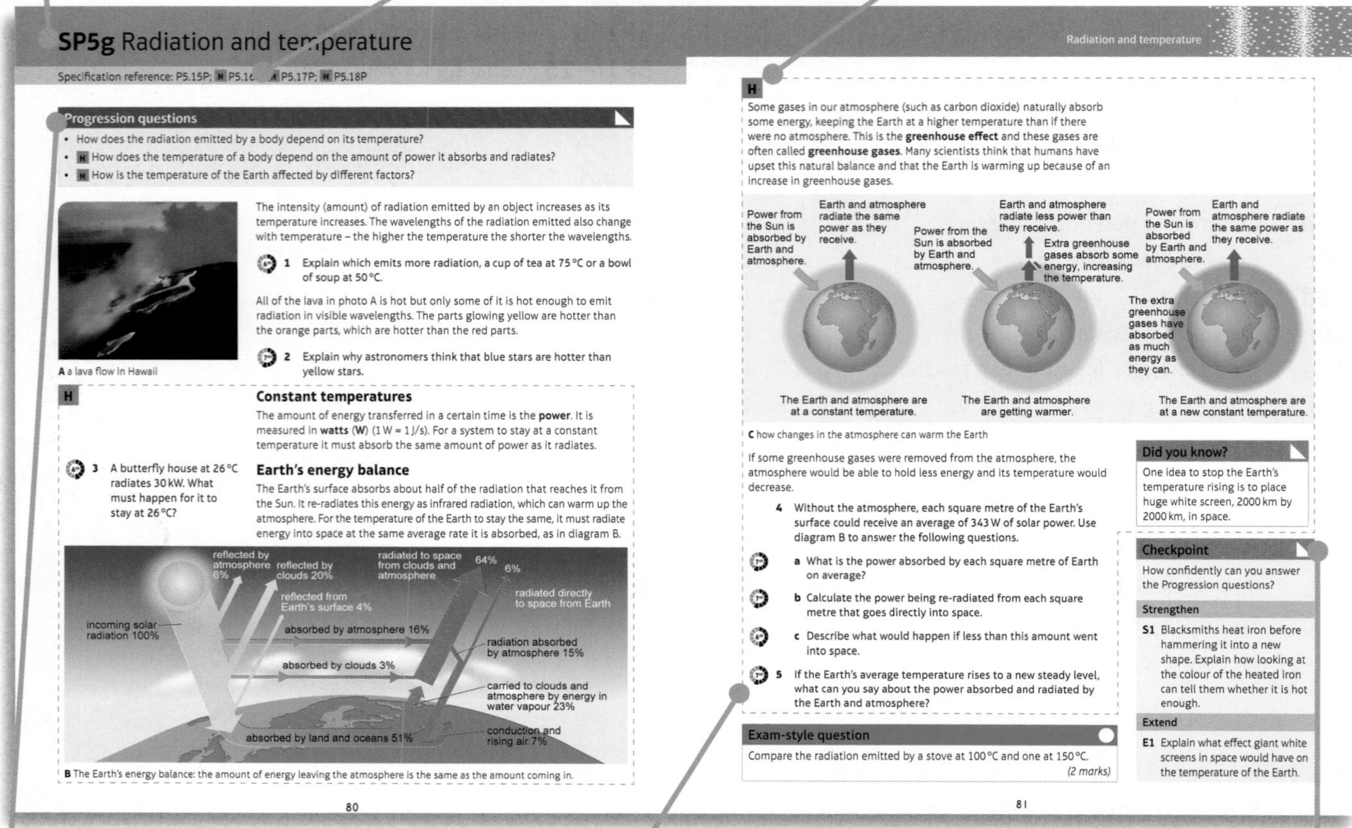

SP5g Radiation and temperature

Specification reference: P5.15P; H P5.16P; H P5.17P; H P5.18P

Progression questions

- How does the radiation emitted by a body depend on its temperature?
- H How does the temperature of a body depend on the amount of power it absorbs and radiates?
- H How is the temperature of the Earth affected by different factors?

A a lava flow in Hawaii

The intensity (amount) of radiation emitted by an object increases as its temperature increases. The wavelengths of the radiation emitted also change with temperature – the higher the temperature the shorter the wavelengths.

1 Explain which emits more radiation, a cup of tea at 75 °C or a bowl of soup at 50 °C.

All of the lava in photo A is hot but only some of it is hot enough to emit radiation in visible wavelengths. The parts glowing yellow are hotter than the orange parts, which are hotter than the red parts.

2 Explain why astronomers think that blue stars are hotter than yellow stars.

H

Constant temperatures

The amount of energy transferred in a certain time is the **power**. It is measured in **watts** (W) (1 W = 1 J/s). For a system to stay at a constant temperature it must absorb the same amount of power as it radiates.

3 A butterfly house at 26 °C radiates 30 kW. What must happen for it to stay at 26 °C?

Earth's energy balance

The Earth's surface absorbs about half of the radiation that reaches it from the Sun. It re-radiates this energy as infrared radiation, which can warm up the atmosphere. For the temperature of the Earth to stay the same, it must radiate energy into space at the same average rate it is absorbed, as in diagram B.

reflected by atmosphere 6%
reflected by clouds 20%
radiated to space from clouds and atmosphere
64%
6%
reflected from Earth's surface 4%
radiated directly to space from Earth
incoming solar radiation 100%
absorbed by atmosphere 16%
radiation absorbed by atmosphere 15%
absorbed by clouds 3%
carried to clouds and atmosphere by energy in water vapour 23%
absorbed by land and oceans 51%
conduction and rising air 7%

B The Earth's energy balance: the amount of energy leaving the atmosphere is the same as the amount coming in.

80

Radiation and temperature

H

Some gases in our atmosphere (such as carbon dioxide) naturally absorb some energy, keeping the Earth at a higher temperature than if there were no atmosphere. This is the **greenhouse effect** and these gases are often called **greenhouse gases**. Many scientists think that humans have upset this natural balance and that the Earth is warming up because of an increase in greenhouse gases.

Power from the Sun is absorbed by Earth and atmosphere.
Earth and atmosphere radiate the same power as they receive.
The Earth and atmosphere are at a constant temperature.
Power from the Sun is absorbed by Earth and atmosphere.
Earth and atmosphere radiate less power than they receive.
Extra greenhouse gases absorb some energy, increasing the temperature.
The Earth and atmosphere are getting warmer.
Power from the Sun is absorbed by Earth and atmosphere.
Earth and atmosphere radiate the same power as they receive.
The extra greenhouse gases have absorbed as much energy as they can.
The Earth and atmosphere are at a new constant temperature.

C how changes in the atmosphere can warm the Earth

If some greenhouse gases were removed from the atmosphere, the atmosphere would be able to hold less energy and its temperature would decrease.

4 Without the atmosphere, each square metre of the Earth's surface could receive an average of 343 W of solar power. Use diagram B to answer the following questions.

a What is the power absorbed by each square metre of Earth on average?

b Calculate the power being re-radiated from each square metre that goes directly into space.

c Describe what would happen if less than this amount went into space.

5 If the Earth's average temperature rises to a new steady level, what can you say about the power absorbed and radiated by the Earth and atmosphere?

Did you know?

One idea to stop the Earth's temperature rising is to place huge white screen, 2000 km by 2000 km, in space.

Exam-style question

Compare the radiation emitted by a stove at 100 °C and one at 150 °C. *(2 marks)*

Checkpoint

How confidently can you answer the Progression questions?

Strengthen

S1 Blacksmiths heat iron before hammering it into a new shape. Explain how looking at the colour of the heated iron can tell them whether it is hot enough.

Extend

E1 Explain what effect giant white screens in space would have on the temperature of the Earth.

81

By the end of the topic you should be able to confidently answer the **Progression questions**. Try to answer them before you start and make a note of your answers. Think about what you know already and what more you need to learn.

Each question has been given a **Pearson Step** from 1 to 12. This tells you how difficult the question is. The higher the step the more challenging the question.

When you've worked through the main student book questions, answer the **Progression questions** again and review your own progress. Decide if you need to reinforce your own learning by answering the **Strengthen question**, or apply, analyse and evaluate your learning in new contexts by tackling the **Extend question**.

Paper 1

SP1 Motion

Penguins cannot climb. They get onto the ice by accelerating to a high speed under the water. As they move upwards out of the water, gravity pulls on them and they slow down. But if they are swimming fast enough they land on the ice before they stop moving.

In this unit you will learn about quantities that have directions (such as forces). You will find out how to calculate speeds and accelerations, and how to represent changes in distance moved and speed on graphs.

The learning journey

Previously you will have learnt at KS3:

- what forces are and the effects of balanced and unbalanced forces
- how average speed, distance and time are related
- how to represent a journey on a distance/time graph.

In this unit you will learn:

- the difference between vector and scalar quantities
- how to calculate speed and acceleration
- how to represent journeys on distance/time and velocity/time graphs
- how to use graphs to calculate speed, acceleration and distance travelled.

SP1a Vectors and scalars

Specification reference: P2.1; P2.2; P2.3; P2.4; P2.5

Progression questions

- What are vector and scalar quantities?
- What are some examples of scalar quantities and their corresponding vector quantities?
- What is the connection between the speed, velocity and acceleration of an object?

A The person in the air stays there because of the force provided by the jets of water.

1 Upthrust is a force that helps objects float. Sketch one of the boats in photo A and add arrows to show two forces on the boat acting in a vertical direction.

2 Describe the differences between mass and weight.

3 Explain why we say that displacement is a vector quantity.

4 Runners in a 400 m race complete one circuit of an athletics track. What is their displacement at the end of the race?

The **force** needed to keep the person in photo A in the air depends on his **weight**. Weight is a force that acts towards the centre of the Earth. All forces have both a **magnitude** (size) and a direction, and are measured in newtons (N).

Quantities that have both size and direction are **vector quantities**. So forces are vectors. Forces are often shown on diagrams using arrows, with longer arrows representing larger forces.

The weight of the person in photo A depends on his **mass**. Mass measures the amount of matter in something and does not have a direction. Quantities that do not have a direction are called **scalar quantities**. Other scalar quantities include **distance**, **speed**, **energy** and time.

Displacement is the distance covered in a straight line, and has a direction. The displacement at the end of a journey is usually less than the distance travelled because of the turns or bends in the journey.

B The bend in the road means that the distance the cyclists cover is greater than their final displacement.

The speed of an object tells you how far it moves in a certain time. **Velocity** is speed in a particular direction. For example a car may have a velocity of 20 m/s northwards.

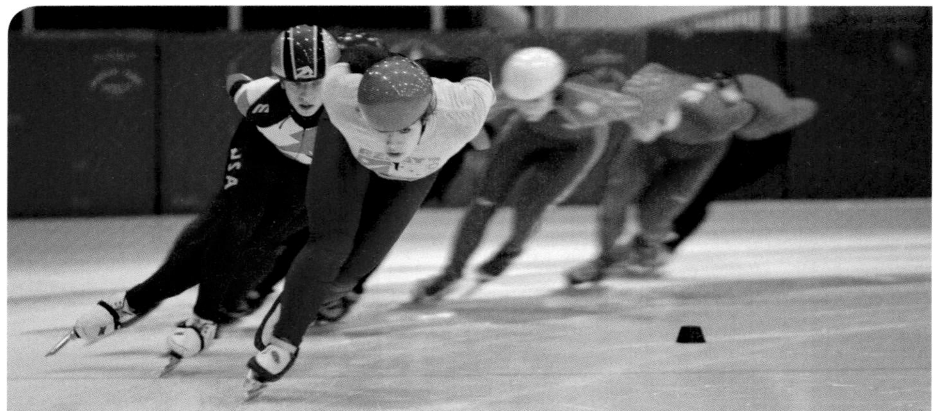

D These skaters maintain a constant speed around the bend, but their velocity is changing.

Other vector quantities include:

- **acceleration** – a measure of how fast velocity is changing
- **momentum** – a combination of mass and velocity.

Exam-style question

Weight and upthrust are both vector quantities.

a Name one other vector quantity that is not a force. *(1 mark)*

b Explain why you do not need to state a direction when describing a weight. *(1 mark)*

Did you know?

Cyclists can achieve speeds of up to 70 mph (that's over 100 km/h and approximately 30 m/s).

5 Look at photo B. Explain why the cyclists' velocity will change even if they maintain the same speed.

6 A student draws the diagram below. Explain what is wrong with it.

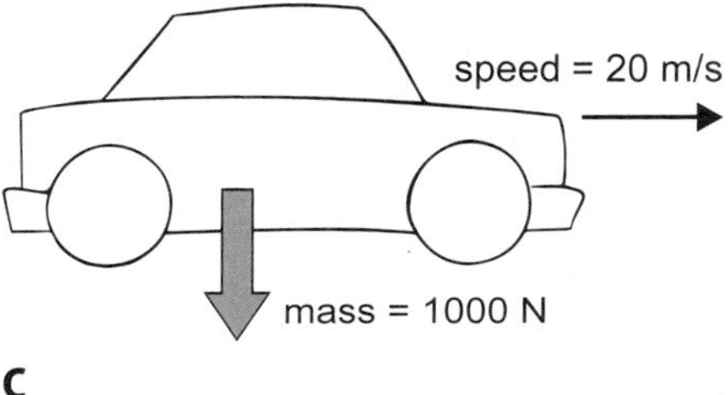

C

Checkpoint

How confidently can you answer the Progression questions?

Strengthen

S1 Sally walks 1 km from her home to school. When she arrives, she tells her science teacher 'My velocity to school this morning was 15 minutes'. What would her teacher say?

S2 Explain the difference between displacement and distance, and between speed and velocity. Give an example of each.

Extend

E1 A car is going around a roundabout. Explain why it is accelerating even if it is moving at a constant speed.

SP1b Distance/time graphs

Specification reference: P2.6; P2.7; P2.11; P2.12

Progression questions

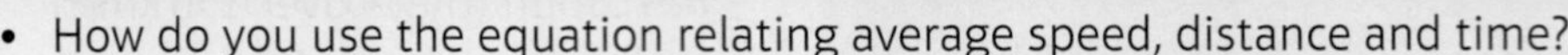

- How do you use the equation relating average speed, distance and time?
- In metres per second, what are the typical speeds that someone might move at during the course of a day?
- How do you represent journeys on a distance/time graph?

A *ThrustSSC* broke the land speed record in 1997 at a speed of 1228 km/h (341 m/s). This was faster than the speed of sound (which is approximately 330 m/s).

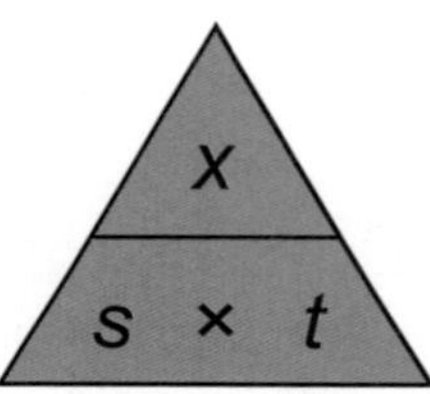

B This equation triangle can help you to rearrange the equation for speed (*s*), where *x* is used to represent distance and *t* represents time. Cover up the quantity you want to calculate and write what you see on the right of the = sign.

Worked example W1

How far would *ThrustSSC* have travelled in 5 seconds during its record-breaking run?

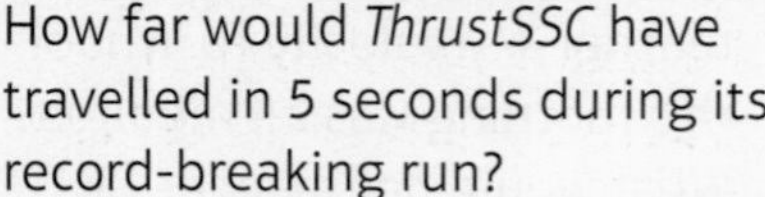

distance = average speed × time

= 341 m/s × 5 s

= 1705 m

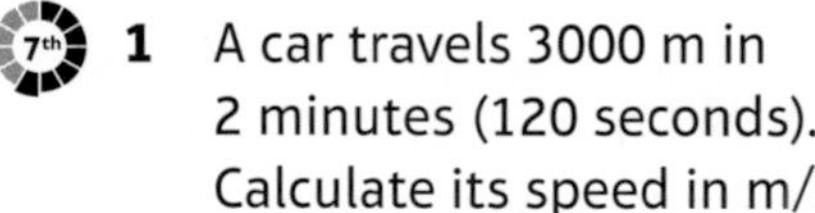

1 A car travels 3000 m in 2 minutes (120 seconds). Calculate its speed in m/s.

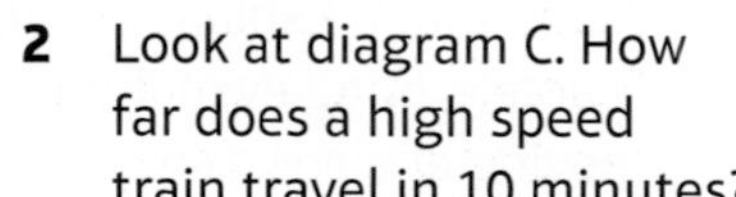

2 Look at diagram C. How far does a high speed train travel in 10 minutes?

The speed of an object tells you how quickly it travels a certain distance. Common units for speed are metres per second (m/s), kilometres per hour (km/h) and miles per hour (mph).

The speed during a journey can change, and the **average speed** is worked out from the total distance travelled and the total time taken. The **instantaneous speed** is the speed at a particular point in a journey.

Speed can be calculated using the following equation:

$$\text{(average) speed (m/s)} = \frac{\text{distance (m)}}{\text{time taken (s)}}$$

The equation can be rearranged to calculate the distance travelled from the speed and the time.

distance travelled = average speed × time
(m) (m/s) (s)

To measure speed in the laboratory you need to measure a distance and a time. For fast-moving objects, using **light gates** to measure time will be more accurate than using a stopwatch.

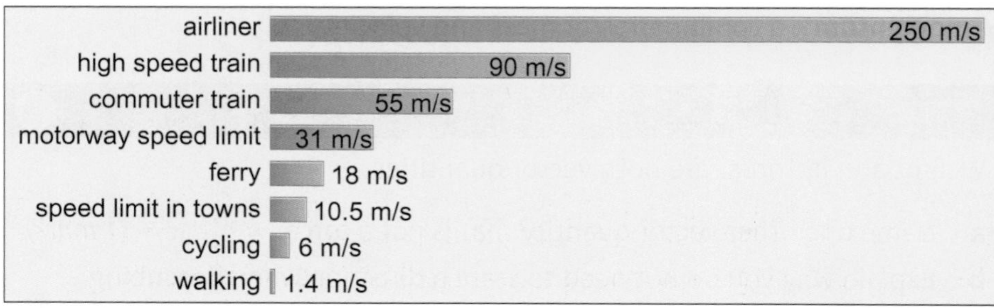

C some typical speeds

Distance/time graphs

A journey can be represented on a **distance/time graph**. Since time and distance are used to calculate speed, the graph can tell us various things about speed:

- horizontal lines mean the object is stationary (its distance from the starting point is not changing)
- straight, sloping lines mean the object is travelling at constant speed
- the steeper the line, the faster the object is travelling
- the speed is calculated from the **gradient** of the line.

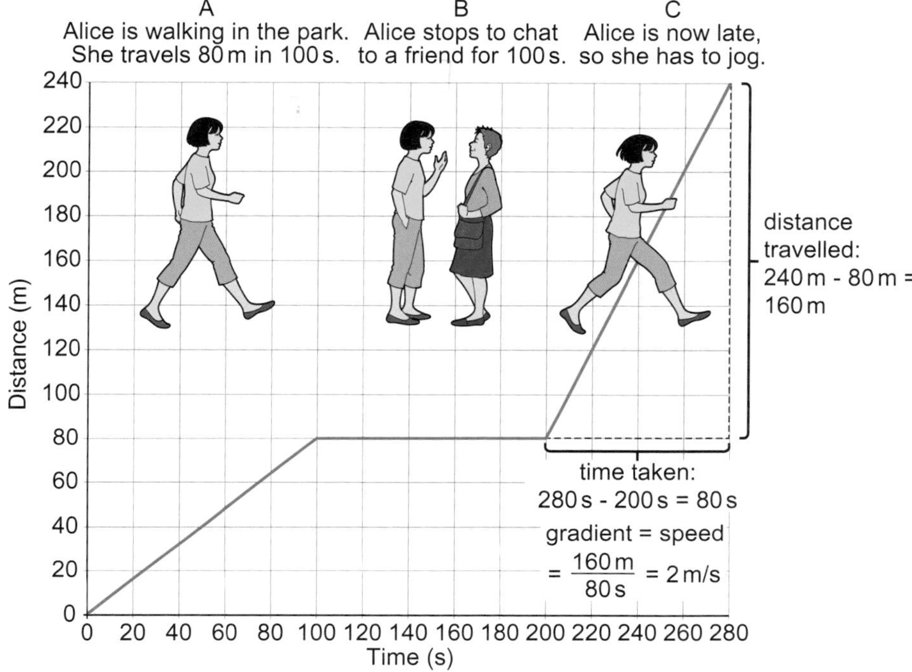
A
Alice is walking in the park. She travels 80 m in 100 s.
B
Alice stops to chat to a friend for 100 s.
C
Alice is now late, so she has to jog.
distance travelled: 240 m - 80 m = 160 m
time taken: 280 s - 200 s = 80 s
gradient = speed
= 160 m / 80 s = 2 m/s
Distance (m)
Time (s)
0
20
40
60
80
100
120
140
160
180
200
220
240
260
280

D The gradient of a distance/time graph gives the speed.

Worked example W2

In graph D, what is Alice's speed for part C of her walk?

$$\text{gradient} = \frac{\text{vertical difference between two points on a graph}}{\text{horizontal difference between the same two points}}$$

$$= \frac{240\text{ m} - 80\text{ m}}{280\text{ s} - 200\text{ s}}$$

Make sure you take the starting value away from the end value each time.

$$\text{speed} = \frac{160\text{ m}}{80\text{ s}}$$

$$\text{speed} = 2\text{ m/s}$$

Exam-style question

A snail travels 300 cm in 4 minutes. Calculate the speed of the snail in m/s. *(3 marks)*

Did you know?

The fastest wind speed recorded was 113 m/s, in Australia in 1996. There may be higher wind speeds than this inside tornados, but they have never been recorded. A gale force wind blows at around 20 m/s.

3 Look at graph D. Calculate Alice's speed for:

 **a** part A on the graph

 **b** part B on the graph.

 **4** If Alice had not stopped to chat but had walked at her initial speed for 280 s, how far would she have travelled?

Checkpoint

How confidently can you answer the Progression questions?

Strengthen

S1 A peregrine falcon flies at 50 m/s for 7 seconds. How far does it fly?

S2 Zahir starts a race fast, then gets a stitch and has to stop. When he starts running again he goes more slowly than before. Sketch a distance/time graph to show Zahir's race if he runs at a constant speed in each section of the race.

Extend

E1 Look at question S2. Zahir's speeds are 3 m/s for 60 seconds, 2 m/s for 90 seconds and his rest lasted for 30 seconds. Plot a distance/time graph on graph paper to show his race.

SP1c Acceleration

Specification reference: P2.8; P2.9; P2.13

Progression questions

- How do you calculate accelerations from a change in velocity and a time?
- How are acceleration, initial velocity, final velocity and distance related?
- What is the acceleration of free fall?

A A fighter plane can accelerate from 0 to 80 m/s (180 mph) in 2 seconds.

Fighter planes taking off from aircraft carriers use a catapult to help them to accelerate to flying speed.

A change in velocity is called acceleration. Acceleration is a vector quantity – it has a size (magnitude) and a direction. If a moving object changes its velocity or direction, then it is accelerating.

The acceleration tells you the change in velocity each second, so the units of acceleration are metres per second per second. This is written as m/s^2 (metres per second squared). An acceleration of 10 m/s^2 means that each second the velocity increases by 10 m/s.

 **1** How are velocity and acceleration connected?

Acceleration is calculated using the following equation:

$$\text{acceleration } (m/s^2) = \frac{\text{change in velocity (m/s)}}{\text{time taken (s)}}$$

This can also be written as:

$$a = \frac{v - u}{t}$$

where a is the acceleration

v is the final velocity

u is the initial velocity

t is the time taken for the change in velocity.

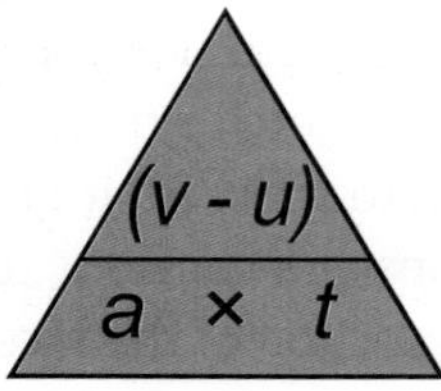

B This triangle can help you to rearrange the equation.

Worked example W1

An airliner's velocity changes from 0 m/s to 60 m/s in 20 seconds. What is its acceleration?

$$a = \frac{v - u}{t}$$

$$= \frac{60 \text{ m/s} - 0 \text{ m/s}}{20 \text{ s}}$$

$$= 3 \text{ m/s}^2$$

 **2** Calculate the take-off acceleration of the fighter plane in photo A.

Acceleration does not always mean getting faster. An acceleration can also cause an object to get slower. This is sometimes called a **deceleration**, and the acceleration will have a negative value.

 **3** A car slows down from 25 m/s to 10 m/s in 5 seconds. Calculate its acceleration.

Acceleration can be related to initial velocity, final velocity and distance travelled by this equation:

$$(\text{final velocity})^2 - (\text{initial velocity})^2 = 2 \times \text{acceleration} \times \text{distance}$$
$$(\text{m/s})^2 \qquad (\text{m/s})^2 \qquad (\text{m/s}^2) \qquad (\text{m})$$

This can also be written as $v^2 - u^2 = 2 \times a \times x$, where x represents distance.

Worked example W2

A car travelling at 15 m/s accelerates at 1.5 m/s² over a distance of 50 m. Calculate its final velocity.

$v^2 = (2 \times a \times x) + u^2$

$\quad = (2 \times 1.5\ \text{m/s}^2 \times 50\ \text{m}) + (15\ \text{m/s} \times 15\ \text{m/s})$

$v^2 = 375\ (\text{m/s})^2$

$v\ = \sqrt{375}\ (\text{m/s})^2$

$\quad = 19.4\ \text{m/s}$

 **4** A cyclist accelerates from 2 m/s to 8 m/s with an acceleration of 1.5 m/s². How far did she travel while she was accelerating? Use the equation $x = \dfrac{v^2 - u^2}{2 \times a}$.

Acceleration due to gravity

An object in free fall is moving downwards because of the force of gravity acting on it. If there are no other forces (such as air resistance), the acceleration due to gravity is 9.8 m/s². This is represented by the symbol g, and is often rounded to 10 m/s² in calculations.

5 Look at photo C.

 **a** Calculate the acceleration on the ejecting pilot in m/s².

 **b** How does this compare to everyday accelerations?

Exam-style question

A cheetah accelerates from rest to 30 m/s in 3 seconds. Calculate the acceleration of the cheetah. *(2 marks)*

Did you know?

Large accelerations are often compared to the acceleration due to gravity (g). The ejector seat in this aircraft can subject the pilot to accelerations of up to 12g or more.

C

Checkpoint

How confidently can you answer the Progression questions?

Strengthen

S1 Explain how positive, negative and zero accelerations change the velocity of a moving object.

S2 A car travelling at 40 m/s comes to a halt in 8 seconds. What is the car's acceleration and how far does it travel while it is stopping?

Extend

E1 A train is travelling at 35 m/s. It slows down with an acceleration of −0.5 m/s². How much time does it take to stop and how far does it travel while it is stopping?

SP1d Velocity/time graphs

Specification reference: P2.10

Progression questions

- How do you compare accelerations on a velocity/time graph?
- How can you calculate acceleration from a velocity/time graph?
- How can you use a velocity/time graph to work out the total distance travelled?

A Top Fuel dragsters can reach velocities of 150 m/s (335 mph) in only 4 seconds.

In a drag race, cars accelerate in a straight line over a short course of only a few hundred metres.

The changing velocity of a dragster during a race can be shown using a **velocity/time graph**.

On a velocity/time graph:

- a horizontal line means the object is travelling at constant velocity
- a sloping line shows that the object is accelerating. The steeper the line, the greater the acceleration. If the line slopes down to the right, the object is decelerating (slowing down). You can find the acceleration of an object from the gradient of the line on a velocity/time graph.
- a negative velocity (a line below the horizontal axis) shows the object moving in the opposite direction.

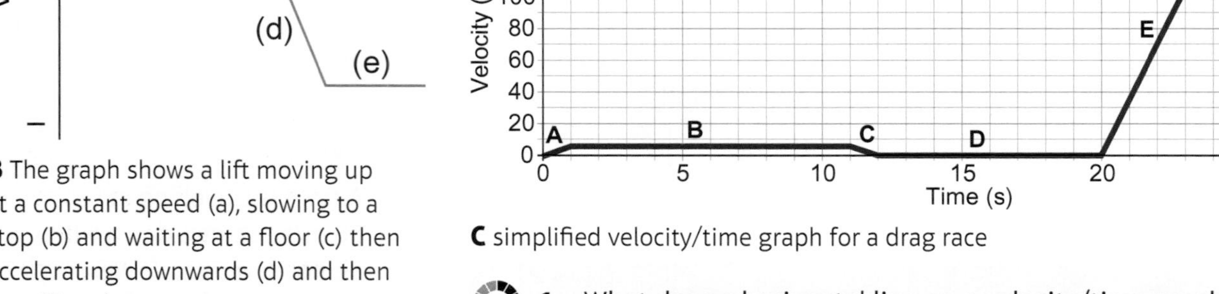

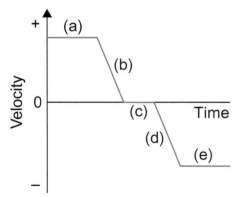

B The graph shows a lift moving up at a constant speed (a), slowing to a stop (b) and waiting at a floor (c) then accelerating downwards (d) and then travelling downwards at a constant speed (e).

Graph C is a simplified velocity/time graph for a dragster. It shows the car driving slowly to the start line, waiting for the signal, and then racing.

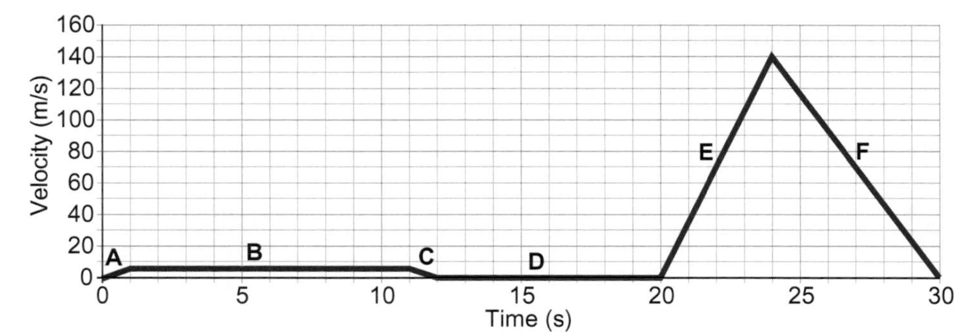

C simplified velocity/time graph for a drag race

6th **1** What does a horizontal line on a velocity/time graph tell you about an object's velocity?

 **2** **a** In which part of graph C is the dragster travelling at a constant velocity?

 **b** In which part of the graph does the dragster have its greatest acceleration?

 **c** Which part(s) of the graph show that the dragster is slowing down?

 **3** Look at graph C. Calculate the acceleration during part F of the journey.

Calculating distance travelled from a graph

The area under a velocity/time graph is the distance the object has travelled (distance is calculated by multiplying a velocity and a time). In graph D, the distance travelled in the first 5 seconds is the area of a rectangle. The distance travelled in the next 5 seconds is found by splitting the shape into a triangle and a rectangle, and finding their areas separately.

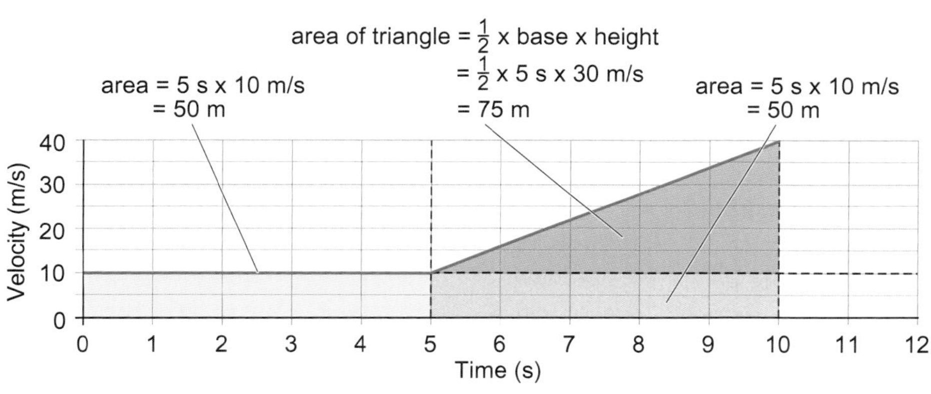

D

The total distance travelled by the object in graph D is the sum of all the areas.

total distance travelled = 50 m + 50 m + 75 m = 175 m

4 Look at graph C. The dragster travels at 5 m/s as it approaches the start line.

 **a** How far does it travel to get to the start line?

 **b** What is the distance travelled by the dragster during the race and slowing down afterwards?

 **5** Mel draws a graph showing a bus journey through town. Explain why this should be called a speed/time graph, not a velocity/time graph.

Exam-style question

Explain why the area under part of a velocity/time graph gives you the distance covered. *(3 marks)*

Checkpoint

How confidently can you answer the Progression questions?

Strengthen

S1 Table E below gives some data for a train journey. Draw a velocity/time graph from this and join the points with straight lines. Label your graph with all the things you can tell from it. Show your working for any calculations you do.

Time (s)	Velocity (m/s)
0	0
20	10
30	30
60	30
120	0

E

Extend

E1 In a fitness test, students run up and down the sports hall. They have to run faster after each time they turn around. Sketch a velocity/time graph for 4 lengths of the hall, if each length is run at a constant speed.

Vectors and scalars

Explain the statements below, giving examples in each case.

- The speed of an object has the same value as its velocity.
- The displacement of an object from its starting point is often not the same as the distance travelled.
- The displacement of an object at the end of a journey can never be greater than the distance travelled. **(6 marks)**

Student answer

Speed is a scalar and velocity is a vector because it has a direction. So if a car goes around a corner at a constant speed, the magnitude [1] *of the velocity does not change but the velocity is changing because the direction is changing* [2].

Displacement is a vector, and is the distance in a straight line between the starting point and finishing point. Displacement has a direction [3]. *Distance is a scalar, and is the actual distance moved during a journey including all the twists and turns* [4].

[1] This is a good use of correct scientific terminology.

[2] This is a good explanation of the first statement, and includes an example.

[3] This is a good description of the meaning of displacement.

[4] Although the student has described distance, there isn't an example that illustrates the difference.

Verdict

This is an acceptable answer. It shows a good understanding of the differences between vectors and scalars. The answer also provides good descriptions of the terms speed, velocity, distance and displacement. The use of scientific language is good and the answer is arranged logically. However, there are some parts of the answer which are missing.

This answer could be improved by including information on the third statement and an explanation for the second statement.

Exam tip

Try to leave enough time to check through your answer. Make sure you have included everything that the question has asked for (in this example, the student should have covered all three statements).

Paper 1

SP2 Motion and Forces

The car crash in this image was staged for a photoshoot. Crashes like this are often used in films and tend to be more spectacular than real road accidents. For example, in films a piston on the road is often used to flip a car into the air. Stunt designers need to carefully calculate the force from a piston to make sure that a car flips in the way that they want it to.

Engineers designing cars also need to know about the forces on cars and about how these forces affect the car and its occupants. This information can help them to design cars that will reduce the harm to occupants in crashes. Knowledge about forces can also help the government to work out what the speed limits should be on different roads, and what safety advice to give drivers.

The learning journey

Previously you will have learnt at KS3:

- what forces are and the effects of balanced and unbalanced forces
- what a resultant force is
- about gravity as a non-contact force
- ways in which energy is stored and transferred.

In this unit you will learn:

- about Newton's Laws of Motion
- how to calculate the weight of an object from its mass
- about the factors that affect the stopping distance of a vehicle
- how to use ideas about energy transfers to calculate braking distances
- about the dangers of large decelerations
- **H** how to calculate momentum, and apply ideas about momentum to collisions.

SP2a Resultant forces

Specification reference: P2.14

Progression questions

- What is the difference between the speed of an object and its velocity?
- How do we represent all the forces acting on an object?
- How do we calculate resultant forces?

A A 'wall of death' is a small arena with almost vertical sides. Motorcyclists ride around the walls, and won't fall as long as they keep moving!

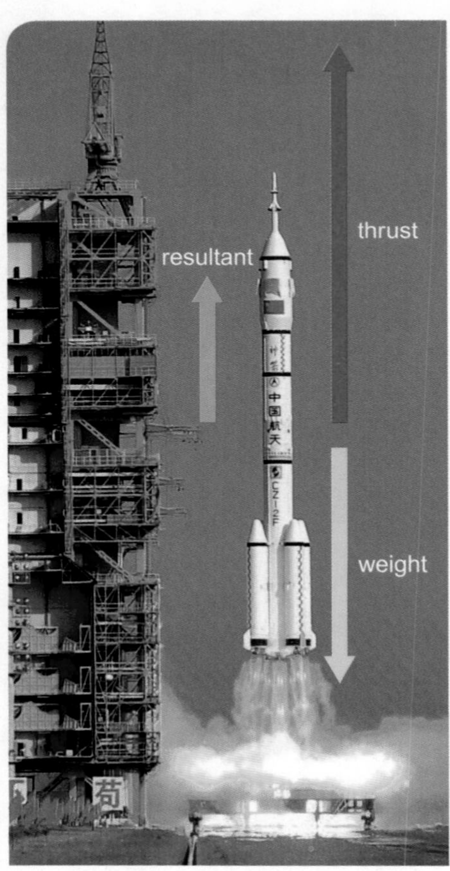

B A Chang Zheng 2F rocket has a take-off weight of approximately 5000 kN, and thrust of about 13 000 kN (1 kN = 1000 N).

Scalars and vectors

The motorcyclist in photo A is moving at a constant **speed** but his **velocity** is changing all the time. This is because velocity is a **vector quantity**. It has a direction as well as a magnitude (size). Speed is a **scalar quantity**. It only has a magnitude.

When an object changes its velocity, it is accelerating. As **acceleration** is a change in a vector quantity (velocity), acceleration is also a vector.

1 A car is driving around a roundabout at 20 km/h. Explain whether or not:

 **a** its speed is changing

 **b** its velocity is changing.

Representing forces

Forces are vector quantities. It is important to know the direction in which a force is acting, as well as how big it is. We can draw diagrams to show the forces on objects to help us to think about the effects the forces will have. The size of the force is represented by the length of the arrows.

The thrust on the rocket in photo B is the upwards force from its engines. You can easily see from the diagram that the thrust is greater than its weight. The weight cancels out part of the thrust, so the overall upwards force is 8000 kN. This is called the **resultant force** on the rocket.

To work out the resultant of two forces:

- if the forces are acting in the same direction, add them
- if they are acting in opposite directions (as in photo B), subtract one from the other.

 **2** **a** A cyclist is riding along a flat road without pedalling. The air resistance is 10 N and **friction** is 5 N. What is the resultant force on the bike?

 **b** What is the resultant force if the cyclist is pedalling with a force of 25 N?

The aeroplane in photo C has two forces acting in the vertical direction and two in the horizontal. We do not have to think about all four forces at one time to work out a resultant, because the two sets of forces are at right angles to each other. We can think about the two sets of forces separately.

C

If the resultant of all the forces on an object is zero, we say the forces are **balanced**. If there is a non-zero resultant force on an object, the forces are **unbalanced**.

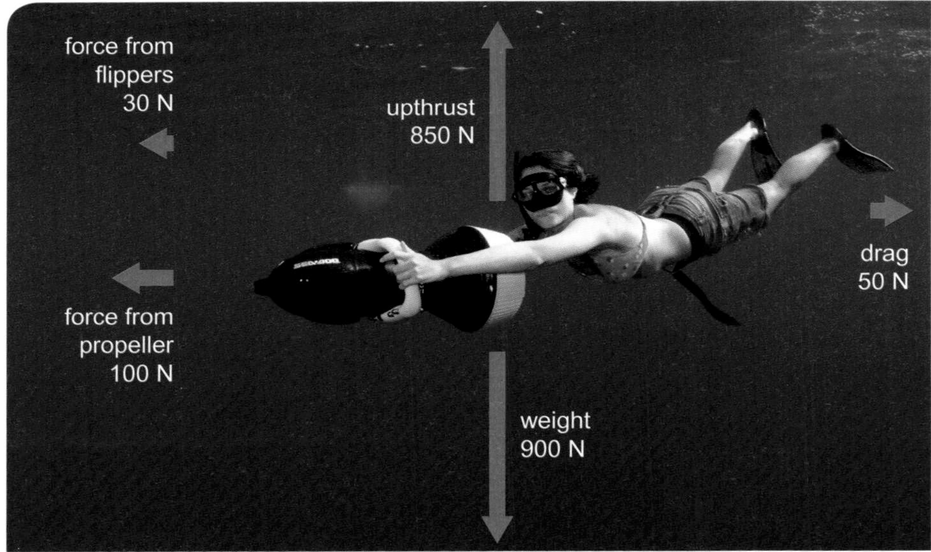

D What are the forces on this diver?

Exam-style question

The forces on a car are balanced. State and explain the size of the resultant force on the car. *(2 marks)*

3 Calculate the resultant force on the aeroplane in photo C in the:

 **a** vertical direction

 **b** horizontal direction.

4 In photo C, are the forces balanced or unbalanced in the:

 **a** vertical direction

b horizontal direction?

Checkpoint

How confidently can you answer the Progression questions?

Strengthen

S1 Draw a concept map to show what you know about forces.

Extend

E1 Calculate the resultant force on the diver in photo D in the vertical direction and in the horizontal direction.

E2 Draw a diagram of the diver and add force arrows to show the two resultant forces.

SP2b Newton's First Law

Specification reference: P2.14; H P2.20; H P2.21

Progression questions

- What happens to the motion of an object when the forces on it are balanced?
- What can happen to the motion of an object when there is a resultant force on it?
- H What is centripetal force?

A Human cannonballs are propelled using unbalanced forces from compressed air or springs – not using explosives!

Sir Isaac Newton (1642–1727) worked out three 'laws' of motion that describe how forces affect the movement of objects.

Newton's First Law of motion can be written as:

- a moving object will continue to move at the same speed and direction unless an external force acts on it
- a stationary object will remain at rest unless an external force acts on it.

It is the overall resultant force that is important when you are looking at how the velocity of an object changes. Balanced forces (zero resultant force) will not change the velocity of an object. Unbalanced forces (non-zero resultant force) will change the speed and/or direction of an object.

1 a What is the resultant force on the human cannonball in the vertical direction when she is flying through the air?

b How will this resultant force affect her velocity?

2 Look at photo C on the previous page again. Explain how the velocity of the aeroplane will change in the:

a vertical direction

b horizontal direction.

The ice yacht in photo B is not changing speed in the vertical direction. Its weight is balanced by an upwards force from the ice.

B An ice yacht can go much faster than a sailing boat in the same wind conditions.

3 A sailing boat has a forwards force of 300 N from the wind in its sails. It is travelling at a constant speed.

7th **a** What is the total force acting backwards on the sailing boat? Explain your answer.

8th **b** The ice yacht in photo B has the same force from its sails. Explain why its velocity will be increasing.

H Circular motion

C This fairground ride is accelerating the people in the chairs.

An object moving in a circle has changing velocity, even though its speed remains the same. The resultant force that causes the change in direction is called the **centripetal force**, and acts towards the centre of the circle. In photo C, the centripetal force is provided by tension in the wires holding the seats. Other types of force that can make objects move in circular paths include friction and gravity.

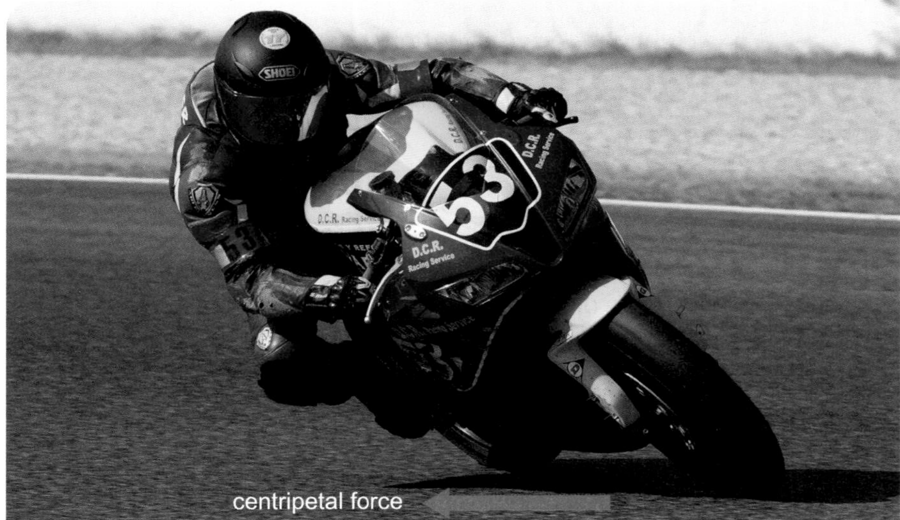

D The centripetal force here is supplied by friction between the tyres and the road.

 **4** A satellite is in a circular orbit around the Earth. Explain how and why its velocity is continuously changing.

Exam-style question

A man is pushing a baby along the pavement in a pushchair. Friction is also acting on the pushchair. The pushchair is slowing down. Compare the horizontal forces on the pushchair. *(2 marks)*

Checkpoint

How confidently can you answer the Progression questions?

Strengthen

S1 You are cycling along a flat road and your speed is increasing. Explain the resultant forces on you in the horizontal and vertical directions.

Extend

E1 Describe the forces on a human cannonball, from just before they are fired to when they land safely in a net, including how these forces affect their motion. You only need to consider forces and motion in the vertical direction.

E2 **H** Explain what a centripetal force is and describe three different kinds of force that can act as centripetal forces.

SP2c Mass and weight

Specification reference: P2.16; P2.17; P2.18

Progression questions

- What is the difference between mass and weight?
- What are the factors that determine the weight of an object?
- How do you calculate weight?

A The Huygens space probe used the air resistance from a parachute to balance its weight when it landed on Titan (one of Saturn's moons).

Mass is the quantity of matter there is in an object, and only changes if the object itself changes. For example, your mass increases when you eat a meal. **Weight** is a measure of the pull of gravity on an object and depends on the strength of gravity. The units for mass are kilograms. Weight is a force, so it is measured in newtons. Weight can be measured using a **force meter**, which has a scale in newtons. Many force meters contain a spring, which stretches as the force on it increases allowing the weight to be read off the scale.

 **1** Suggest one way in which you can decrease your mass.

On Earth the **gravitational field strength** has a value of about 10 newtons per kilogram (N/kg). This means that each kilogram is pulled down with a force of 10 N. The gravitational field strength is different on other planets and moons.

The weight of an object can be calculated using the following equation:

weight = mass × gravitational field strength
(N) (kg) (N/kg)

This is often written as: $W = m \times g$

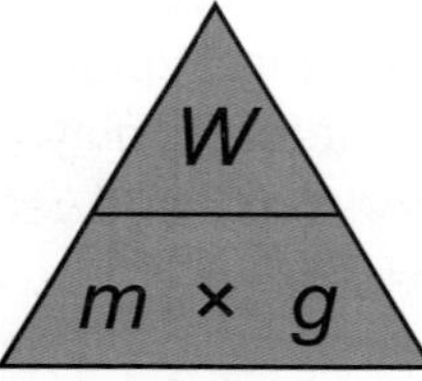

B This triangle can help you to change the subject of the equation. Cover up the quantity you want to find, and what you can see is the equation you need to use.

Worked examples

What is the weight of a 90 kg astronaut on the surface of the Earth?

$W = m \times g$

$W = 90\text{ kg} \times 10\text{ N/kg}$

$= 900\text{ N}$

A space probe has a weight of 3000 N on the Earth. What is its mass?

$m = \frac{W}{g}$

$= \frac{3000\text{ N}}{10\text{ N/kg}}$

$= 300\text{ kg}$

Did you know?

Gravity is not the same everywhere on the Earth. Your weight is greater standing at the North Pole than it would be standing at the equator.

2 A 300 kg space probe lands on Titan, where the gravitational field strength is 1.4 N/kg.

 **a** What is its mass on Titan? Explain your answer.

 **b** What is its weight on Titan?

 **3** A Mars rover has a mass of 185 kg. Its weight on Mars is 685 N. What is the gravitational field strength on Mars?

Forces on falling bodies

On Earth, a falling object has a force of air resistance on it as well as its weight. Figure C shows how the forces on a skydiver change during her fall.

0.5 seconds after jumping, speed = 5 m/s
resultant
Air resistance increases with speed, so just after jumping the air resistance is much smaller than her weight. The large resultant force makes her accelerate downwards.

3 seconds after jumping, speed = 25 m/s
resultant
Her air resistance is larger but her weight stays the same. The resultant force is smaller, so she is still accelerating, but not as much.

12 seconds after jumping, speed = 55 m/s
resultant = 0
She is moving so fast that the air resistance balances her weight. She continues to fall at the same speed.

C

 **4** Explain why the weight of the skydiver stays the same throughout the jump.

 **5** When the skydiver opens her parachute her air resistance increases very suddenly. Explain how this affects the resultant force and her velocity.

Exam-style question

The mass of a spanner on Earth is 0.2 kg, and its weight is 2 N.
The spanner is taken to the Moon as part of an astronaut's tool kit.

Compare the mass and weight of the spanner on the Moon and on the Earth. *(4 marks)*

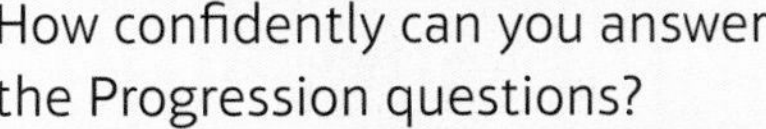

Checkpoint

How confidently can you answer the Progression questions?

Strengthen

S1 Write glossary entries for 'mass' and 'weight'.

S2 A cat has a mass of 2 kg. Calculate its weight on Earth.

Extend

E1 Write an encyclopaedia entry that explains the difference between mass and weight.

E2 A space probe on Titan has a weight of 280 N. Calculate its mass and its weight on Earth.

SP2d Newton's Second Law

Specification reference: P2.15; H P2.22

Progression questions

- What are the factors that affect the acceleration of an object?
- How do you calculate the different factors that affect acceleration?
- H What is inertial mass and how is it defined?

A The safety rules for building Formula 1® cars include a limit to the engine force and a minimum mass for the car.

The acceleration of an object is a measure of how much its velocity changes in a certain time. Sir Isaac Newton's Second Law of Motion describes the factors that affect the acceleration of an object.

The acceleration in the direction of a resultant force depends on:

- the size of the force (for the same mass, the bigger the force the bigger the acceleration)
- the mass of the object (for the same force, the more massive the object the smaller the acceleration).

1 The resultant force on a ball is not zero. What will happen to the ball?

2 **a** The same force is used to accelerate a small car and a lorry. What will be different about their motions? Explain your answer.

b If you wanted to make the same two vehicles accelerate at the same rate, what can you say about the forces needed to do this? Explain your answer.

Did you know?

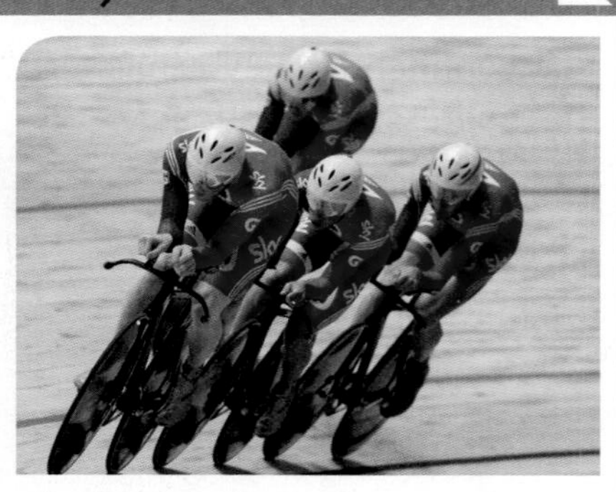

B A bike with low mass is so important for track racers that their bikes do not even have brakes! Racing rules state that these bikes must have a minimum mass of 6.8 kg.

Calculating forces

The force needed to accelerate a particular object can be calculated using the equation:

force = mass × acceleration
(N) (kg) (m/s^2)

This is often written as $F = m \times a$

1 newton is the force needed to accelerate a mass of 1 kg by 1 m/s^2.

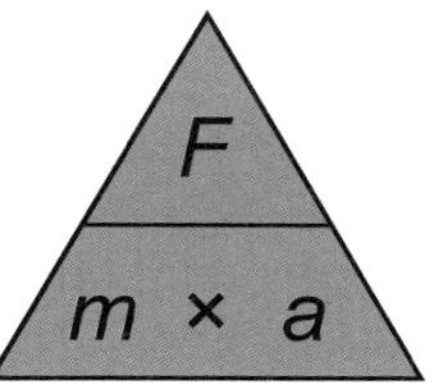

C This triangle can help you to change the subject of the equation. Cover up the quantity you want to find, and what you can see is the equation you need to use.

Worked example

A motorcycle has a mass of 200 kg. What force is needed to give it an acceleration of 7 m/s^2?

$F = m \times a$
$= 200 \text{ kg} \times 7 \text{ m/s}^2$
$= 1400 \text{ N}$

 **3** A car has a mass of 1500 kg. What force is needed to give it an acceleration of 4 m/s^2?

 **4** A force of 800 N accelerates the car in question 3. What is its acceleration?

D What do you need to know to work out whether the car or aeroplane has the greater acceleration?

H Inertial mass

The more massive an object is, the more force is needed to change its velocity (either to make it start moving or to change the velocity of a moving object). We define the **inertial mass** of an object as the force on it divided by the acceleration that force produces.

Calculating an object's inertial mass from values of force and acceleration gives the same mass value as that found by measuring the force of gravity on it.

 **5** A force of 160 N on a bicycle produces an acceleration of 2 m/s^2. What is the total inertial mass of the bicycle and its rider?

Exam-style question

A car accelerates at 2 m/s^2. The resultant force acting on the car is 3000 N.

Calculate the mass of the car. *(3 marks)*

Checkpoint

How confidently can you answer the Progression questions?

Strengthen

S1 Look at photo A. Explain whether the Formula 1® rules are designed to set an upper or lower limit to the accelerations of the cars.

S2 Calculate the force needed to accelerate a 250 kg motorbike at 5 m/s^2.

Extend

E1 Look at photo D. Explain what you need to know to help you to work out which vehicle will have the greater acceleration.

E2 Explain why two objects dropped on the Moon will accelerate at the same rate, even when they have different masses.

SP2d Core practical – Investigating acceleration

Specification reference: P2.19

Aim

Investigate the relationship between force, mass and acceleration by varying the masses added to trolleys.

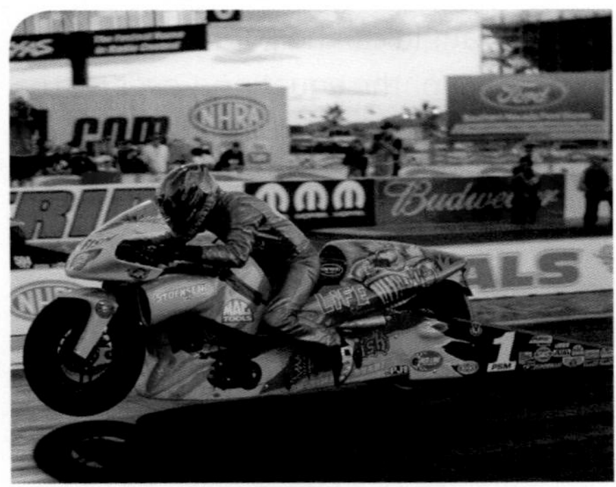

A

In drag racing, the aim is to get to the end of a straight track as quickly as possible, and so the most important feature of the bike is its acceleration. Drag racers can try to improve the performance of their bikes by changing the force produced by the engine and the tyres or by changing the mass of the bike.

Your task

You are going to use trolleys as a model of a motorbike to investigate the effects that mass and force have on acceleration. The force will be provided by masses hanging on a string. You can vary the mass of the trolley by adding stacking masses to it.

Method

A Prop up one end of the ramp. Place a trolley on the ramp. Adjust the slope of the ramp until the trolley just starts to move on its own. Keep the ramp at this slope for the whole investigation. Set up the light gates and the pulley and string as shown in diagram B.

B Stick a piece of card to the top of the trolley. Measure the length of the card and write it down.

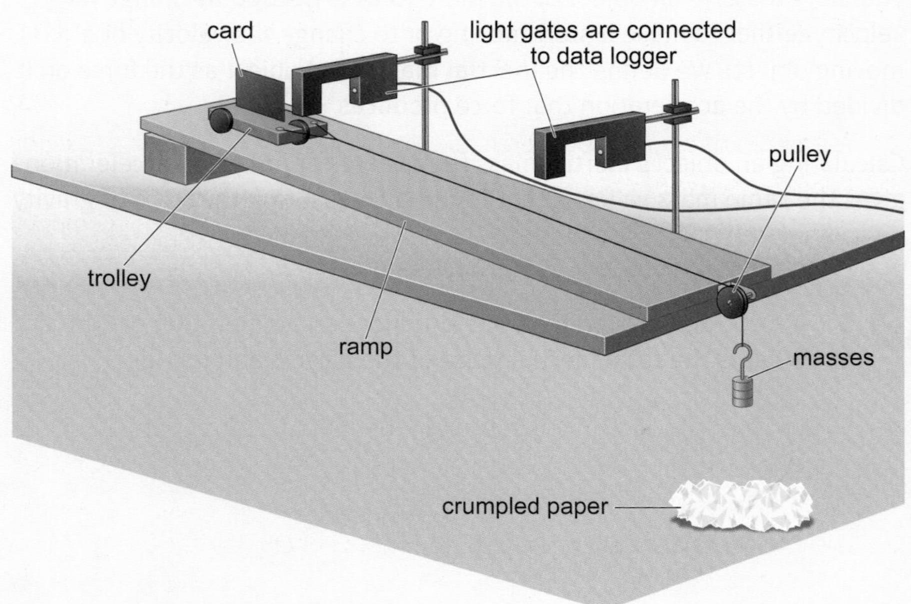

B apparatus for investigating acceleration

C Find the mass of the trolley and write it down.

D Put a mass on the end of the string. You will keep this mass the same for all your tests. You will have to decide what mass to use.

E Release the trolley from the top of the ramp and write down the speed of the trolley (from the datalogger) as it passes through each light gate. Also write down the time it takes for the trolley to go from one light gate to the other.

F Put a mass on top of the trolley. Keep the masses on the end of the string as they were before. Repeat step E.

G Repeat step E for other masses on top of the trolley. You will have to decide what masses to use, how many different masses you are going to test, and whether you need to repeat any of your tests.

H The steps above are investigating how the mass of the trolley affects the acceleration. If you wish to investigate the effect of force on acceleration, you need to keep the mass the same. However, the masses on the end of the string are also accelerating, along with the trolley, and it is the overall mass that you need to keep the same. You can do this by starting with a stack of masses on the trolley. Take one mass off the trolley and hang it on the end of the string. Then follow step E to measure the acceleration.

I Now transfer another mass from the trolley to the end of the string and find the acceleration again. Keep doing this until all the masses that started on the trolley have been transferred to the end of the string.

Exam-style questions

1 The light gates and datalogger record the speed of the trolley at the top of the ramp and at the bottom of the ramp, and also record the time the trolley takes to move between the two light gates. Describe how this information can be used to calculate the acceleration. *(2 marks)*

2 State the difference between mass and weight. *(1 mark)*

3 Explain one way in which you would stay safe while doing the experiments. *(2 marks)*

4 Make a list of the apparatus you need to carry out the method. *(2 marks)*

5 Explain why a light gate is needed at the top of the ramp as well as at the bottom. *(2 marks)*

6 Look at steps H and I. Explain why all the masses to be used in investigating the effect of the force on acceleration have to be on the trolley to start with. *(3 marks)*

7 Use the results shown in graph C to draw a conclusion for this part of the investigation. *(1 mark)*

8 Look at graph D.

a Use graph D to draw a conclusion for this part of the investigation. *(1 mark)*

b Explain how you would present the data in graph D to allow you to draw a better conclusion. *(2 marks)*

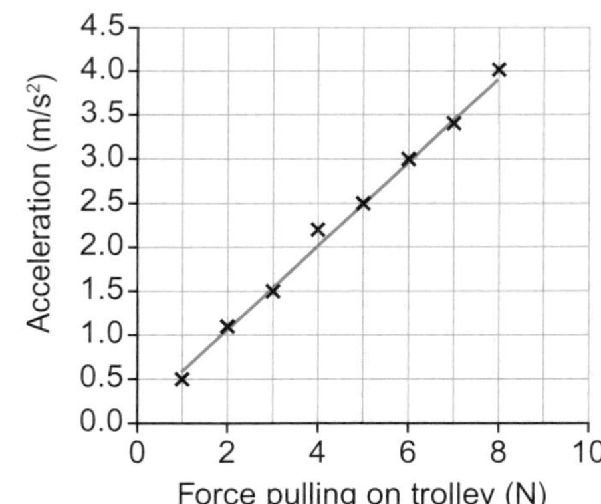

C

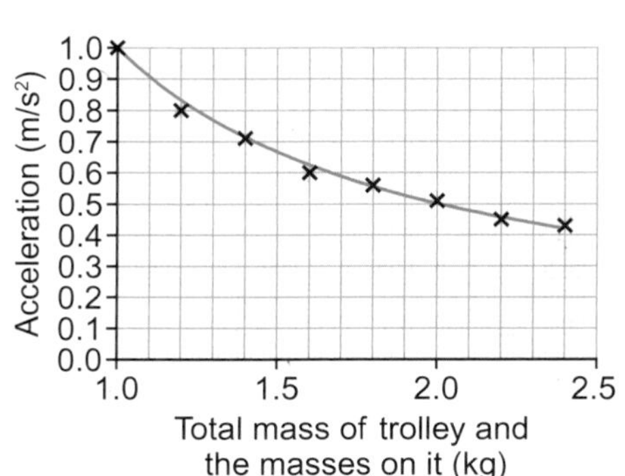

D

SP2e Newton's Third Law

Specification reference: P2.23; H P2.23

Progression questions

- What does Newton's Third Law tell us?
- How does Newton's Third Law apply to stationary objects?
- H How do objects affect each other when they collide?

A The dog is not moving. What are the forces acting here?

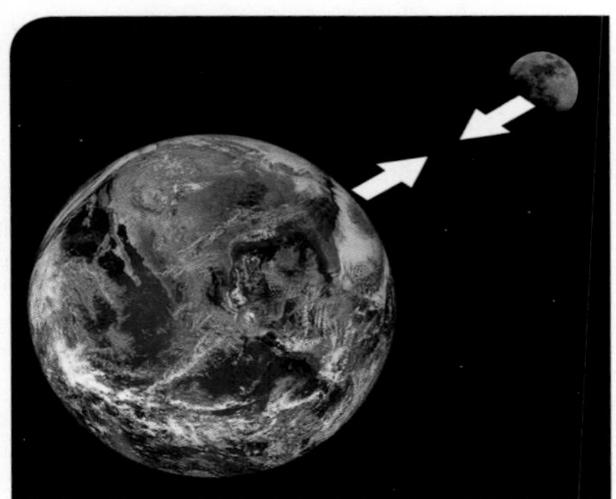

B The Earth attracts the Moon with the same force as the Moon attracts the Earth.

2 You are standing leaning on a wall. Draw a sketch to show this (a stick man will do) and add arrows to show an action–reaction pair of forces acting in the:

 **a** vertical direction

 **b** horizontal direction.

7th **3** For the situation in question 2, describe the balanced forces on you acting in the vertical direction.

Newton's Third Law is about the forces on two different objects when they interact with each other. This interaction can happen:

- when objects touch, such as when you sit on a chair
- at a distance, such as the gravitational attraction between the Earth and the Moon.

There is a pair of forces acting on the two interactive objects, often called **action–reaction forces**. The two forces are always the same size and in opposite directions. They are also the same type of force. In photo A the rope and the dog are both exerting pulling forces on each other. In photo B the two forces are both gravitational forces.

Photo A shows an **equilibrium** situation, because nothing is moving. A force in the rope is pulling on the dog but the dog is also pulling on the rope.

 **1** **Think about the vertical forces in photo A. One force is the weight of the dog pushing down on the ground. What is the other force in this pair?**

The weight of a dog on the ground is equal to the force pushing up on the dog from the ground. In photo A there is another pair of action–reaction forces acting on the rope and the dog – there is a force from the dog on the rope and a force from the rope on the dog.

Action–reaction forces are not the same as balanced forces. In both cases the sizes of the forces are equal and act in opposite directions, but:

- action–reaction forces act *on different objects.*
- balanced forces all act *on the same object*

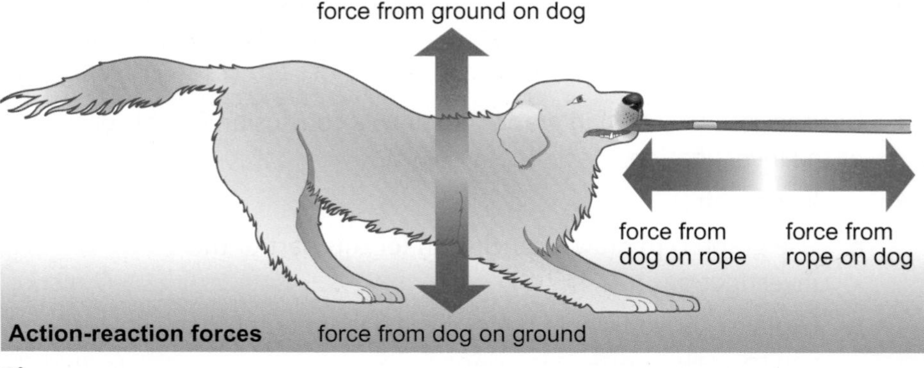

Ci

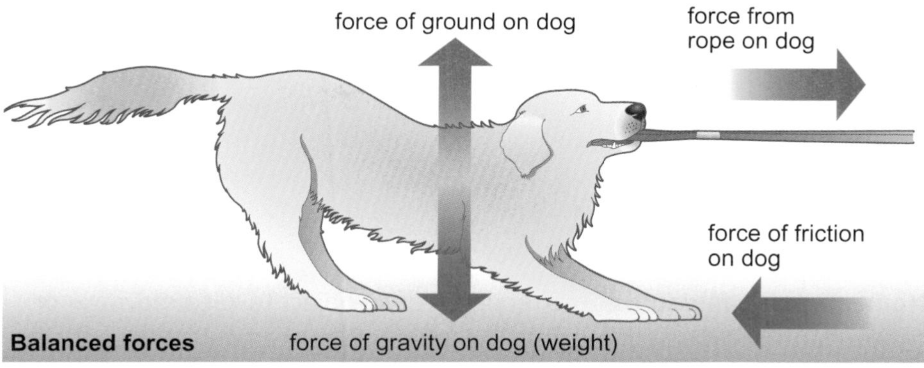

Cii

H Collisions

We can apply the idea of action–reaction forces to what happens when things collide. In photo D, the ball will bounce off the footballer's head. His head exerts a force on the ball, but the ball also exerts a force on his head, as you can feel if you have ever tried heading a ball!

The action and reaction forces that occur during the collision are the same size, but they do not necessarily have the same effects on the two objects, because the objects have different masses.

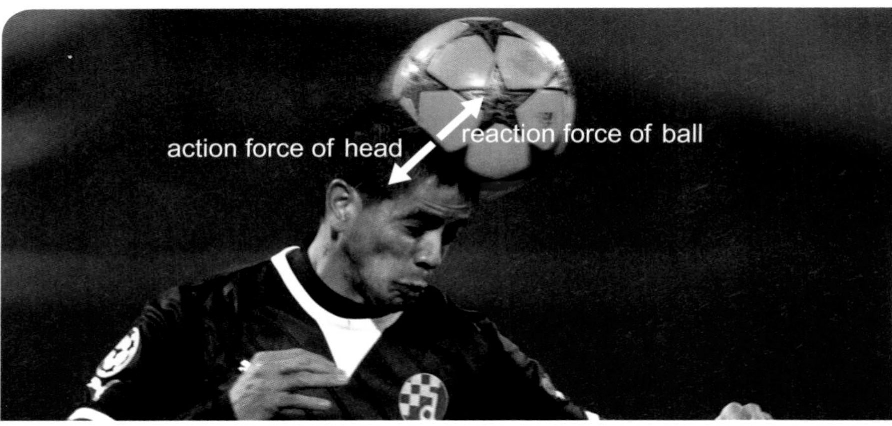

D Action–reaction forces during a collision.

 **4** Describe the action–reaction forces when a ball bounces on the ground.

 **5** Look at photo D. The player's head and the ball both change velocity during the collision. Describe the effects on the two objects and explain why the effects are different.

Exam-style question

A student is using a force meter to find the weight of an apple. Describe an action-reaction pair of forces in this situation. *(1 mark)*

Checkpoint

How confidently can you answer the Progression questions?

Strengthen

S1 Describe the action–reaction forces when you sit in a chair. Describe how these forces are different to a pair of balanced forces acting on you.

Extend

E1 Two teams are having a tug-of-war. Make a sketch and add labelled arrows to show three action–reaction force pairs and three pairs of balanced forces.

E2 **H** A ball is released a metre above the surface of the Earth. Describe the action–reaction forces due to gravity on the ball and the Earth. Describe the forces when the ball and the Earth collide. Explain how the effects on the two objects are different.

SP2f Momentum

Specification reference: H P2.23; H P2.24; H P2.25; H P2.26

Progression questions

- H How is momentum calculated?
- H How is momentum related to force and acceleration?
- H What happens to momentum in collisions?

Momentum is a measure of the tendency of an object to keep moving – or of how hard it is to stop it moving. The momentum of an object depends on its mass and its velocity. Momentum depends on a vector quantity (velocity), and is also a vector.

Momentum is calculated using this equation:

momentum = mass × velocity
(kg m/s) (kg) (m/s)

This can also be written as $p = m \times v$, where p stands for momentum.

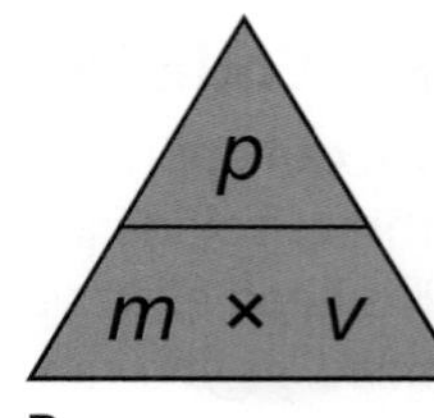

B

A The damage caused by a wrecking ball depends on its mass and how fast it is moving when it hits.

8th **1** Explain why a motorcycle travelling at 30 m/s has less momentum than a car travelling at the same velocity.

9th **2** A 500 kg wrecking ball is moving at 10 m/s when it hits a building. What is its momentum?

10th **3** The same ball at a different time has a momentum of 1500 kg m/s. What is its velocity?

Momentum and acceleration

Table C shows two equations involving acceleration. These can be combined to give:

$$\text{force} = \frac{\text{mass} \times \text{change in velocity}}{\text{time}} \quad \text{or} \quad \frac{m(v-u)}{t}$$

where v is the final velocity and u is the starting velocity.

As mass × velocity is the momentum of an object, this equation can also be written as:

$$\text{force} = \frac{\text{change in momentum}}{\text{time}} \quad \text{or} \quad \frac{mv - mu}{t}$$

force, mass and acceleration	$F = m \times a$
change in velocity and time	$a = \frac{v-u}{t}$

C equations involving acceleration

Did you know?

The largest ships are oil tankers. It can take several miles for a moving oil tanker to come to a stop.

Worked example

A 2000 kg car accelerates from 10 m/s to 25 m/s in 10 seconds. What resultant force produced this acceleration?

$$\text{force} = \frac{mv - mu}{t}$$

$$= \frac{2000\text{ kg} \times 25\text{ m/s} - 2000\text{ kg} \times 10\text{ m/s}}{10\text{ s}}$$

$$= \frac{50\,000\text{ kg m/s} - 20\,000\text{ kg m/s}}{10\text{ s}}$$

$$= 3000\text{ N}$$

4 A 1000 kg car accelerates from rest to 15 m/s in 15 seconds. What resultant force caused this?

H Momentum and collisions

When moving objects collide the total momentum of both objects is the same before the collision as it is after the collision, as long as there are no external forces acting. This is known as **conservation of momentum**. Remember, momentum is a vector so you need to consider direction when you add the quantities together. If two objects are moving in opposite directions, we give the momentum of one object a positive sign and the other a negative sign.

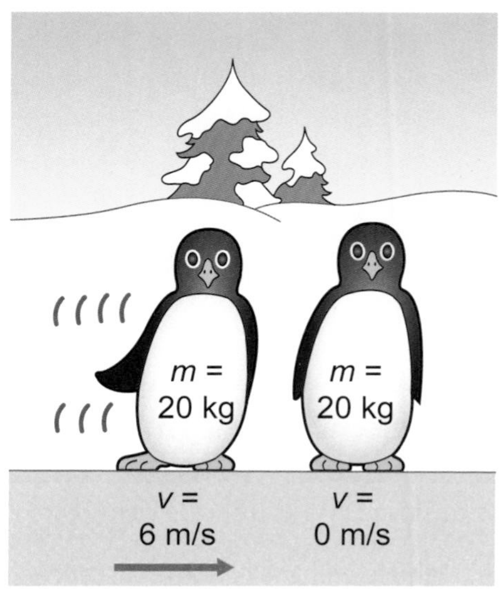

before collision

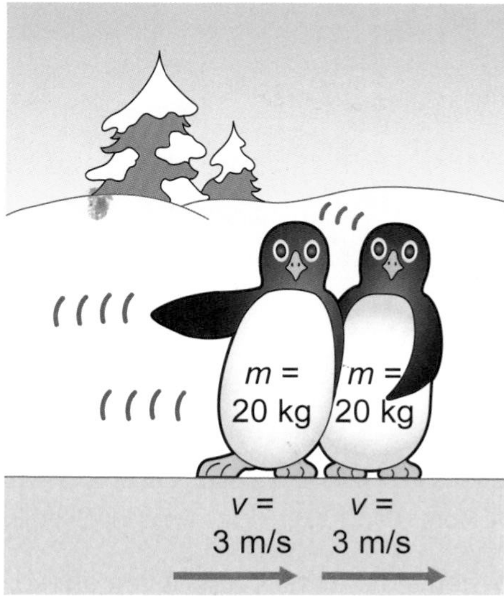

after collision

E

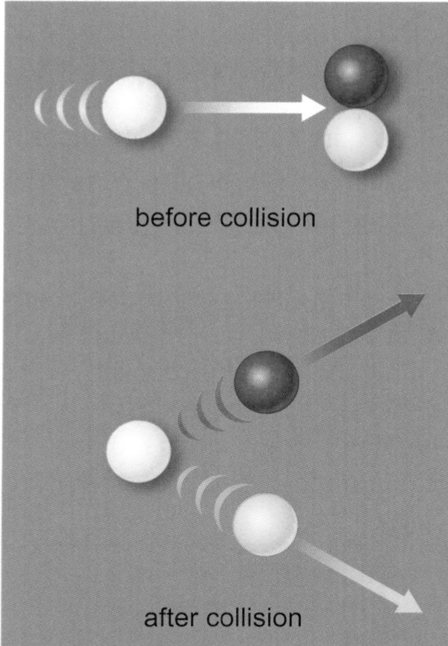

D The total momentum of the two coloured balls will be the same as the momentum of the white ball that hit them.

5 Look at diagram E.

 **a** Calculate the momentum of each penguin before they collide.

 **b** Calculate the total momentum before the penguins collide.

 **c** In which direction is the total momentum before the collision?

 **d** What is the total momentum after the collision and in which direction?

 **e** Explain whether momentum has been conserved.

Exam-style question

A car has a mass of 1800 kg. It is moving with a velocity of 35 m/s.

Calculate the momentum of the car. *(3 marks)*

Checkpoint

How confidently can you answer the Progression questions?

Strengthen

S1 Two 5000 kg railway trucks are travelling at 5 m/s in opposite directions when they collide. After the collision they are stationary. Show that momentum is conserved.

Extend

E1 A 1 g bullet is travelling at 300 m/s when it enters a stationary 1 kg block of wood. The impact of the bullet makes the wood move. What is the speed of the block immediately after the impact? Explain how you worked out your answer.

SP2g Stopping distances

Specification reference: P2.27; P2.28; P2.29; P2.30

Progression questions

- How are human reaction times measured?
- What are typical human reaction times?
- What are the factors that affect the stopping distance of a vehicle?

Did you know?

Until 1896 all 'self-propelled' vehicles had to have a man walking in front with a red flag, to warn other road users that it was coming.

A More than 130 vehicles were involved in this crash and over 200 people were injured.

When a driver sees a problem ahead, their vehicle will travel some distance while the driver reacts to the situation. This is called the **thinking distance**. The vehicle will then go some distance further while the brakes are working to bring it to a halt. This is called the **braking distance**. The overall **stopping distance** for any road vehicle is the sum of the thinking and braking distances.

stopping distance = thinking distance + braking distance

5th **1** Why is it important for drivers to know their stopping distances?

6th **2** For a thinking distance of 5 m and a braking distance of 12 m, what is the overall stopping distance?

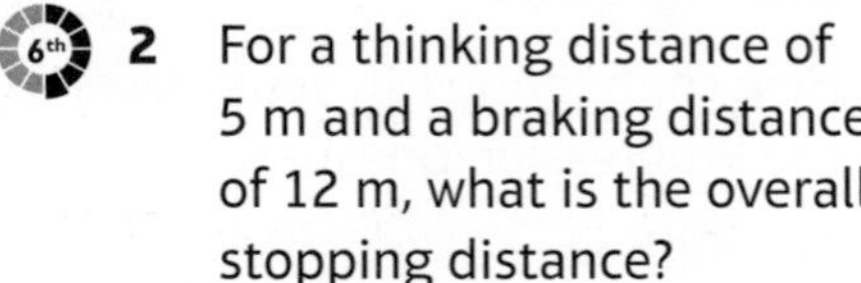

Reaction times

A **reaction time** is the time between a person detecting a **stimulus** (such as a flashing light or a sound) and their **response** (such as pressing a button or applying the brakes in a car). Response times can be measured using computers or electric circuits that measure the time between a stimulus and a response.

The typical reaction time to a visual stimulus, such as a computer screen changing colour, is about 0.25 seconds. However this time can be much longer if the person is tired, ill or has been taking drugs or drinking alcohol. Distractions, such as using a mobile phone, can also increase reaction times.

B A driving simulator can be used to test reaction times in a realistic situation.

6th **3** Explain why the thinking distance depends on:

a the driver's reaction time **b** the speed of the car.

6th **4** Suggest why the reaction time measured in a driving simulator might be longer than the time measured using a test on a computer.

6th **5** Explain why there are legal limits for the amount of alcohol drivers are allowed in their blood.

Braking distances

Car brakes use friction to slow the car down. If the brakes are worn, they create less friction and do not slow the vehicle as effectively. Friction between the tyres and road is also important. If the road is wet or has loose gravel on it, or if the tyres are worn, there is less friction and the braking distance is increased.

If a vehicle has more mass, more force is needed to decelerate it. So if the same amount of friction is used to stop a vehicle, a heavier vehicle will travel further than a lighter one (it has a greater braking distance).

 **6** Why are the overall stopping distances for cars less than for lorries?

 **7** Look at photo D. Suggest why there are two separate speed limits.

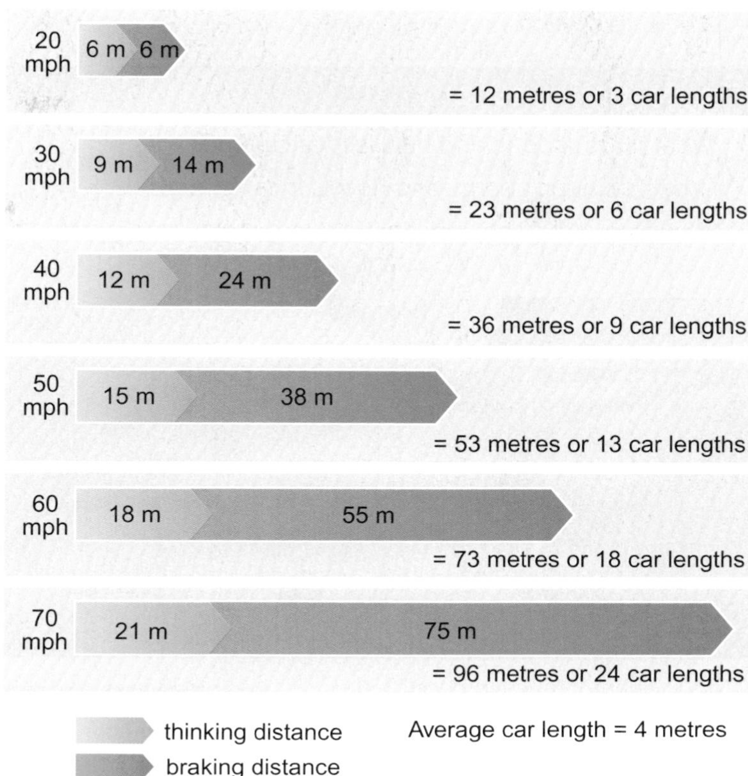

C The Highway Code shows typical stopping distances for a family car.

D This sign is from an autoroute (motorway) in France. The speed limits are in km/h.

Exam-style question

Explain how thinking distance and braking distance depend on the speed of the vehicle. *(4 marks)*

Checkpoint

How confidently can you answer the Progression questions?

Strengthen

S1 List the factors that affect stopping distance. State whether each factor affects the thinking distance or the braking distance, and how they affect this distance.

Extend

E1 The crash in photo A happened in a sudden patch of fog. Write a paragraph for a road-safety website to explain why fog can be a hazard on the roads, and what drivers can do to avoid crashing in foggy conditions.

SP2h Braking distance and energy

Specification reference: P2.32P; P2.33P

Progression questions

- What is work done and how is it calculated?
- What is kinetic energy and how is it calculated?
- How are work done and kinetic energy related to braking distances?

A This accident happened because the runway was not long enough to allow the aeroplane to stop in the icy conditions at the time.

The force used to accelerate an object transfers energy to it. The amount of energy transferred depends on the size of the force and how far the object moves while the force is pushing it. The energy transferred by a force acting over a distance is called **work done**. We calculate the work done using this equation:

$$\underset{(\text{J})}{\text{work done}} = \underset{(\text{N})}{\text{force}} \times \underset{(\text{m})}{\text{distance moved in the direction of the force}}$$

Worked example W1

A 25 N force pushes an object for 10 metres, accelerating it from rest (0 m/s) to 10 m/s. Calculate the work done.

work done = force × distance moved in the direction of the force

$= 25\,\text{N} \times 10\,\text{m}$

$= 250\,\text{J}$

1 A 5 N force pushes a trolley for 5 metres. Calculate the work done.

The energy stored in a moving object is called **kinetic energy**. The amount of kinetic energy depends on the mass of an object and its velocity.

$$\underset{(\text{J})}{\text{kinetic energy}} = \frac{1}{2} \times \underset{(\text{kg})}{\text{mass}} \times \underset{(\text{m/s})^2}{(\text{speed})^2}$$

Worked example W2

The object in Worked example W1 has a mass of 5 kg. Calculate its kinetic energy when it is moving at 10 m/s.

kinetic energy $= \frac{1}{2} \times \text{mass} \times \text{velocity}^2$

$= \frac{1}{2} \times 5\,\text{kg} \times (10\,\text{m/s})^2$

$= 250\,\text{J}$

2 The object in question 1 is accelerated by the 5 N force. Explain what its final kinetic energy is.

3 A ball with mass 0.5 kg is moving at 20 m/s. Calculate its kinetic energy.

The answers to Worked examples W1 and W2 show that the kinetic energy of the moving object is the same as the energy transferred to accelerate it.

When a vehicle stops, the kinetic energy is transferred to other **energy** stores by the braking force. We can use this idea to calculate the braking distance of a vehicle.

B The energy transferred by the braking force increases the temperature of the brakes and surroundings. The brake discs on this car are glowing red.

Worked example W3

A 1500 kg car is travelling at 10 m/s. The driver applies a braking force of 10 000 N. How far does the car travel before it comes to a stop?

$$\text{kinetic energy} = \frac{1}{2} \times \text{mass} \times \text{velocity}^2$$

$$= \frac{1}{2} \times 1500\,\text{kg} \times (10\,\text{m/s})^2$$

$$= 75\,000\,\text{J}$$

Work done to stop the car is 75 000 J.

$$\text{distance} = \frac{\text{work done}}{\text{force}}$$

$$= \frac{75\,000\,\text{J}}{10\,000\,\text{N}}$$

$$= 7.5\,\text{m}$$

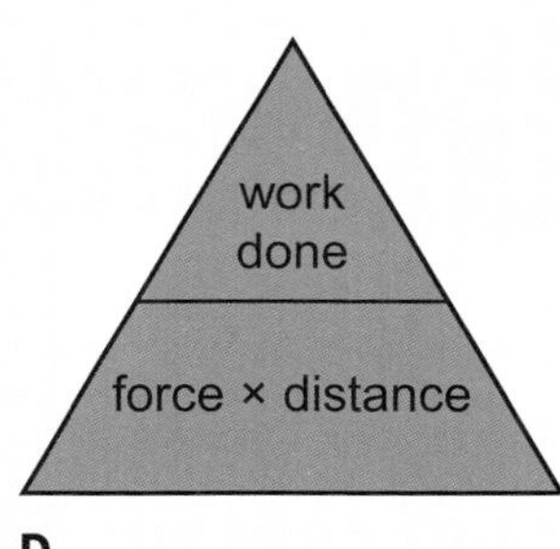

D

The braking distance of a car depends on its kinetic energy, and so it depends on the square of its velocity. This means that if the velocity doubles, the braking distance is multiplied by 2^2 which is 4.

4 Calculate the braking distance of the car in Worked example W3 if it has a velocity of:

a 20 m/s **b** 40 m/s.

5 Look at Worked example W3 and your answers to question 4. When the velocity of the car doubles, how does the braking distance change?

Exam-style question

A car is that is moving at 10 m/s travels 10 metres while braking to a stop. Explain what its braking distance would be if it were travelling at 20 m/s. Include any assumptions you have made in your answer. *(4 marks)*

Did you know?

Aircraft landing on aircraft carriers use a dangling hook to catch a wire across the deck. This brings the aircraft to a stop quickly enough to prevent them falling into the sea.

C

Checkpoint

How confidently can you answer the Progression questions?

Strengthen

S1 A car travels at 15 m/s for one part of a journey and at 30 m/s for another part of the journey. The driver brakes to a stop using a braking force of 9000 N each time. Explain how its braking distance will be different at the two speeds.

Extend

E1 The brakes on lorries exert a lot more force than car brakes. Lorries also often have lower speed limits on motorways. Use ideas about work and kinetic energy to explain these facts. Include a calculation to illustrate your answer.

SP2i Crash hazards

Specification reference: H P2.26; P2.31; H P2.31

Progression questions

- What are the dangers caused by large decelerations?
- How can the hazards of large decelerations be reduced?
- H How can you use momentum to calculate the forces involved in crashes?

A The amount of damage caused by a collision depends on the mass of the lorry and on how fast it was travelling.

B The large forces in road collisions injure or kill people and damage cars.

In a car crash, the vehicles involved come to a stop very quickly. Slowing down is a **deceleration** (or a negative acceleration). The force needed for any kind of acceleration depends on the size of the acceleration and on the mass of the object.

1 Explain why the force on a vehicle in a crash is larger:

a if the vehicle is moving faster before the crash

b for a lorry than for a car travelling at the same speed.

Modern cars have lots of safety features built into them to help to reduce the forces on the occupants in a collision. **Crumple zones** are built into the front (and sometimes the back) of cars. If the car hits something it takes a little time for this crumpling to happen, so the deceleration of the car is less and the force on the car is also less than if it had a more solid front.

C Forces on humans can result from hitting the dashboard or steering wheel, or if other passengers hit them.

Photo C also shows that the passengers do not stop moving when the car stops! Seat belts hold the passengers into the car, so the effect of the crumple zone reduces the forces on the passengers as well as on the car. Airbags increase the time it takes for a person's head to stop in a collision.

D Airbags were used to help the Mars Pathfinder to land safely by increasing the time for the probe to hit the ground, and so reducing the force on it..

 **2** Look at photo C. Explain why front *and* back seat passengers in a car should wear seat belts.

 **3** Bubble wrap is a plastic covering with many air bubbles. How do you think bubble wrap protects fragile items?

H

The force in a road collision depends on the change of momentum as the car comes to a stop. You have seen (in Topic *SP2f Momentum*) that we can use the equation below to calculate the force:

$F = \frac{mv - mu}{t}$, where u is initial velocity and v is final velocity.

Worked example

A 1500 kg car is travelling at 15 m/s (just over 30 mph) when it hits a wall. It comes to a stop in 0.07 seconds. What is the force acting on the car?

$$\text{force} = \frac{1500\,\text{kg} \times 0\,\text{m/s} - 1500\,\text{kg} \times 15\,\text{m/s}}{0.07\,\text{s}}$$

$$= \frac{-22\,500}{0.07}$$

$$= -321\,429\,\text{N}$$

The negative sign shows that the force is in the opposite direction to the original motion.

 **4 a** **H** An 1800 kg car travelling at 20 m/s stops in 0.03 seconds when it hits a wall. What is the force on it?

 **b** **H** Explain why this force is different to the one in the worked example. Give as many reasons as you can.

Exam-style question

Explain why a crumple zone only reduces the force on vehicle passengers if they are wearing seat belts. *(4 marks)*

Did you know?

Many people have survived falls from aeroplanes without parachutes. In 2004, skydiver Christine McKenzie's parachute failed – she escaped with only a cracked pelvis because she fell into some power lines which slowed her down before she hit the ground.

Checkpoint

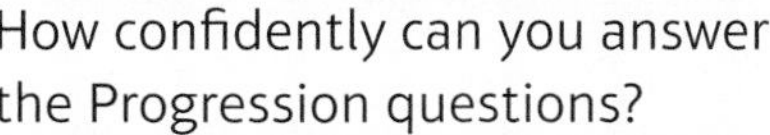

How confidently can you answer the Progression questions?

Strengthen

S1 Describe two ways of reducing the forces in a collision, and explain how they work.

Extend

E1 **H** Use calculations to show the effects of velocity, mass and crumple zones on the forces acting in a road collision. You will need to find or estimate values for speed, mass and the change in length of a crumple zone in a crash.

Braking distances

The braking distance of a vehicle is affected by its mass, its speed and the state of its brakes.

Explain how these factors affect the braking distance, using ideas about forces and acceleration. **(6 marks)**

Student answer

The braking distance is longer if the car is going faster or has a larger mass. If its brakes are working well, the braking distance is shorter [1]. *The deceleration of the car depends on the force applied, which depends on how good the brakes are, so if the brakes are good the deceleration is more and it can stop sooner* [2]. *The deceleration also depends on the weight* [3], *so if the brakes are the same* [4] *the deceleration will be less with a heavier car so it will go further while it is stopping.*

[1] This part of the answer describes how each factor affects the braking distance.

[2] This explains how the state of the brakes affects the braking distance.

[3] The student should refer to mass here, not weight.

[4] This would have been clearer if the student had referred to the *force* from the brakes being the same.

Verdict

This is an acceptable answer. It shows a good understanding of some of the factors that affect braking distances. The answer clearly states how each factor affects the braking distance, and provides good explanations for two of the factors. The use of scientific language is generally good, although with a mistake about weight and mass.

This answer could be improved by including an explanation of how the speed of a vehicle affects the braking distance, and making sure that correct scientific words are used.

Exam tip

In this example the command word is 'explain' so you need to say how *and why* each factor affects the braking distance.

Paper 1

SP3 Conservation of Energy

This is a solar power station at Sanlúcar la Mayor in Seville, Spain. Rings of mirrors focus energy from the Sun into a central furnace where water is heated to make steam. The steam is used to turn turbines, which drive generators to make electricity in a similar way to a normal power station.

In this unit you will learn about the ways in which energy can be transferred and stored, how to reduce energy transfers, and the renewable and non-renewable resources we use in everyday life.

The learning journey

Previously you will have learnt at KS3:

- that temperature differences lead to energy transfers
- how energy can be transferred by conduction, convection and radiation
- ways of reducing energy transferred by heating
- that energy is conserved
- ways in which energy can be stored and transferred.

In this unit you will learn:

- how energy is stored and transferred
- how to represent energy transfers using diagrams
- how to calculate efficiency
- how to reduce transfers of wasted energy
- how to calculate the amount of gravitational potential energy or kinetic energy stored in objects
- about the different renewable and non-renewable resources we use to make electricity, for heating and cooking, and for transport.

SP3a Energy stores and transfers

Specification reference: P3.3; P3.4; P3.5; P3.6; P3.8

Progression questions

- How is energy transferred between different stores?
- How can we represent energy transfers in diagrams?
- What happens to the total amount of energy when energy is transferred?

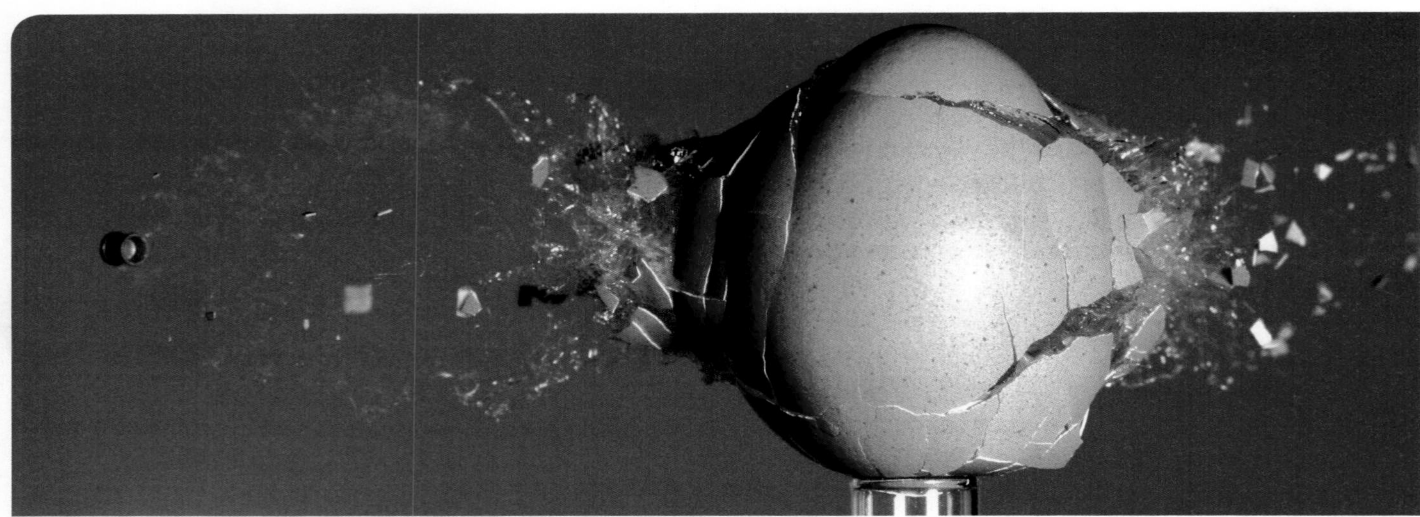

A What energy stores and transfers are involved as the bullet goes through the egg?

Energy is stored in different ways. Energy stored in food, fuel and batteries is often called **chemical energy**. Energy can also be stored in moving objects (**kinetic energy**), hot objects (**thermal energy**), in stretched, squashed or twisted materials (**strain energy** or **elastic potential energy**) and in objects in high positions (**gravitational potential energy**). Energy stored inside atoms is called **atomic energy** or **nuclear energy**.

Energy can be transferred between different stores. In photo A, some of the kinetic energy stored in the moving bullet is transferred to the egg by forces. Some of this energy is stored in the moving fragments of the egg, and some will heat up the egg.

When an electrical kettle is used to heat water, energy transferred to the kettle by electricity ends up as a store of thermal energy in the hot water. As the hot water is at a higher temperature than the kettle and the surroundings, some energy is transferred to these things by heating. Energy can also be transferred by light and sound.

Conservation of energy

In physics, a **system** describes something in which we are studying changes. An electrical kettle and its surroundings form a simple system. Energy cannot be created or destroyed. It can only be transferred from one store to another. This is called the **law of conservation of energy**. This means that the total energy transferred by a system is the same as the energy put into the system. The units for measuring energy are **joules** (J).

5th **1** Describe the changes in energy stores when a car accelerates.

6th **2** A ball thrown upwards has a store of kinetic energy as it leaves the person's hand. Describe the changes in energy stores as the ball rises and then falls again.

5th **3** List the ways in which energy can be transferred that are mentioned on this page.

Although energy is always conserved, it is not always transferred into forms that are useful. Think of the kettle – the energy stored in the hot water is useful, but the energy stored in the kettle itself and in the surroundings is not.

Energy diagrams

We represent energy stores and transfers using diagrams, such as diagram B.

energy stored in moving car (kinetic energy)	energy transferred by forces during braking →	energy stored in hot brakes (thermal energy)

B A flow diagram showing the energy transfers when a car brakes.

Did you know?

The energy transferred by the braking force can increase the temperature of brakes so much that they glow red.

C

A **Sankey diagram** shows the amount of energy transferred. The width of the arrows represents the amount of energy in joules.

D energy transfers in a light bulb

 **7** Sketch a Sankey diagram for a kettle using the energy values given in question 4.

Exam-style question

a Draw a diagram to represent the energy transfers in a television. *(3 marks)*

b Identify the useful energy transfers in the television. *(1 mark)*

 **4** 1000 J of energy is transferred to a kettle. 850 J ends up stored in the hot water in the kettle. Explain how much energy is transferred to the kettle itself and its surroundings.

 **5** Look at diagram B. The car's brakes do not stay hot. Describe the final energy store in this system.

 **6** **a** Draw an energy transfer diagram similar to diagram B for the bullet and egg in photo A.

 **b** Describe how your diagram would be different if it were a car hitting a wall.

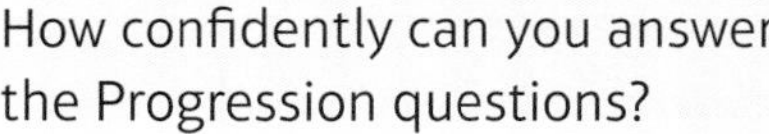

Checkpoint

How confidently can you answer the Progression questions?

Strengthen

S1 Describe the energy stores and transfers when you climb up to a high diving board and then jump into a swimming pool.

Extend

E1 A light bulb has 25 J of energy transferred to it every second, and 10 J of energy are transferred to the surroundings by light. Draw a Sankey diagram to show the energy transfers and explain how you worked out the amount of energy transferred by heating.

SP3b Energy efficiency

Specification reference: P3.7; P3.9; P3.11; H P3.12

Progression questions

- What does efficiency mean?
- How do we calculate the efficiency of an energy transfer?
- How can we reduce unwanted energy transfers in machines?

A Oiling the chain on a bicycle makes pedalling it much easier.

When a light bulb is switched on, most of the energy supplied to it by electricity is transferred to the surroundings by heating. This energy is **dissipated** (it spreads out) and cannot be used for other useful energy transfers – it is wasted.

Most machines waste energy when they get hot. Whenever two moving parts touch each other, friction causes them to heat up. The thermal energy stored in the hot machine is transferred to the surroundings by heating, which dissipates the energy. This energy is wasted energy.

Friction between moving parts can be reduced by **lubrication**. Oil or other liquids, and sometimes even gases, can be used as lubricants.

Did you know?

When two sticks are rubbed together the temperature rise can be used to start a fire. In this case the energy transferred by heating is useful!

B

1 **a** Explain why it is harder to pedal a bicycle if the chain needs oiling.

b When you pedal a bicycle, how is wasted energy transferred to the surroundings?

Efficiency

Efficiency is a way of describing how good a machine is at transferring energy into useful forms. The efficiency of a machine is given as a number between 0 and 1. The higher the number, the more efficient the machine. This is shown in diagram C.

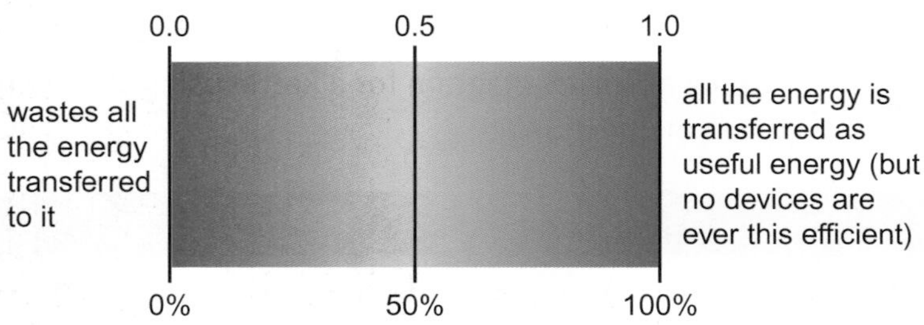

C The efficiency of a machine is sometimes given as a percentage.

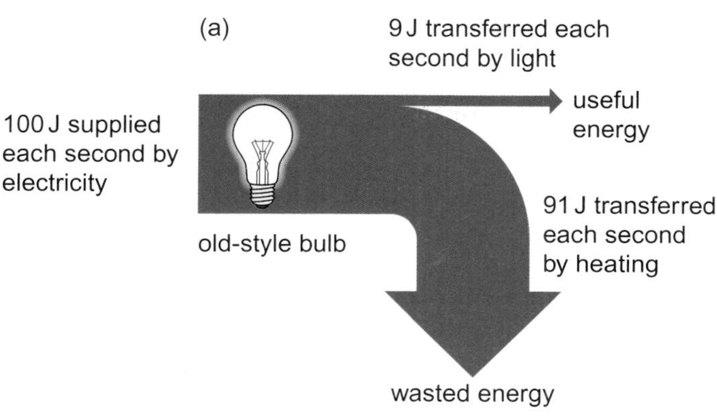

(b)
45 J transferred each second by light
useful energy
100 J supplied each second by electricity
low-energy bulb
55 J transferred each second by heating
wasted energy

D Modern low-energy bulbs are more efficient than old-style bulbs.

 **2** Look at diagram D. How can you tell from these diagrams that modern low-energy bulbs are more efficient?

The efficiency of a device can be calculated using this equation:

$$\text{efficiency} = \frac{\text{(useful energy transferred by the device)}}{\text{(total energy supplied to the device)}}$$

Worked example

Calculate the efficiency of the old-style bulb shown in diagram D.

$$\text{efficiency} = \frac{\text{(useful energy transferred by the device)}}{\text{(total energy supplied to the device)}}$$

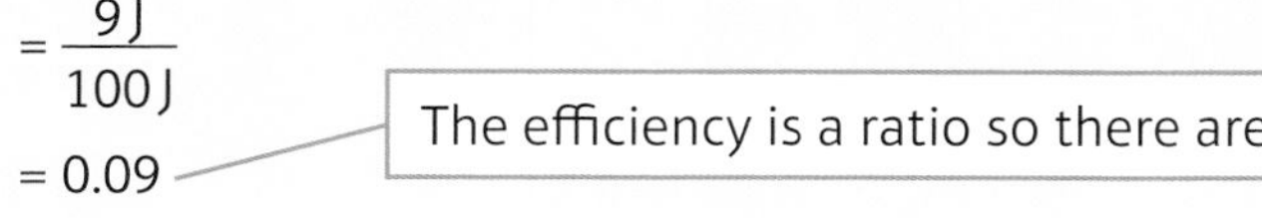

$$= \frac{9\,\text{J}}{100\,\text{J}}$$

$$= 0.09$$

The efficiency is a ratio so there are no units.

 **3** Calculate the efficiency of the low-energy light bulb shown in diagram D.

 **4** Look at diagram D in *SP3a Energy stores and transfers*. Calculate the efficiency of the light bulb.

H

Reducing the amount of wasted energy can increase the efficiency of a device or a process. For mechanical processes, such as engines, this can mean reducing friction. It can also mean finding ways to make sure all the fuel going into an engine is burned, or finding a way of using the energy transferred by heating that would otherwise be wasted.

 **5** Explain how the efficiency of a bicycle can be increased.

Exam-style question

Kettle A transfers 200 J of energy to boil some water. Kettle B transfers 250 J to boil the same volume of water. Explain which kettle is more efficient. *(3 marks)*

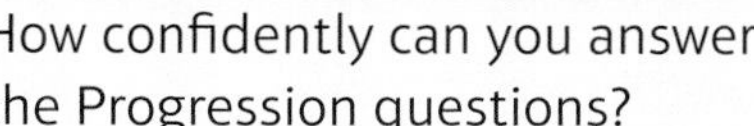

Checkpoint

How confidently can you answer the Progression questions?

Strengthen

S1 Explain why adding oil to door hinges makes the door easier and quieter to open.

S2 A radio is supplied with 50 J of energy and transfers 5 J of this by sound. Explain what happens to the rest of the energy and calculate the efficiency of the radio.

Extend

E1 A coal-fired power station has an efficiency of 0.4. Some of the energy wasted is stored in hot water that needs to be cooled down. The efficiency can be doubled if this hot water is used to heat nearby buildings. Explain how much useful energy is now transferred by electricity and how much is transferred by heating for each 1000 J of energy stored in the coal. Suggest what causes the remaining wasted energy and what happens to it.

SP3c Keeping warm

Specification reference: P3.9; P3.10

Progression questions

- What does thermal conductivity mean?
- What affects the rate at which buildings cool?
- How can insulation reduce unwanted energy transfers?

Did you know?

The best insulator is aerogel. This is a solid made from a silica or carbon framework with air trapped inside it. Aerogel is 99.8% air, and so it has an extremely low density. It has many uses, including insulation for space suits.

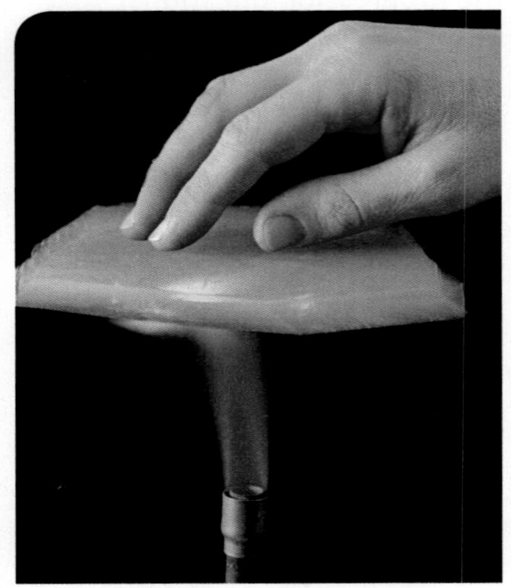

B The aerogel is protecting the hand from the heat of the flame. (Safety note: Do not try this.)

It costs money to keep our houses warm. **Insulation** slows down the rate at which energy is transferred out of a house by heating.

A This house is being built from straw bales. Straw is around 10 times better as an insulator than bricks.

Energy can be transferred by heating in different ways.

- In **conduction** vibrations are passed on between particles in a solid. Metals are good **thermal conductors** and materials such as wood are poor thermal conductors (good **thermal insulators**).
- In **convection** part of a **fluid** that is warmer than the rest rises and sets up a convection current.
- **Radiation** is the only way in which energy can be transferred through a vacuum. **Infrared radiation** can also pass through gases and some solid materials. Infrared radiation is **absorbed** and **emitted** easily by dull, dark surfaces, and is absorbed and emitted poorly by light, shiny surfaces.

The straw bales in photo A have a low **thermal conductivity**. This means that energy is not transferred through them very easily by heating. Materials that contain air are good insulators because air has a very low thermal conductivity. When air is trapped it cannot form convection currents and so does not transfer much energy.

1 A pan of water is heated on a cooker. Describe how:

a energy is transferred from the cooker to the water

b the energy spreads out through the water.

2 Explain why bubble wrapping is a good insulating material.

The rate at which energy is transferred through a material by heating depends on its thickness, on its thermal conductivity and also on the temperature difference across it. The rate of energy transfer is reduced by increasing thickness, decreasing thermal conductivity and decreasing temperature difference.

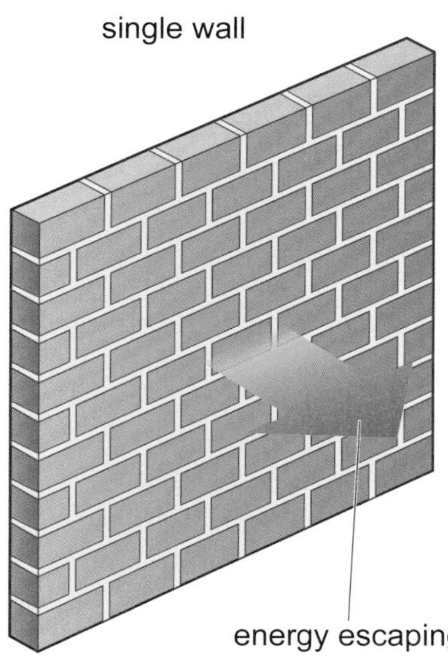

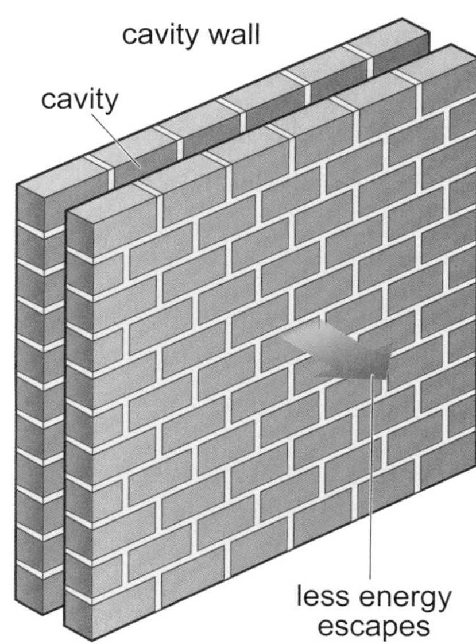

C Modern brick walls are built from two layers with a cavity (air gap) between them, which helps to insulate a house.

6 The container in diagram D keeps hot drinks hot but also keeps cold drinks cold. Explain how it can do this.

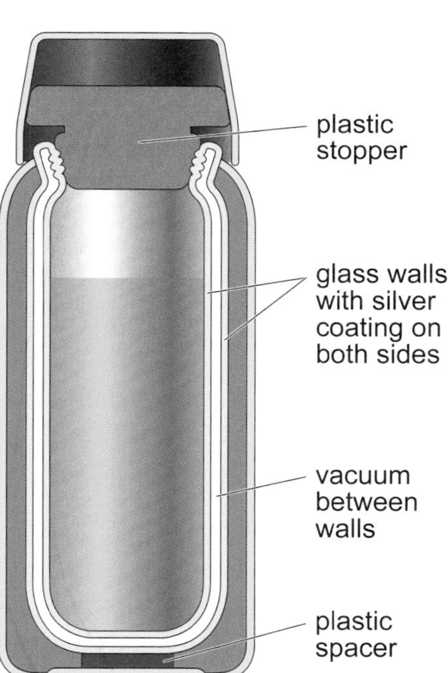

D A 'vacuum flask' is often used to store hot or cold liquids and uses a combination of different materials to reduce energy transfer by heating.

Exam-style question

Energy is needed to keep homes warm in the winter. Explain how the thermal conductivity of the walls in a house affects the energy needed to keep a house warm. *(2 marks)*

3 a Look at photo B. Do you think the aerogel has a higher or lower thermal conductivity than straw bales?

b Explain how you worked out your answer to part a.

4 Straw bale houses have very thick walls. Give two reasons why the walls in the house in photo A are better insulators than normal brick walls.

5 Look at diagram C.

a Give two reasons why the cavity wall keeps a house warmer than a single wall.

b Suggest why modern buildings have the cavity filled with foam or a similar material.

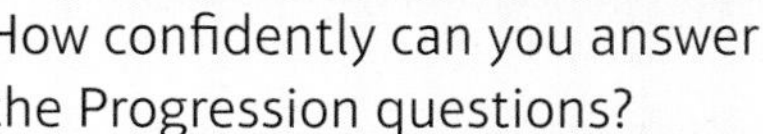

Checkpoint

How confidently can you answer the Progression questions?

Strengthen

S1 Explain two ways in which walls can be built to keep a house warmer.

S2 Explain two ways in which insulation is used at home to reduce energy transfers.

Extend

E1 Look at diagram D. Explain which features of the flask reduce energy transfer by radiation, by conduction and by convection.

SP3d Stored energies

Specification reference: P3.1; P3.2

Progression questions

- What factors affect the gravitational potential energy stored in an object?
- How do you calculate gravitational potential energy?
- How do you calculate the amount of kinetic energy stored in a moving object?

A There are three heavy 'weights' on steel cables inside the Big Ben clock tower. They are lifted up three times a week to store the energy needed to drive the clock and the bells.

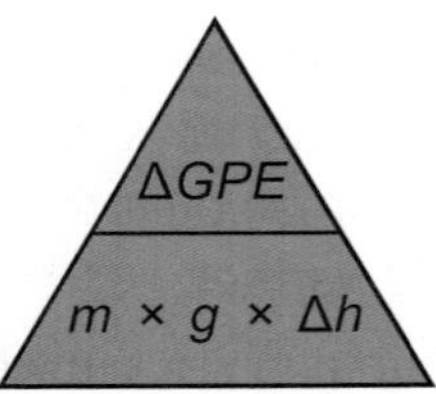

B This equation triangle can help you to change the subject of the equation. Δ is the Greek letter delta and stands for 'change in'.

Gravitational potential energy (GPE) is energy that is stored because of an object's position in a gravitational field. Any object that is above the surface of the Earth contains a store of gravitational potential energy. Every time something is moved upwards, it stores more gravitational potential energy.

The amount of GPE stored depends on the mass of the object, the strength of gravity and how far the object is moved upwards. It can be calculated using this equation:

$$\text{change in gravitational potential energy (J)} = \text{mass (kg)} \times \text{gravitational field strength (N/kg)} \times \text{change in vertical height (m)}$$

This can be written as: $\Delta GPE = m \times g \times \Delta h$

where ΔGPE represents the change in gravitational potential energy

m represents mass

g represents gravitational field strength

Δh represents change in vertical height.

The value for gravitational field strength on Earth is approximately 10 N/kg.

Worked example W1

A 5 kg box stores an extra 25 J of GPE when it is lifted onto a shelf. Calculate the distance it was lifted.

$$\Delta h = \frac{\Delta GPE}{m \times g}$$

$$= \frac{25\text{ J}}{5\text{ kg} \times 10\text{ N/kg}}$$

$$= 0.5\text{ m}$$

1 The gravitational field strength on the Moon is about 1.6 N/kg. Explain why the GPE stored by an object lifted 1 metre above the Moon's surface is less than when it is lifted by 1 m on the Earth.

2 One 'weight' in the Big Ben clock tower has a mass of 100 kg.

a Calculate the change in the GPE when it is raised by 5 metres.

b How far does this 'weight' have to be lifted to store 3000 J of energy?

Did you know?

Some birds crack open the shells of nuts or animals they are going to eat by dropping them onto stones. The birds store gravitational potential energy in the shells by lifting them into the air.

Kinetic energy

Energy is stored in moving objects. We call this kinetic energy. The amount of kinetic energy stored in a moving object depends on its mass and its speed.

Kinetic energy can be calculated using this equation:

$$\underset{(J)}{\text{kinetic energy}} = \frac{1}{2} \times \underset{(kg)}{\text{mass}} \times \underset{(m/s)^2}{(\text{speed})^2}$$

This can be written as:

$$KE = \frac{1}{2} \times m \times v^2$$

where KE represents kinetic energy

m represents mass.

v represents speed.

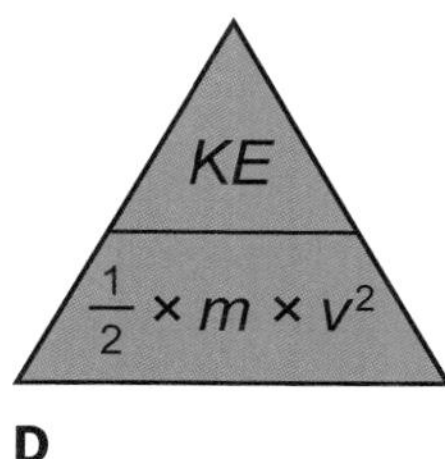

D

C The heavy disc at the bottom of this potter's wheel stores energy while it is spinning.

Worked example W2

A cricket ball with a mass of 160 g is bowled at a speed of 30 m/s. How much kinetic energy is stored in the moving ball?

160 g = 0.16 kg

$KE = \frac{1}{2} \times m \times v^2$

$= \frac{1}{2} \times 0.16\text{ kg} \times (30\text{ m/s})^2$

$= 72\text{ J}$

3 Calculate the kinetic energy stored in the following.

a A 2 kg toy robot dog walking at 2 m/s.

b A boy on a bike riding at 8 m/s. The mass of the boy and his bike is 70 kg.

4 A whale swimming at 7 m/s stores 98 000 J of kinetic energy. Calculate the mass of the whale.

Exam-style question

Explain why a car moving at 20 m/s is storing more kinetic energy than a cyclist moving at 2 m/s. *(2 marks)*

Checkpoint

How confidently can you answer the Progression questions?

Strengthen

S1 A missile is flying at 220 m/s at 100 m above the sea. Its mass is 1000 kg. Calculate its:

a gravitational potential energy

b kinetic energy.

Extend

E1 A crow drops a 15 g walnut from 5 m above the ground. Calculate the amount of GPE stored in the nut just before it fell.

E2 Assume all the GPE stored in the walnut was transferred to kinetic energy just before it reached the ground. How fast was it moving when it hit the ground?

SP3e Non-renewable resources

Specification reference: P3.13; P3.14

Progression questions

- What non-renewable energy resources can we use?
- How are the different non-renewable resources used?
- How is the use of non-renewable energy resources changing?

A The New Horizons space probe was launched in 2006 and flew past Pluto in 2015. Power for the spacecraft is provided by energy stored in a radioactive nuclear fuel.

Nuclear fuels such as **uranium** store a lot of energy in a small piece of material. This makes nuclear fuels very useful for spacecraft, where the mass of the fuel is important.

Most of the electricity used in the UK is generated using nuclear fuels or **fossil fuels** such as coal, oil and natural gas. These are all **non-renewable** energy resources, which means they will run out one day.

Petrol and diesel are fossil fuels made from oil. They are used in most vehicles, aeroplanes and ships because they store a lot of energy and they are easy to store and to use in engines. Another fossil fuel, natural gas, is burnt to heat homes or for cooking.

 **1** Describe two reasons why fuels made from oil are used in vehicles.

Burning fossil fuels release carbon dioxide and other gases. Carbon dioxide emissions contribute to **climate change**. Other emissions from power stations and vehicles cause further pollution problems. There are various ways of reducing this pollution, but these cost money.

Burning natural gas causes less pollution than burning coal. Natural gas power stations also emit less carbon dioxide than other fossil-fuelled power stations producing the same amount of electricity.

 **2** Describe two advantages of using natural gas instead of coal to generate electricity.

Nuclear power stations do not emit any carbon dioxide or other gases. However, the waste they produce is radioactive and some of it will stay radioactive for millions of years. This is expensive to dispose of safely. It is also very expensive to decommission (dismantle safely) a nuclear power station at the end of its life. It costs a lot more to build and to decommission a nuclear power station than a fossil-fuelled one.

There are not many accidents in nuclear power stations and the stations are designed to contain any radioactive leaks. However, if a major accident occurs it can have very serious consequences.

B Accidents with oil rigs or oil tankers can pollute large areas, harming wildlife.

C In 2011 a tsunami swept over a nuclear power station in Japan and damaged it. Radioactive materials were released into the atmosphere. Some of this polluted the sea nearby and then spread across the Pacific Ocean.

Most countries in the world are trying to cut down the use of fossil fuels. This will reduce pollution and also help to make supplies of the fuels last longer. **Renewable** resources are energy resources such as solar or wind energy that will not run out. Most renewable resources do not emit polluting gases.

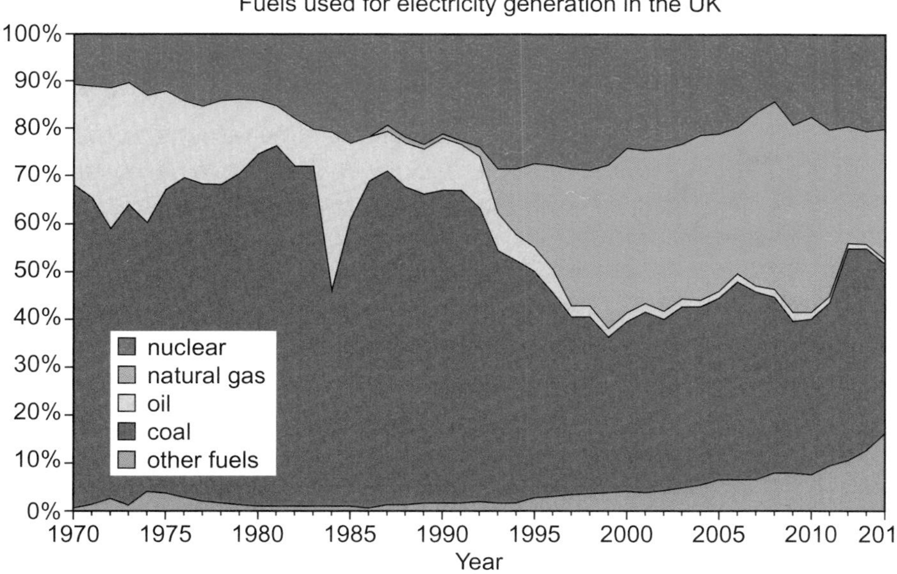

D 'Other fuels' on this graph include renewable fuels.

 **6** Look at graph D. Describe how the energy resources the UK uses for generating electricity have changed since 1970.

Exam-style question

State two reasons why many countries are trying to reduce the amount of non-renewable fuel they use. *(2 marks)*

 **3** **a** Write down one advantage of nuclear power over fossil-fuelled power stations.

 **b** Write down two disadvantages.

 **4** **a** Why do you think a nuclear power station costs more to build?

 **b** Why is decommissioning it properly very important?

 **5** How can a nuclear accident affect people in many different parts of the world?

Checkpoint

How confidently can you answer the Progression questions?

Strengthen

S1 Name four different non-renewable fuels and describe how they are used.

S2 Suggest why the use of renewable energy resources has been increasing in the UK in recent years.

Extend

E1 List three different ways in which non-renewable fuels are used. Describe the advantages and disadvantages of each fuel for each use.

SP3f Renewable resources

Specification reference: P3.13; P3.14

Progression questions

- What renewable energy resources can we use?
- How are the different renewable resources used?
- How is the use of renewable energy resources changing?

A In a solar chimney power station, solar energy heats the air under the glass. The hot air rises up the tower, turning turbines as it moves.

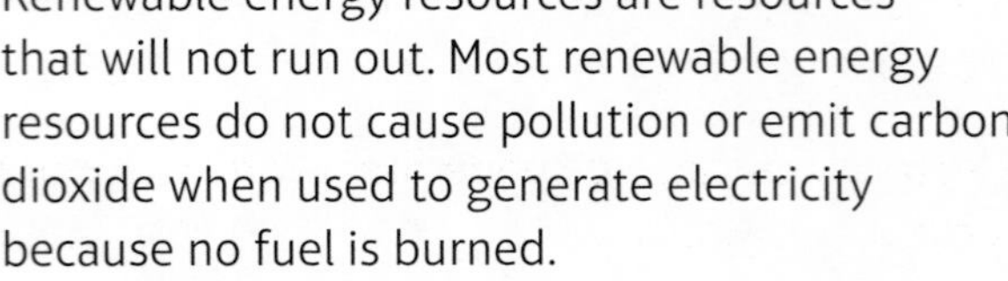

Renewable energy resources are resources that will not run out. Most renewable energy resources do not cause pollution or emit carbon dioxide when used to generate electricity because no fuel is burned.

1 What advantage do almost all renewable energy resources have over fossil fuels for generating electricity?

Solar cells convert **solar energy** directly into electrical energy, in 'solar farms' or on house roof-tops. Solar energy can also be used in power stations such as the one in diagram A or the one on the opening page of this unit. Solar energy can also be used to heat water for use in homes. Solar energy is not available all the time.

2 Describe two ways in which electricity can be produced using energy from the Sun.

Hydroelectricity is generated by falling water in places where water can be trapped in high reservoirs. It is available at any time (as long as the reservoir does not dry up). A hydroelectric power station can be started and stopped very quickly unlike fossil fuel power stations.

Wind turbines can be used to generate electricity as long as the wind speed is not too slow or too fast. A lot of wind turbines are needed to produce the same amount of energy as a fossil-fuelled power station and some people think they spoil the landscape.

B artist's impression of tidal stream turbines

Tidal power can generate electricity when turbines in a huge barrage (dam) across a river estuary turn as the tides flow in and out. Tidal power is not available all the time but is available at predictable times. There are not many places in the UK that are suitable for barrages and they may affect birds and other wildlife that live or feed on tidal mudflats. Underwater turbines can be placed in water currents in the sea to generate electricity, as seen in Figure B.

3 Describe two ways in which electricity can be produced using tides.

Bio-fuels can be used in the same ways as fossil fuels. They are made from animal wastes or from plants. Bio-fuels can be made from waste wood or the parts of plants that are not used for food, but some crops are grown specifically to be made into bio-fuels. Bio-fuels are called **carbon neutral** because when they burn, they release the same amount of carbon dioxide that they took from the atmosphere when the plants grew. However, energy is also needed to grow and harvest the crops and to turn them into fuel, so most bio-fuels are not really carbon-neutral.

Electricity can also be generated from waves or from hot rocks underground.

C Some people object to growing crops for fuel because this reduces the land available for farming food and can increase food prices.

We cannot only use renewable resources to generate electricity because most are not available all the time. It also takes a lot of land to obtain energy from bio-fuels or other renewable resources such as solar farms.

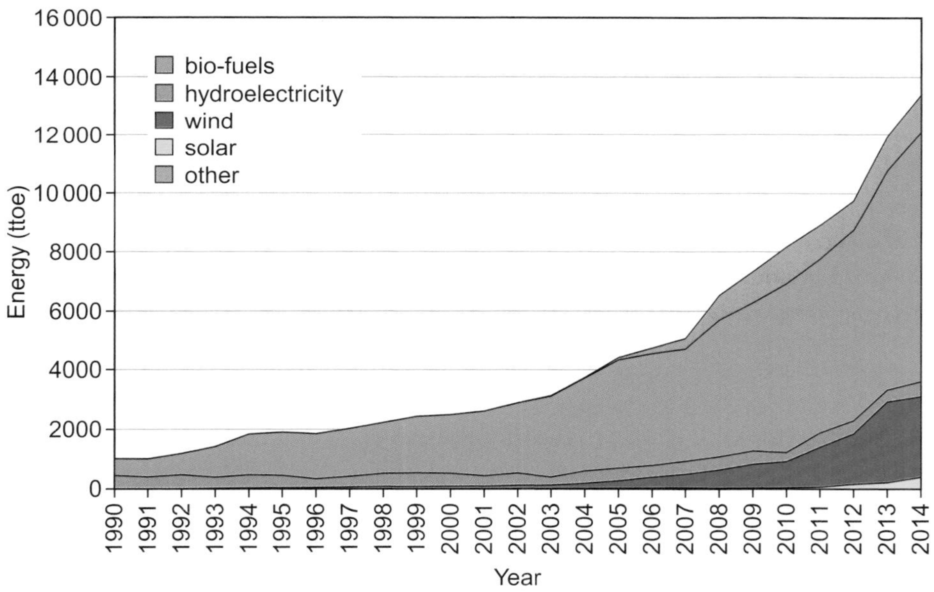

D On the graph, 'other' includes wave, tidal, geothermal and biofuel resources. The unit 'ttoe' (thousand tonnes of oil equivalent) is used to make fair comparisons between energy resources.

5 Look at graph D.

 **a** Describe how the use of renewable energy has changed since 1990.

 **b** Suggest two reasons for the changes you have described.

Exam-style question

State one advantage and one disadvantage of using renewable resources for generating electricity. *(2 marks)*

4 State which renewable resources:

 **a** are available all the time

 **b** are available at predictable times

 **c** can only be used in certain places

 **d** depend on the weather.

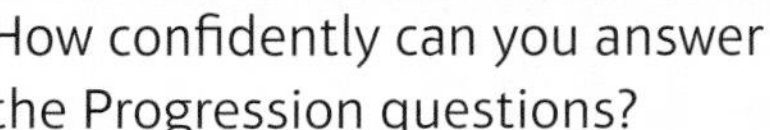

Checkpoint

How confidently can you answer the Progression questions?

Strengthen

S1 Describe five renewable energy resources and how they are used.

S2 State two advantages of a hydroelectric power station compared with a natural gas power station.

Extend

E1 Explain why we would need an efficient way of storing electricity before we could generate all our electricity from renewable resources. Give examples to support your answer.

Generating electricity

Hydroelectric power and solar power can be used as alternatives to fossil fuels.

Assess hydroelectric power and solar power as energy resources for the large-scale generation of electricity in the UK. **(6 marks)**

Student answer

Both these resources are renewable, and they do not produce gases that harm the environment. Using these resources would help to reduce the gases put into the air by burning fossil fuels [1]. *Hydroelectricity is available at any time, but solar energy is only available during the day and if the clouds are not too thick* [2]. *Not many places in the UK are suitable for building the reservoirs for hydroelectricity, and they can affect wildlife and habitats. Solar panels need to cover large areas of land to produce the same amount of electricity as a fossil-fuelled power station, and there is not enough sunshine in the UK to produce a lot of electricity* [3]. *So neither resource is really good enough to produce enough electricity to replace fossil fuels for generating electricity* [4].

[1] The student has given some advantages that apply to both resources compared to using fossil fuels.

[2] This sentence considers the availability of hydroelectricity and solar power.

[3] This section explains the limitations of using hydroelectricity and solar power in the UK.

[4] The last sentence provides a final conclusion, and links this to the use of fossil fuels as mentioned in the question.

Verdict

This is a strong answer. It clearly lists the factors that affect the use of the two renewable resources and provides a conclusion. The ideas are presented clearly and in a logical order. The answer links scientific ideas together, for example, that burning fossil fuels produces gases and that these gases can harm the environment. The use of scientific language is good.

Exam tip

If a question asks you to 'assess' something, you need to consider all the factors that apply, and consider which are the most important. Most 'assess' questions require you to draw a conclusion.

Paper 1

SP4 Waves

This photo was taken as an aeroplane flew in front of the Sun. The dark lines show shock waves in the air made by the aircraft. The waves in the air cause light waves from the Sun to be refracted so we see brighter and darker areas.

The learning journey

Previously you will have learnt at KS3:

- about light waves and sound waves, and how they can be described
- how sound waves are produced and how they are detected by our ears
- some uses of sound waves
- how light can be absorbed, scattered and reflected
- different colours of light.

In this unit you will learn:

- that waves transfer energy and information
- how to describe the characteristics of waves
- how the speed of a wave is related to its frequency and wavelength, and to the time it takes to travel a certain distance
- how waves are refracted at boundaries between different materials
- what happens when waves are reflected, refracted, transmitted or absorbed by different materials
- more about how our ears work
- about the uses of ultrasound and infrasound.

SP4a Describing waves

Specification reference: P4.1; P4.2; P4.3; P4.4; P4.5

Progression questions

- What do waves transfer?
- How can we describe waves?
- What is the difference between a longitudinal wave and a transverse wave?

A Energy from waves demolished part of the railway line and road at Dawlish in Devon in 2014 .

Sea **waves** transfer energy to the shore. When waves hit the land, the energy is transferred to the land and can wear it away.

Waves on the surface of water are **transverse** waves. Particles in the water move up and down as a wave passes – the particles are not carried to the shore.

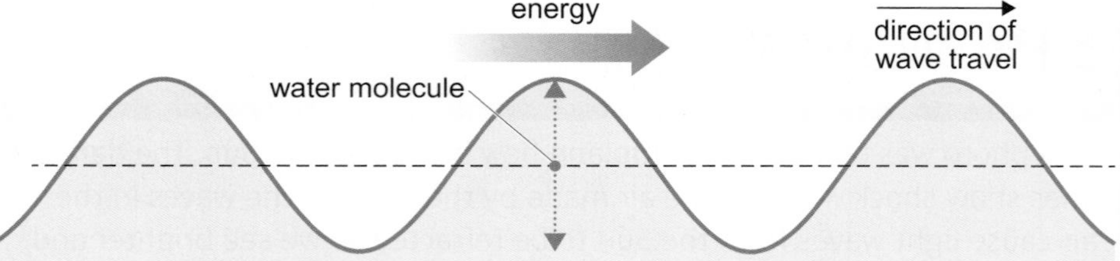

B In a transverse wave the particles move up and down at right angles to the direction the wave is moving.

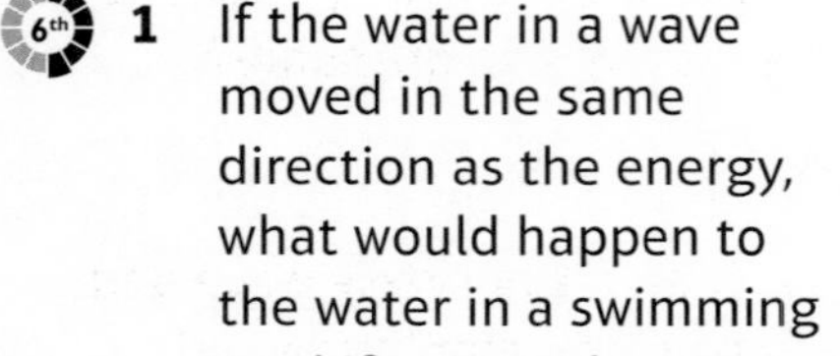

1 If the water in a wave moved in the same direction as the energy, what would happen to the water in a swimming pool if you made waves at one end?

2 Particles in a sound wave move in the same direction as the wave is travelling. Explain why loudspeakers do not move all the air in a room away from them.

Sound waves also transfer energy. Sound waves are **longitudinal** waves. Particles in the material through which the wave is travelling move backwards and forwards as the wave passes.

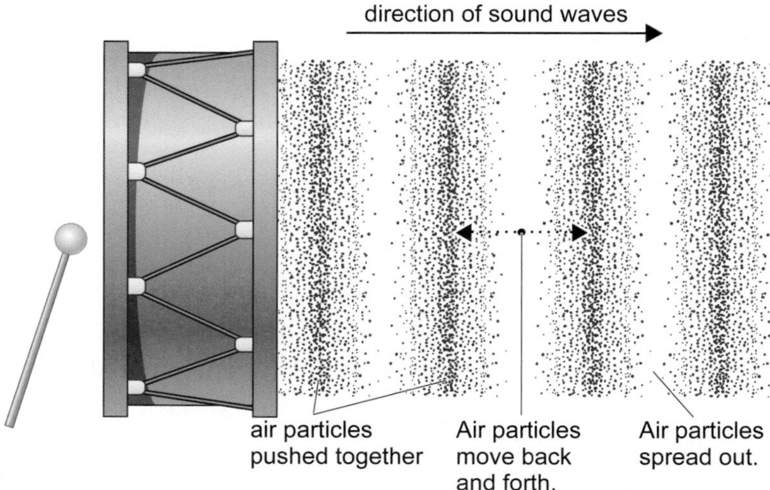

C Sound waves are longitudinal waves. The particles move back and forth in the same direction as the wave is travelling.

Earthquakes and explosions produce **seismic waves** that travel through the Earth. Solid rock material can be pushed and pulled (longitudinal seismic waves) or moved up and down, or side to side (transverse seismic waves).

Electromagnetic waves (such as light, radio waves, microwaves) are transverse waves and do not need a **medium** (material) through which to travel.

Describing waves

Wave **frequency** is the number of waves passing a point each second. It is measured in **hertz** (**Hz**). A frequency of 1 hertz means 1 wave passing per second. For sound, the wave frequency determines the pitch (how high or low it sounds) and for light the frequency determines the colour.

The **period** is the length of time it takes one wave to pass a given point.

The **wavelength** of a wave is the distance from a point on one wave to a point in the same position on the next wave, measured in metres.

The **amplitude** of a wave is the maximum distance of a point on the wave away from its rest position, measured in metres. The greater the amplitude of a sound wave, the louder the sound.

The **velocity** of a wave is the speed of the wave in the direction it is travelling. Waves travel at different speeds in different materials.

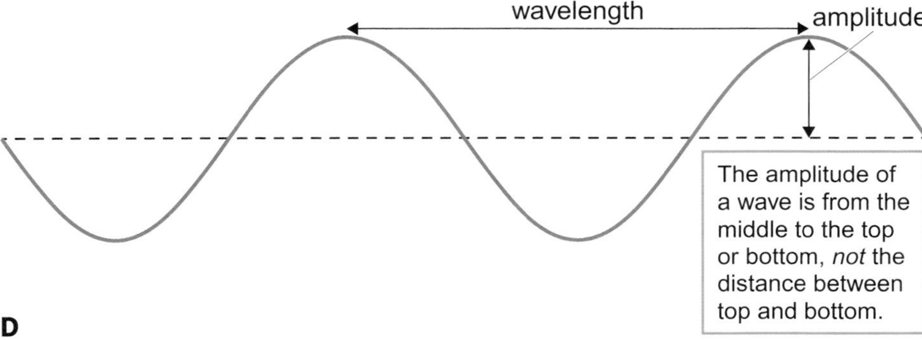
wavelength
amplitude
The amplitude of a wave is from the middle to the top or bottom, *not* the distance between top and bottom.

D

Changes in the frequency, wavelength or amplitude of a wave can be used to transfer information from one place to another. For example, when you listen to FM radio, the music is sent by variations in the frequency of the radio waves.

5 The tops of sea waves pass a stick twice every second.

 **a** What is the frequency?

 **b** What is the period?

 **6** Write down two things that waves transfer and give an example of each.

Exam-style question

Compare and contrast the way particles move in a sound wave and in a wave on the surface of water. *(4 marks)*

3 List two types of wave that are:

 **a** transverse waves

 **b** longitudinal waves.

Did you know?

The Sun has 'sunquakes'. Huge explosions of gas, called solar flares, cause waves to spread through the Sun in a similar way to Earth movements causing earthquakes.

 **4** Suggest how we see light change when the amplitude of light waves varies.

Checkpoint

How confidently can you answer the Progression questions?

Strengthen

S1 Draw a transverse wave and label the amplitude and wavelength.

S2 Describe the similarities and differences between longitudinal and transverse waves.

Extend

E1 Write glossary entries for the different terms used to describe waves, including examples of different types of wave.

E2 Explain the differences between waves on the surface of water and sound waves, in terms of what they transfer and the characteristics of the waves.

SP4b Wave speeds

Specification reference: P4.6; P4.7

Progression questions

- How can we calculate the speed (or velocity) of a wave?
- How can we measure the speed of sound in air?
- How can we measure the speed of waves on water?

A The sound waves from this volcano plume took time to reach the photographer. This time can be used to calculate how far away the lightning was.

The speed of a wave can be calculated from the distance it travels in a certain time. This is the same equation we use for calculating the speed of moving objects.

$$\text{speed (m/s)} = \frac{\text{distance (m)}}{\text{time (s)}}$$

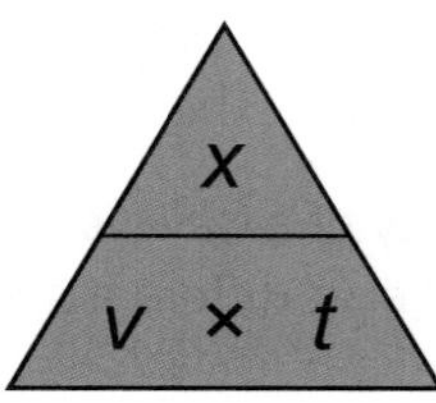

B You can rearrange the equation for speed using this triangle. *v* stands for speed and *x* stands for distance.

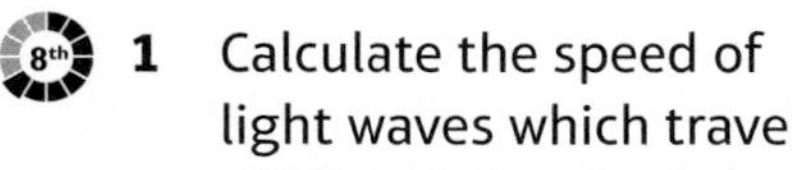

1 Calculate the speed of light waves which travel 900 000 000 m in 3 s.

2 You hear thunder 5 s after you see lightning.

a Sound travels at 330 m/s in air. How far away was the lightning strike?

b Explain what assumption you made in your answer.

Worked example W1

A surfer travels 52 m on the front of a wave in 8 s. Calculate the wave speed.

$$\text{wave speed} = \frac{\text{distance}}{\text{time}}$$

$$\text{wave speed} = \frac{52\text{ m}}{8\text{ s}}$$

$$= 6.5\text{ m/s}$$

Did you know?

Waves on the surface of water get slower as the water gets shallower. This is what causes waves to break as they reach the shore.

C

The wave speed is linked to the wave frequency and wavelength by this equation.

wave speed (m/s) = frequency (Hz) × wavelength (m)

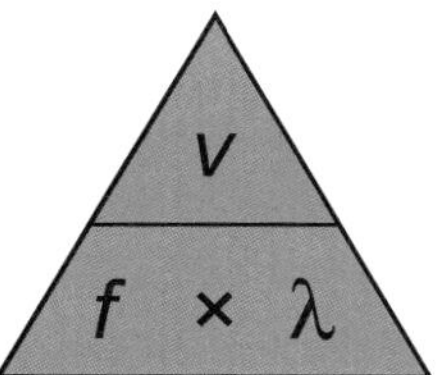

D You can rearrange the equation for wave speed using this triangle. v stands for speed and f stands for frequency. λ is the Greek letter lambda and represents wavelength.

Worked example W2

Some waves have a wavelength of 13 m and a frequency of 0.5 Hz. Calculate their speed.

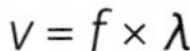

$v = f \times \lambda$

$= 0.5\text{ Hz} \times 13\text{ m}$

$= 6.5\text{ m/s}$

3 Calculate the speed of sound waves that have a wavelength of 2 m and a frequency of 170 Hz.

4 Calculate the wavelength of seismic waves that travel at 5000 m/s and have a frequency of 100 Hz.

The speed of a wave depends on the medium through which it is travelling. Light always travels at 300 000 000 m/s in a vacuum but it travels more slowly in glass or water. When light goes from air into water its wavelength also reduces.

5 When light travels from air into water, its frequency does not change. Explain why its wavelength decreases.

Measuring the speed of waves

You can find the speed of sound by measuring the time it takes for a sound to travel a certain distance. For example, if you stand in front of a large wall you can measure the time it takes for an echo of a loud sound to reach you. The speed can be calculated using the speed, time, distance equation.

One way of measuring the speed of waves on water is to measure the time it takes for a wave to travel between two fixed points such as buoys. The speed can be calculated from the time and the distance between the points.

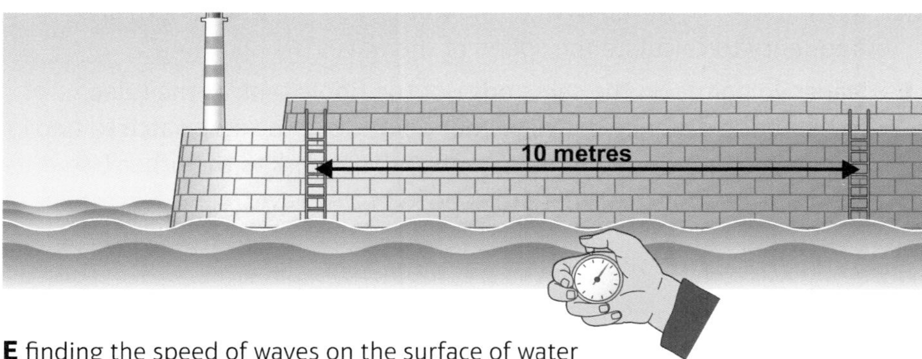

E finding the speed of waves on the surface of water

6 Look at diagram E. It takes 7 s for a wave to move from one ladder to the other. Calculate the speed of the wave.

Exam-style question

Humans can hear sounds with a wavelength of 16 m. The speed of sound in air is 330 m/s. Calculate the frequency of these sounds. *(3 marks)*

Checkpoint

How confidently can you answer the Progression questions?

Strengthen

S1 An underwater sound wave travels 2000 m in 1.3 s. Calculate its speed.

S2 The frequency of the sound wave in **S1** is 3000 Hz. Calculate the wavelength of the sound wave.

Extend

E1 You are asked to find the speed of ripples on water using the equation linking speed, frequency and wavelength. Describe how to take the measurements you need and how you would work out the speed.

SP4b Core practical – Investigating waves

Specification reference: P4.17

Aim

Investigate the suitability of equipment to measure the speed, frequency and wavelength of a wave in a solid and a fluid.

A This photo shows a detailed image of the USS *Monitor* made using frequency sonar. The USS *Monitor* was an iron-hulled steamship that sank in 1862.

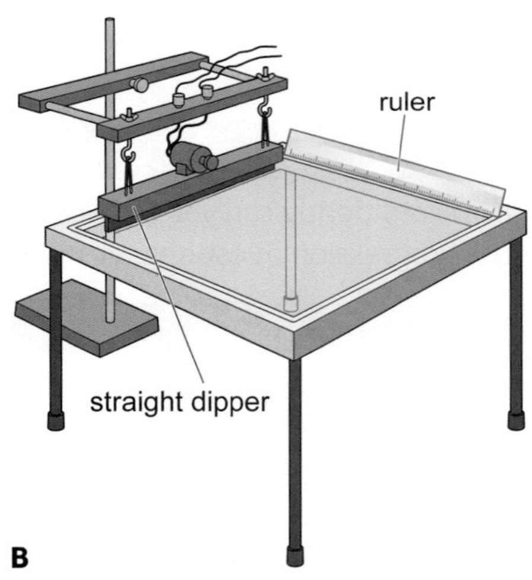

B

Light waves do not travel very far through sea water before being absorbed by the water or reflected by tiny particles in the water. This makes it impossible to take pictures of things that are deep down on the sea bed. Scientists and explorers use sonar equipment to send sound waves into the water and detect the echoes. The depth can be worked out from the speed of sound in the water and the time it takes for the echo to return.

Your task

You are going to use different pieces of equipment to measure the speed and wavelength of waves on the surface of water, and the speed and frequency of sound waves in solids.

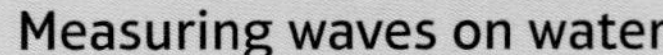

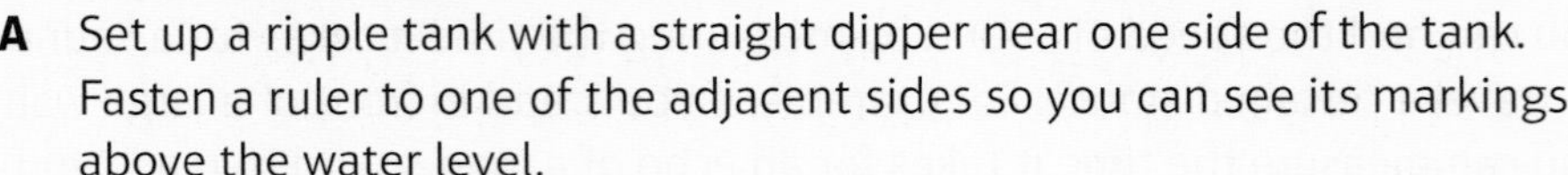

Method

Measuring waves on water

A Set up a ripple tank with a straight dipper near one side of the tank. Fasten a ruler to one of the adjacent sides so you can see its markings above the water level.

B Vary the voltage to the motor until you get waves with a wavelength about half as long as the ripple tank (so you can always see two waves).

C Count how many waves are formed in 10 seconds and write it down.

D Look at the waves against the ruler. Use the markings on the ruler to estimate the wavelength of the waves. Use the wavelength and frequency to calculate the speed of the waves.

E Mark two points on the same edge of the ripple tank as the ruler. Measure the distance between your points. Use the stopwatch to find out how long it takes a wave to go from one mark to the other. Use this information to calculate the speed of the waves.

Measuring waves in solids

F Suspend a metal rod horizontally using clamp stands and rubber bands.

G Hit one end of the rod with a hammer. Hold a smartphone with a frequency app near the rod and note down the peak frequency.

H Measure the length of the rod and write it down. The wavelength will be twice the length of the rod.

I Use the frequency and wavelength to calculate the speed of sound in the rod.

Exam-style questions

1 A sound wave in air travels 660 metres in two seconds. Calculate the speed of the sound wave. *(2 marks)*

2 A sound wave travelling in water has a frequency of 100 Hz. The speed of sound in water is 1482 m/s. Calculate the wavelength of the wave. *(2 marks)*

3 Describe how to find the frequency of the waves in the ripple tank using the method in step C. *(2 marks)*

4 Luke estimated the wavelength of the waves in the ripple tank using the method described in step D. Emily took a photo of the waves in the ripple tank and estimated the wavelength using the photo.

Explain which method was likely to give the more accurate result. *(2 marks)*

5 Adanna is watching waves on the sea go past two buoys. She knows the buoys are 20 metres apart. Describe how she can find the speed of the waves. *(2 marks)*

6 Liwei measured the frequency and wavelength of waves in a ripple tank and calculated their speed as 0.4 m/s. Using the method in step E, she calculated the speed as 0.2 m/s.

Explain which result is likely to be more accurate. *(2 marks)*

7 The speed of sound in air can be measured by finding the time it takes for a sound to echo from a nearby wall, and measuring the distance to the wall.

Hitting the end of a metal rod with a hammer causes sound waves to travel along the rod. They reflect from the far end of the rod and continue to move up and down the rod until the energy dissipates.

Give a reason why the method used for finding the speed of sound in air cannot be used for finding the speed of sound in a metal. *(2 marks)*

8 Gina used the method described in step G to measure the frequency of sound in an aluminium rod 0.8 metres in length. She recorded a peak frequency of 4000 Hz. Sound inside a metal rod has a wavelength that is twice the length of the rod.

Use Gina's results to calculate the speed of sound in aluminium. *(3 marks)*

SP4c Refraction

Specification reference: P4.10; H P4.10

Progression questions

- What happens when waves refract?
- When does refraction occur?
- H How does a change in the speed of a wave affect its direction?

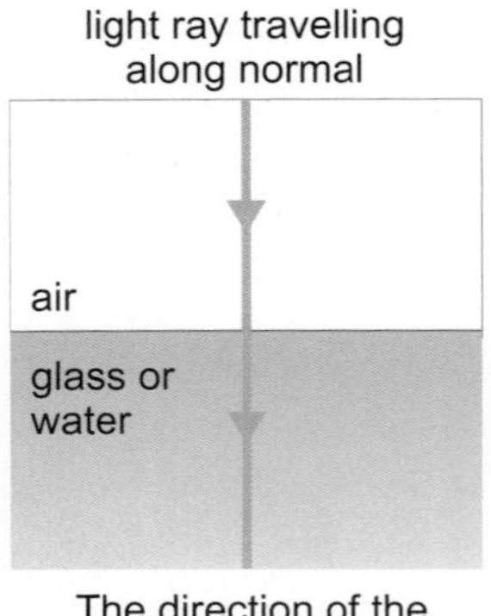

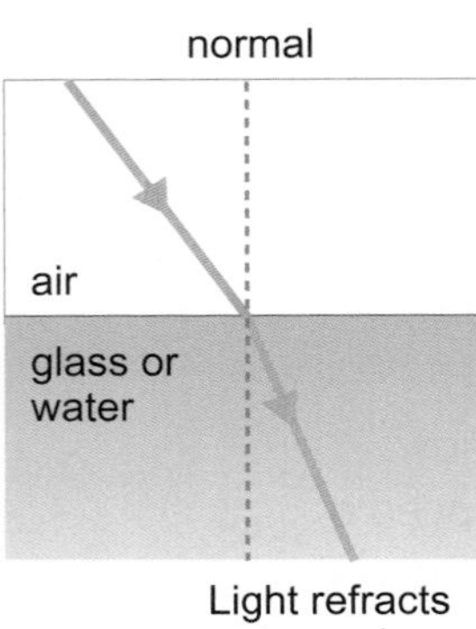

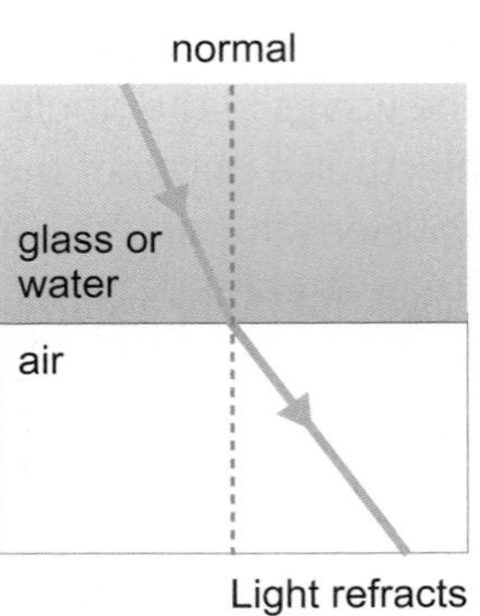

A Light is refracted when it goes from one material to another.

Most waves travel outwards from their source in straight lines. However, waves can change direction when they move into a different medium. The change in direction is called **refraction** and happens at the **interface** (boundary) between the two media. A line at right angles to the interface is called the **normal** line. Light travelling along the normal does not change direction when it goes into a different medium.

We see things when light reflected from them reaches our eyes. An object on the bottom of a swimming pool looks closer than it really is because light reflected by it changes direction when it leaves the water.

1 Describe how the direction of a light wave changes when it moves:

a from glass into air

b from air into glass.

2 a Explain why archerfish (photo B) have to learn to compensate for refraction when aiming at insects.

b Draw a diagram to show how light reflected by the insect reaches the fish's eyes.

B Archerfish knock insects into the water by spitting at them. The fish have to learn to compensate for refraction when aiming at insects.

H

Waves can travel through many different media but with different speeds. For example, light travels faster in air than it does in glass or water. As light passes the interface between one medium and another it changes speed. This change in speed causes the direction of the light to change.

The bend depends on how fast the light travels in the two media and the angle of the light hitting the interface. The greater the difference in speed between the two media, the more the light is bent. The light bends towards the normal when it slows down.

We can use waves on water as a model to help us to understand what happens with light waves. The speed of waves on water depends on how deep the water is. Waves moving from deep water into shallow water slow down and change direction (diagram D). Lines on ray diagrams (diagram A) show the direction in which the waves are moving, not the waves themselves.

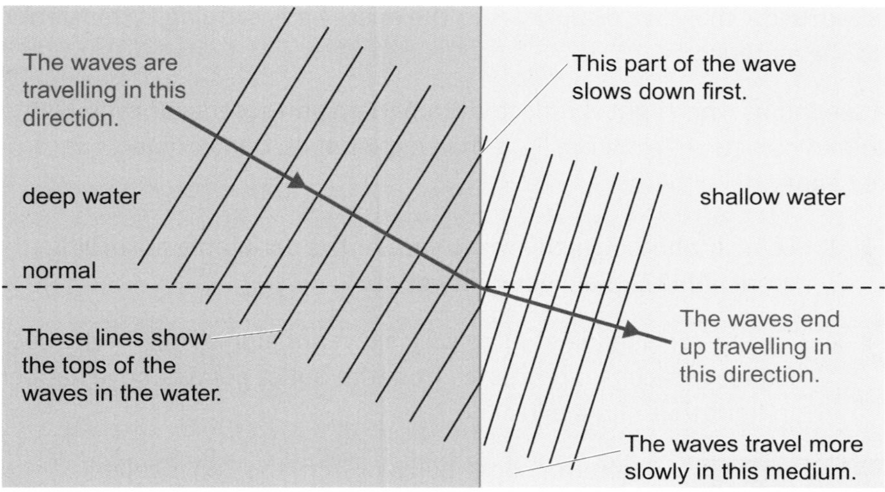

The waves are travelling in this direction.
This part of the wave slows down first.
deep water
shallow water
normal
These lines show the tops of the waves in the water.
The waves end up travelling in this direction.
The waves travel more slowly in this medium.

D Water waves change direction when the depth changes.

3 Look at diagram D. Explain what happens to the waves when they move into shallow water.

4 Explain why the waves do not change direction when they are travelling at right angles to the interface.

5 Explain what happens to waves on the surface of water when they cross an interface from shallow water into deeper water.

6 Explain how diagram A shows that light travels more slowly in glass and water than it does in air.

Exam-style question

Lane markings on the bottom of a swimming pool are straight lines. Explain why they do not usually look straight when you look at them from above the water. *(2 marks)*

Did you know?

Refraction can also happen when the properties of a material change gradually. A mirage occurs when air near the ground is hotter than air higher up. In this photo, refraction is distorting the path of the light from the sky, making it appear to come from the ground and giving the impression of a puddle of water in the road.

C

Checkpoint

How confidently can you answer the Progression questions?

Strengthen

S1 Describe how the direction of a light ray changes as it goes from air into water, and when it goes from water into air. Use the word 'normal' in your answer.

Extend

E1 **H** A ray of light shines through a thick piece of glass. Explain why the light ray emerges from the glass travelling in the same direction as originally, but not along the same line. You may use a diagram to help you to explain.

SP4d Waves crossing boundaries

Specification reference: P4.9P; P4.16P

Progression questions

- What happens when waves are reflected or refracted?
- What happens when waves are transmitted or absorbed?
- How are changes in velocity, frequency and wavelength related?

A The materials in this room reflect and absorb different amounts of light.

When a wave reaches an interface (boundary) between different materials it can be:

- reflected – the wave 'bounces' off
- refracted – the wave passes into the new material but changes the direction in which it is travelling
- **transmitted** – the wave passes through the material and is not absorbed or reflected
- **absorbed** – the wave disappears as the energy it is carrying is transferred to the material.

We see things when light is reflected from them and enters our eyes. Lighter coloured objects reflect more light than darker ones. Darker objects absorb more light.

1 Look at photo A. Is the white vase or the dark table absorbing more light? Explain your answer.

2 The mirror is made of glass with a layer of aluminium on the back. Explain which two processes from the bullet points above happen to light that hits a mirror.

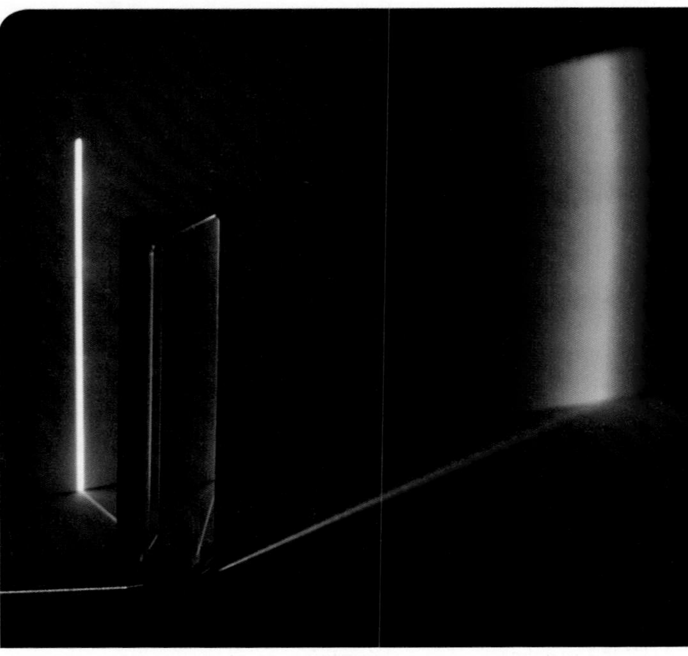

B A prism can separate the colours in white light.

Light from light bulbs or from the Sun is called white light, and is made up of a mixture of different frequencies of light. We see these different frequencies as different colours. The different colours in light change speed by different amounts when they travel from air to glass (or from glass to air). This means they are refracted through varying angles, which is why a prism can be used to split up visible light into the colours of the rainbow.

3 Look at photo B.

a How can you tell that the prism is transmitting and refracting light?

b How does the photo show that different frequencies are refracted by different amounts?

Sound waves can be affected in the same way as light waves. We hear echoes when sound is reflected by a hard surface. Some materials absorb sound well and some transmit it well. Sound is also refracted when it goes into different materials but this is much harder to observe.

C An anechoic chamber is designed to remove all echoes. Anechoic chambers are used to test loudspeakers and other audio equipment.

4 Look at photo C above. Suggest how well the material of the walls reflects and absorbs sound waves. Explain your answer.

Sound waves travel at different speeds in different materials. Wave velocity is equal to the frequency multiplied by the wavelength, so if the velocity changes, either the frequency or wavelength (or both) must also change. When we hear sounds at different pitches, our ears are detecting the different frequencies of sound (not the wavelengths). A sound has the same pitch if it reaches us through air or through a solid, so when a sound wave enters a different material it is the wavelength and velocity that change, not the frequency.

5 Sound travels at approximately 330 m/s in air. A sound wave has a frequency of 10 kHz.

a Calculate its wavelength.

b What will the wavelength be if the sound wave passes into water where the speed is 1500 m/s?

Exam-style question

Compare and contrast the reflection and refraction of waves at a boundary. *(2 marks)*

Did you know?

Before radar was invented, acoustic mirrors like this were used to help observers listen for enemy aircraft approaching. They helped to focus the sound waves.

D an acoustic mirror

Checkpoint

How confidently can you answer the Progression questions?

Strengthen

S1 Compare and contrast a pair of prescription glasses (to help you to see more clearly) with a pair of sunglasses, using the terms in the bullet points near the top of the previous page.

Extend

E1 Sound travels at approximately 4170 m/s in brick. Describe what happens when a sound wave reaches a brick wall and some of it forms an echo. Use the three Progression questions to structure your answer.

SP4e Ears and hearing

Specification reference: H P4.12P

Progression questions

- H How do our ears work?
- H How are sound waves converted to waves in solids?
- H How does the frequency affect the energy transferred to a solid?

H

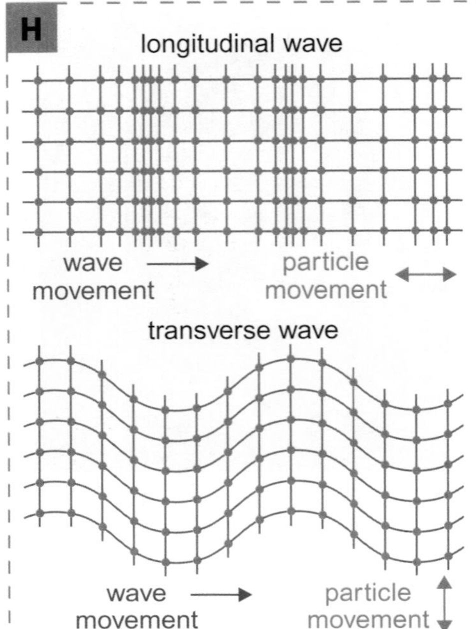

longitudinal wave
wave movement
particle movement
transverse wave
wave movement
particle movement

A longitudinal and transverse waves in a solid

Reflecting and transmitting sound

Sound waves are longitudinal waves. Particles in a gas or liquid vibrate backwards and forwards as a sound wave passes. When the sound wave reaches a solid, some of the energy it is transferring is reflected and some is transmitted through the solid or absorbed by it.

A sound wave causes changes in pressure on the surface of a solid. This causes particles in the solid to vibrate and so the disturbance is passed from the air to the solid. The vibrations in the solid can be passed on both as longitudinal waves and transverse waves.

The shape and properties of a solid (such as density and stiffness) determine how vibrations of different frequencies will affect it.

Human ears

We detect sound waves using our ears. The part of the ear on the outside of our heads helps to channel sound waves into our heads. The vibrations caused by the sound waves are passed on through various parts of the ear until they are detected and converted to electrical impulses that travel to the brain.

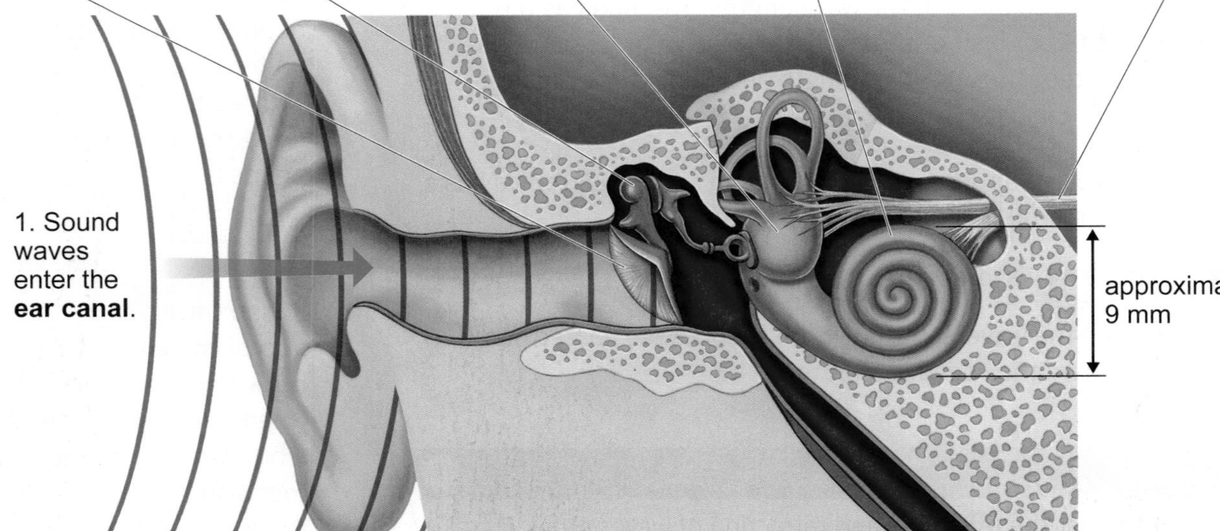

2. The **eardrum** is a thin membrane. Sound waves make it vibrate.
3. Vibrations are passed on to tiny bones which **amplify** the vibrations (make them bigger).
4. Vibrations are passed on to the liquid inside the **cochlea**.
5. Tiny hairs inside the cochlea detect these vibrations and create electrical signals called **impulses**.
6. Impulses travel along neurones in the **auditory nerve** to reach the brain.
1. Sound waves enter the **ear canal**.
approximately 9 mm

B the internal structure of a human ear

H

 **1** List the parts of the ear in the order in which vibrations affect them.

 **2** Which part of the ear converts vibrations into nerve impulses?

3 In which parts of the ear are the vibrations occurring in a:

 **a** solid **b** liquid **c** gas?

How the cochlea works

The cochlea is a coiled tube containing a liquid. It can detect the different frequencies of sound reaching the ear. Human ears can detect sounds from 20 Hz to 20 000 Hz.

Diagram D shows what the coiled tube of the cochlea would look like if it were unwound. The membrane in the middle of the tube is thicker and stiffer at the base and thinner at the apex. The part of the membrane that vibrates depends on the frequency of the sound waves in the liquid inside the cochlea, as different thicknesses of the membrane vibrate best at different frequencies. There are thousands of hair cells along the membrane, which detect its vibrations. Each hair cell is connected to a **neurone** that sends impulses to the brain. The brain interprets signals from different neurones as different pitches of sound.

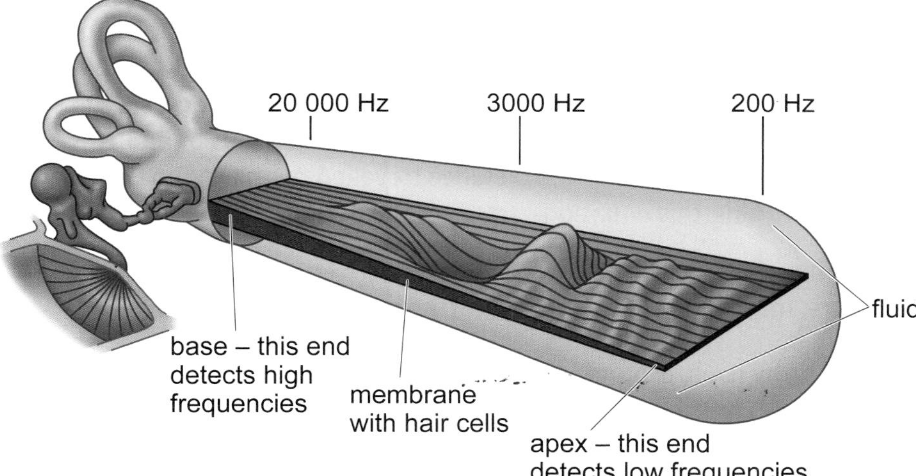

D the membrane in an unrolled human cochlea

5 Look at diagram D.

 **a** Does the thin or thick part of the membrane in the cochlea detect high frequencies?

 **b** Why is it important that different parts of the membrane in the cochlea have different stiffnesses?

Exam-style question

Describe how sound waves in a fluid are converted to vibrations in a solid. *(2 marks)*

Did you know?

Our bones can conduct vibrations to our cochleas. Bone conduction headphones like these allow joggers and cyclists to listen to music without blocking sounds from things around them.

C

4 Describe the functions of the following parts of the cochlea.

 **a** fluid

 **b** membrane

 **c** hair cells

Checkpoint

How confidently can you answer the Progression questions?

Strengthen

S1 Draw a flow chart to describe how we hear sounds. Include how the vibrations are passed on inside the ear, and how the sounds are detected and converted to nerve impulses.

Extend

E1 Like all birds, pigeons have straight cochleas. They have a hearing range of 200–7500 Hz. Suggest ways in which the structure of the pigeon cochlea is different to that of humans.

SP4f Ultrasound

Specification reference: H P4.8P; H P4.13P; H P4.15P

Progression questions

- H What is ultrasound?
- H How is ultrasound used in sonar systems?
- H How is ultrasound used to look inside our bodies?

H

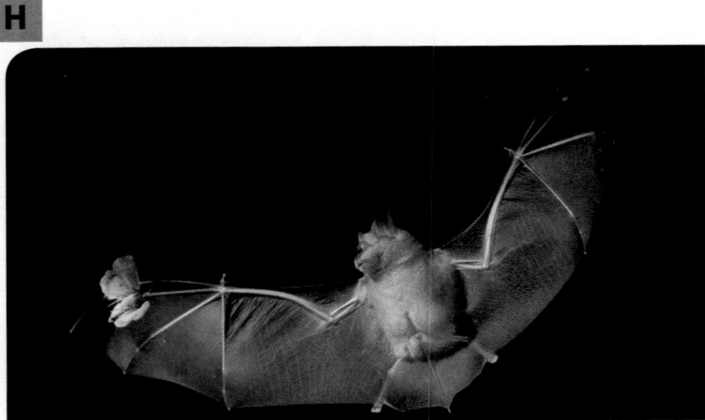

A Bats use echolocation to catch insects.

Humans can detect sound waves up to 20 000 Hz (or 20 kHz). Sounds made by waves with higher frequencies than this are called **ultrasound**. Some animals, such as mice, use ultrasound to communicate with each other. Other animals, such as dolphins and bats, can detect objects around them using ultrasound waves. The ultrasound waves they make are reflected by things around them and the animals listen for the echoes.

1 What is ultrasound?

2 Why do biologists need to use special equipment to detect the noises that bats make?

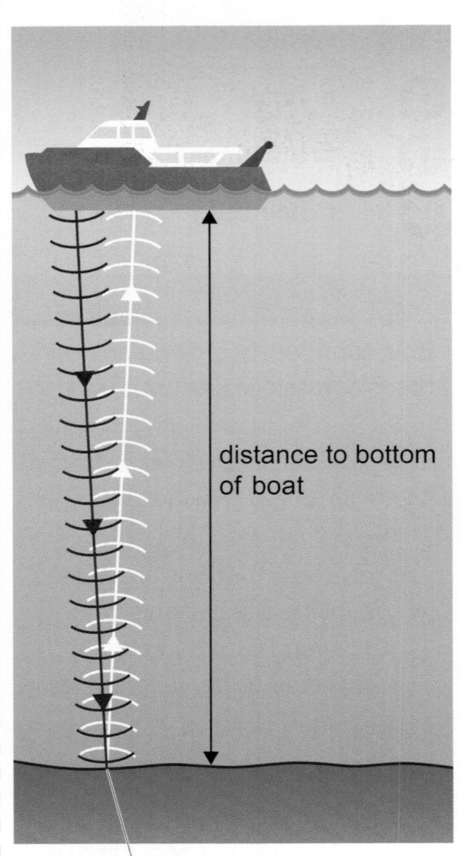

B Sonar equipment can map the shape of the sea bed.

Sonar equipment carried on ships or submarines uses a similar method to find the depth of the sea or to detect fish. A loudspeaker on the ship emits a pulse of ultrasound. This spreads out through the water and some of it is reflected by the sea bed. A special microphone on the ship detects the echo, and the sonar equipment measures the time between the sound being sent out and the echo returning. The distance travelled by the sound wave can be worked out using this equation.

distance =	speed	×	time
(metre, m)	(metre/second, m/s)		(second, s)

3 Why is ultrasound used for investigating the sea floor rather than light?

Worked example

The speed of sound in sea water is 1500 m/s. The sonar equipment on a boat receives an echo 0.01 s after the ultrasound pulse was sent out. Calculate the depth of the water.

distance = speed × time = 1500 m/s × 0.01 s

= 15 m

The sound has travelled 15 m down to the sea bed and back up again, so the depth of the water is half of this distance.

depth = $\frac{15}{2}$ m

= 7.5 m

4 A ship detects a sonar pulse 3 seconds after it was emitted. The speed of sound in sea water is 1500 m/s.

a How far has the sound travelled?

b How deep is the water?

C The US Navy uses dolphins to find mines, including ones buried in sand. This dolphin is carrying a marker that it will place on a mine it has found.

Ultrasound scans

Ultrasound can also be used to make images of things inside the body. One common use is to make detailed images of unborn babies so that doctors can monitor how well the fetus is developing.

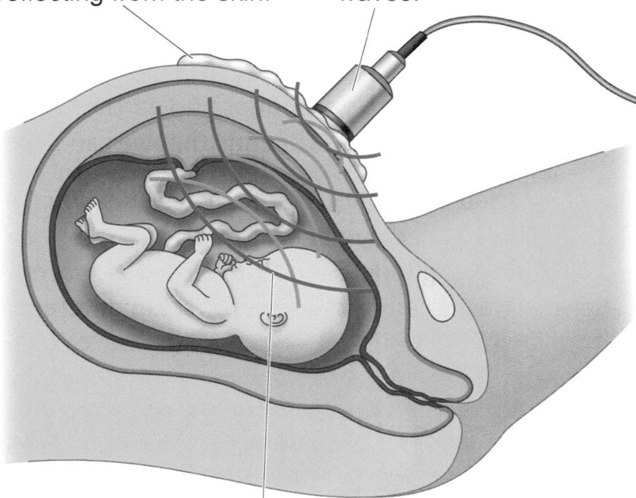

A gel is used to stop the ultrasound just reflecting from the skin.
The probe emits and receives ultrasound waves.
Some sound is reflected when the ultrasound waves pass into a different medium, such as fat or bone.

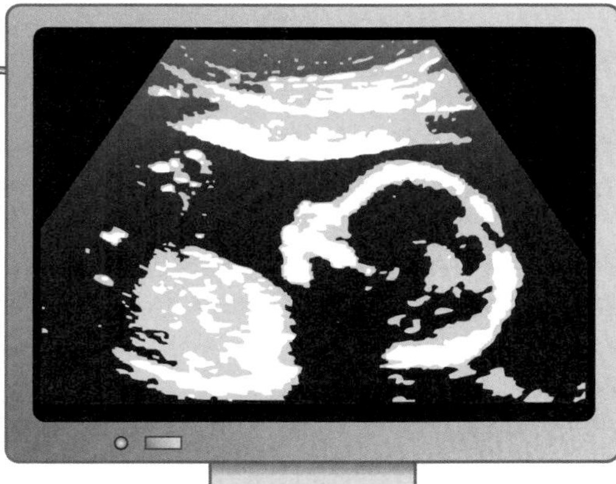

The ultrasound machine detects the time between sending the pulse out and receiving the echo. The display shows where the echoes come from.
The further down the screen, the longer the echo took to get back to the machine.

D When an **ultrasound scan** is made, some of the ultrasound waves are reflected each time the waves pass into a different medium.

Did you know?

Ultrasound can be used to clean things – even your teeth! Most ultrasonic cleaners work by placing delicate items in a solvent. The pressure caused by the ultrasound waves makes millions of tiny bubbles. The forces produced when these bubbles collapse knocks dirt off the item being cleaned.

5 Suggest why ultrasound is used in medical scans rather than visible light.

Exam-style question

Dolphins are used by some navies to find mines. Explain why dolphins are better than human divers at finding mines. *(2 marks)*

Checkpoint

How confidently can you answer the Progression questions?

Strengthen

S1 Design a labelled diagram for the manufacturers of 'FishFinder' sonar systems to explain how sonar works and how it can be used to detect fish. Include an example calculation on your diagram.

Extend

E1 Explain how an ultrasound scanner works.

SP4g Infrasound

Specification reference: **H** P4.14P; **H** P4.15P

Progression questions

- **H** What is infrasound?
- **H** How does infrasound travel through the Earth?
- **H** How can infrasound tell us about the inside of the Earth?

H

A The vibrations of seismic waves can make wet soil turn into a liquid. Buildings can just sink into the ground.

Sounds with a frequency less than 20 Hz are too low for humans to hear. These sounds are called **infrasounds**. Infrasound waves travel further than higher frequency sound waves before they become too faint to detect. Animals such as elephants can hear infrasounds.

1 A vibration has a frequency of 200 Hz. Is it an infrasound? Explain your answer.

Natural events such as volcanic eruptions and earthquakes create infrasound waves as well as sounds that we can hear.

The vibrations caused by earthquakes are called **seismic waves.** The energy released by an earthquake can travel through the Earth as longitudinal **P waves** and as transverse **S waves**. Seismic waves can be detected by **seismometers**.

2 What is:

a a seismic wave **b** a seismometer?

3 Explain whether P waves or S waves are infrasound waves.

Longitudinal waves can be transmitted through solids, liquids and gases. However, transverse waves that need a medium to travel through can only be transmitted by solids. The waves produced by an earthquake can be detected by seismometers all over the world.

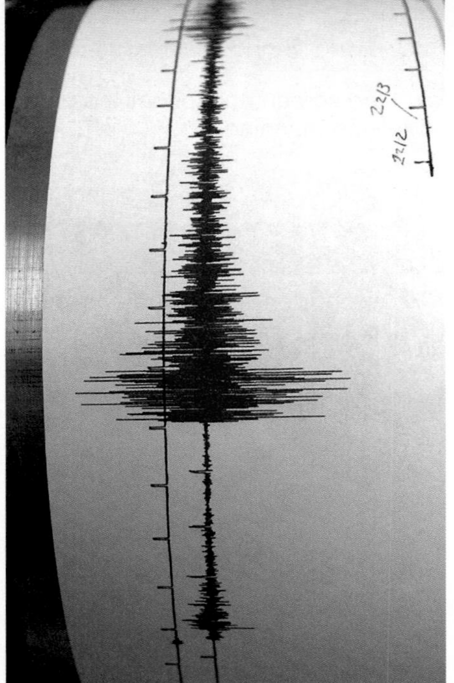

B The vibrations detected by seismometers are recorded digitally. In the past, paper drums like this were used to record seismic waves.

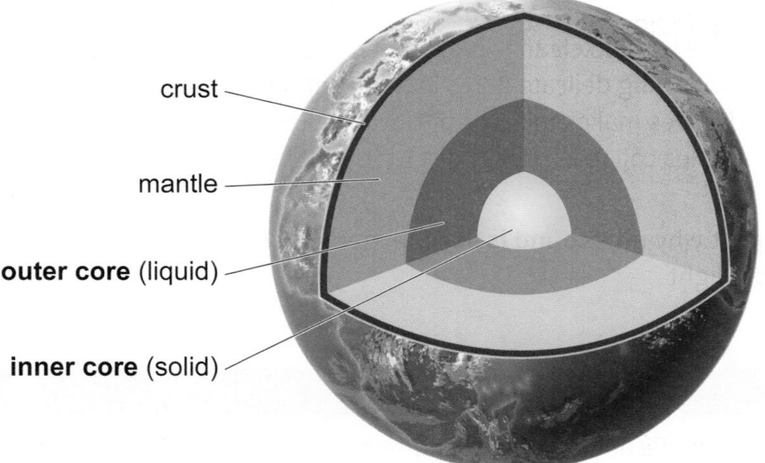

C Information from seismic waves has been used to develop this model of the Earth.

H

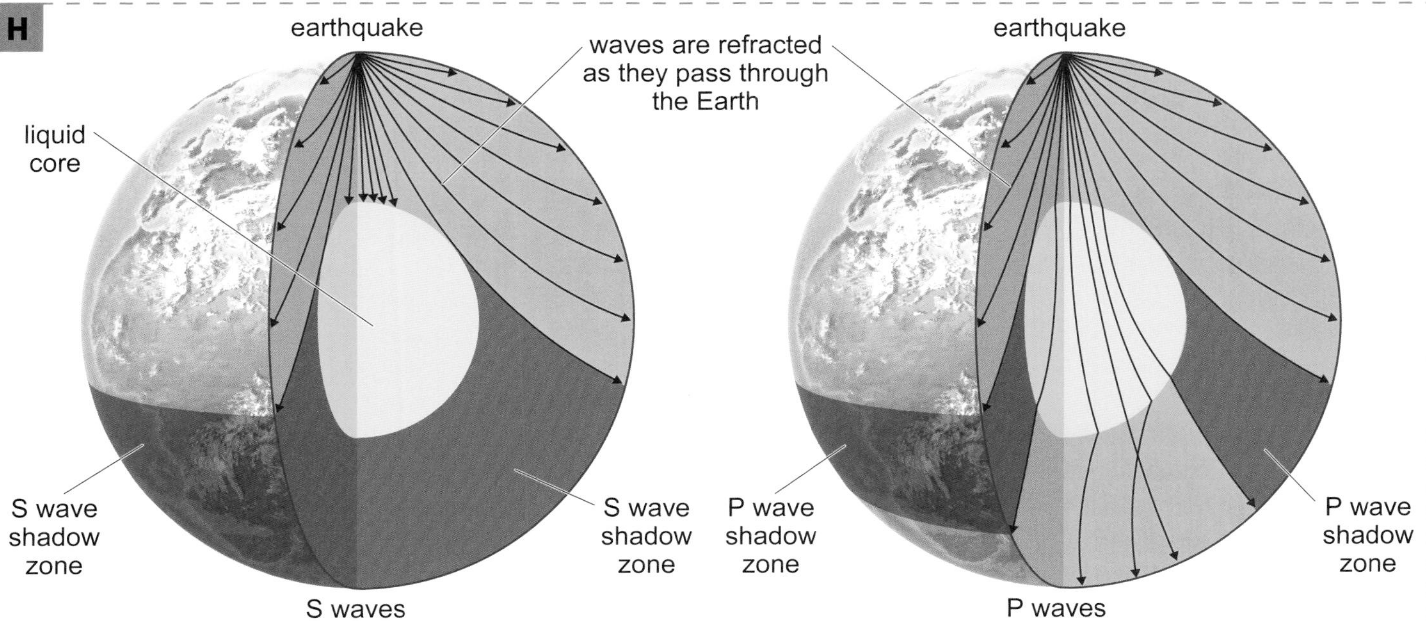

D P waves and S waves travel through the Earth in different ways. The same patterns are produced wherever the earthquake occurs.

Scientists use information about the time the waves arrive in different places and the speed of the waves in different types of rocks to model the paths the waves have taken through the Earth.

The places where the waves are detected depend on where an earthquake occurs, but there is always a large area of the Earth on the opposite side to the earthquake were no S waves are detected. This is called the S wave **shadow zone** and occurs because part of the interior of the Earth is liquid. There is also a band around the Earth called the P wave shadow zone.

After the model shown in diagram D was developed, it was discovered that there were a few, weak P waves arriving in the P wave shadow zone. These could only occur if waves in the liquid core had been reflected by something solid. The current model includes a liquid outer core and a solid inner core.

Did you know?

The solid inner core was first suggested by the Danish scientist Inge Lehmann (1888–1993) in 1936, based on P wave arrivals. Many scientists accepted her model but it was not confirmed until 1971 when computer modelling was used to check her interpretation.

4 Why does the S wave shadow zone suggest that part of the Earth must be liquid?

5 Seismic waves follow curved paths through the Earth. What does this tell you about:

a the speed of the waves within the Earth

b the properties of the rocks in the Earth?

6 Look at diagram D. Explain why there is a P wave shadow zone.

Exam-style question

a State the meaning of infrasound. *(1 mark)*

b Give one use for it. *(1 mark)*

Checkpoint

How confidently can you answer the Progression questions?

Strengthen

S1 Explain why infrasound is useful for studying the Earth's core.

Extend

E1 Explain how the P wave and S wave shadow zones show the nature of the Earth's core.

Waves

Waves on water in a ripple tank are often used as a model to help students to understand sound waves and light waves.

Compare and contrast waves on water, sound waves and light waves. **(6 marks)**

Student answer

All waves move energy around [1]. *We get big water waves on the sea and they can damage things in storms* [2]. *Light waves don't have particals, but waves on water and sound waves both have moving particals that make the waves* [3]. *Light can go through vacumes. We can use light waves and sound waves to send information, but we don't use waves on water for sending information* [4].

[1] This is one similarity between all the types of waves mentioned.

[2] The statement about waves on the sea does not add any further scientific information.

[3] This is a difference between light waves and the other two types of waves.

[4] This is a difference between waves on water and the other two types of waves.

Verdict

This is an acceptable answer. The student has given one similarity between all the waves and pointed out some differences.

The answer could be improved by including more comparisons, such as whether the waves are transverse or longitudinal, or commenting on how fast they travel. A really good answer would also link facts together with scientific ideas – for example, by making it clear that light waves can travel through a vacuum *because* these waves do not need particles to pass them on.

The answer could also be improved by correcting the spelling of the scientific words.

Exam tip

When a question asks you to 'compare and contrast', you need to mention at least one similarity *and* at least one difference between *all* the things mentioned in the question.

Paper 1

SP5 Light and the Electromagnetic Spectrum

We see things when a form of radiation that we call visible light enters our eyes. But there are other forms of radiation similar to light that we cannot see.

Our skin can detect infrared radiation. All objects emit infrared radiation – the hotter the object the more infrared it emits. The thermogram of the penguins is an image made using a special camera that detects infrared radiation. White shows the warmest parts of the image, then red, orange and yellow, with green and blue showing the coldest areas.

In this unit you will learn about different forms of radiation that we cannot see, their uses and dangers.

The learning journey

Previously you will have learnt at KS3:

- that light transfers energy
- about colours and how different colours are absorbed and reflected differently.

In this unit you will learn:

- how to use ray diagrams to explain reflection, refraction and total internal reflection
- how to make coloured light and why some objects appear coloured
- how lenses work and some things they can be used for
- that light is part of a family of waves called the electromagnetic spectrum, which all have some properties in common
- about some uses of waves in different parts of the electromagnetic spectrum
- about some of the harmful effects of waves in different parts of the electromagnetic spectrum
- about some of the factors that affect the temperature of the Earth.

SP5a Ray diagrams

Specification reference: P5.1P

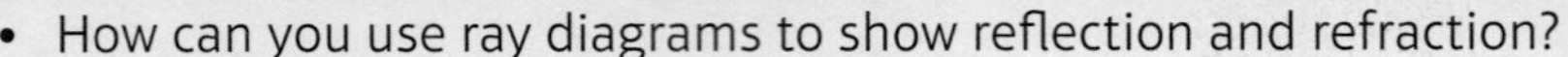

Progression questions

- How can you use ray diagrams to show reflection and refraction?
- What is the law of reflection?
- What is total internal reflection?

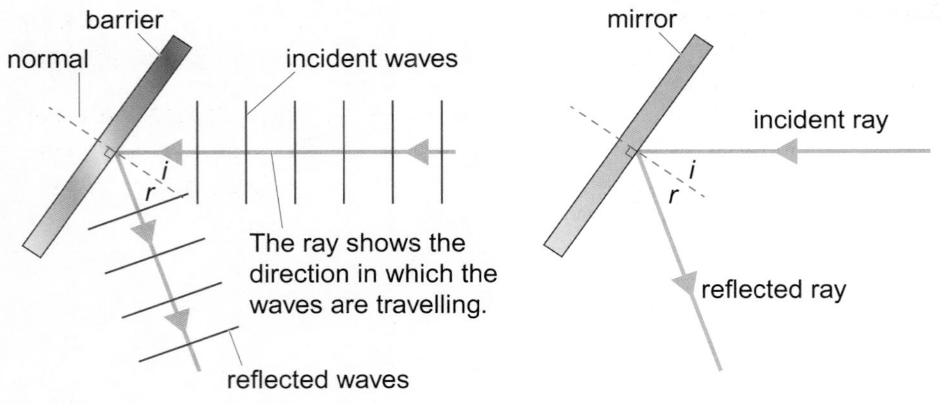

A water waves reflected by a barrier and light waves reflected by a mirror

A **ray diagram** is a way of modelling what happens when light is **reflected** or **refracted**. You can also use waves on water as a model for what happens to light. Diagram A shows how water and light waves are reflected. The rays are lines that show the direction the waves are travelling. A **normal** is a line drawn at right angles to the barrier or mirror. The angles of the **incident ray** and **reflected ray** are always measured from the normal.

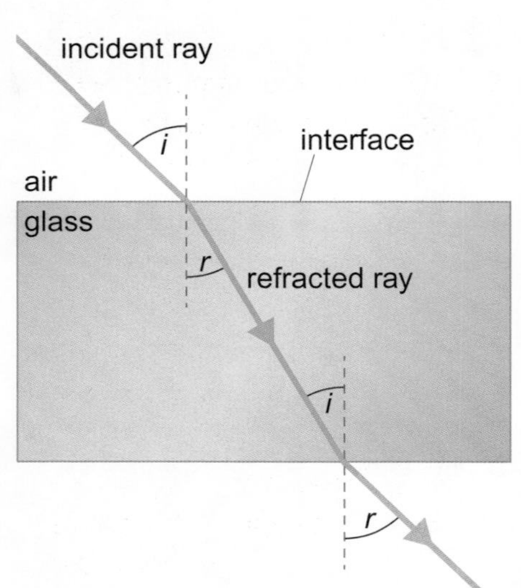

B Light bends towards the normal if it goes into a medium where it travels more slowly. It bends away from the normal if it goes into a medium where it travels faster.

When waves are reflected, the angle of reflection is equal to the angle of incidence. This is called the **law of reflection.**

1 Light hits a mirror with an angle of incidence of 30°. What is the angle of reflection?

Light travels at different speeds in different materials. It travels faster in air than it does in water or glass. When a ray of light moves into a material where it travels at a different speed, it usually changes direction. This is called **refraction**. The angle of incidence (*i*) and **angle of refraction** (*r*) are both measured from the normal. When light meets the **interface** (boundary) at right angles to it (i.e. along the normal) there is no change in direction.

2 Describe what happens to the direction of a light ray when it goes from water into air.

Total internal reflection

When light passes from water or glass into air with small angles of incidence, most of the light passes through the interface but a little is reflected (diagram C, part a). As the angle of incidence increases, the angle of refraction also increases until the refracted light passes along the interface (diagram C, part b). If the angle of incidence increases further, the light is completely reflected inside the glass. This is called **total internal reflection** and the angle of incidence at which this starts to happen is called the **critical angle**.

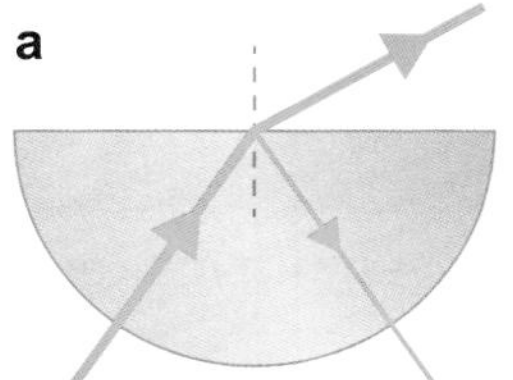

a A small amount of light is reflected, but most is refracted.

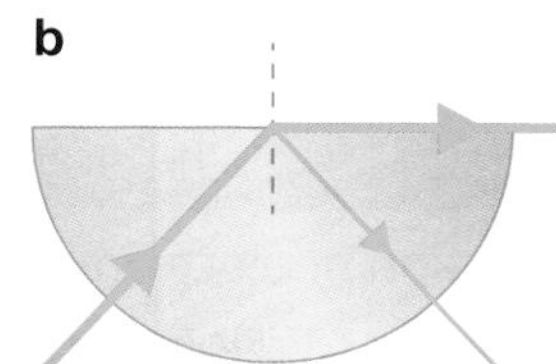

b When the angle of incidence equals the critical angle, the refracted light passes along the interface (boundary) of the glass block.

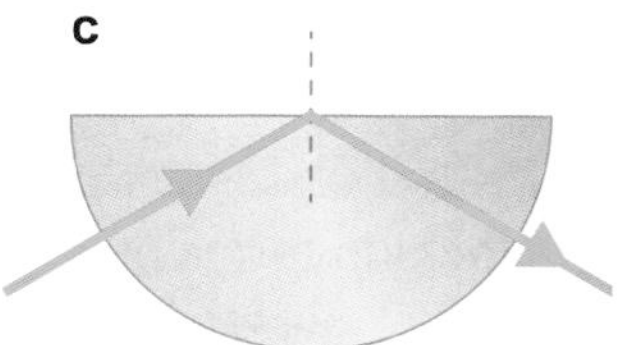

c At angles of incidence greater than the critical angle, the light is completely reflected inside the block.

C light passing through a semi-circular glass block, showing total internal reflection and the critical angle

Did you know?

If you swim underwater, you may be able to see reflections on the underside of the water. These are caused by total internal reflection.

D

5 Look at photo D. Draw a ray diagram to show how light from the manatee reaches the camera.

6 The critical angle for glass is 42°. Use diagrams to help you to explain what happens when light leaves a glass block with the following angles of incidence.

a 35° **b** 45°

Exam-style question

Compare and contrast reflection and total internal reflection. *(3 marks)*

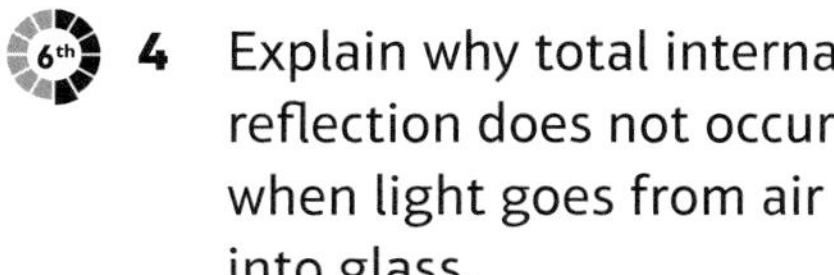

3 You can see your reflection in a window when it is dark outside. Explain why this happens.

4 Explain why total internal reflection does not occur when light goes from air into glass.

Checkpoint

How confidently can you answer the Progression questions?

Strengthen

S1 Draw a ray diagram to show light being reflected by a mirror when the angle of incidence is 45°.

S2 The critical angle for glass is 42°. Explain what this means.

Extend

E1 A triangular glass prism has one 90° angle and two 45° angles. Draw a ray diagram to show what happens to light as it enters one of the short sides with these angles of incidence.

a 60° **b** 90°

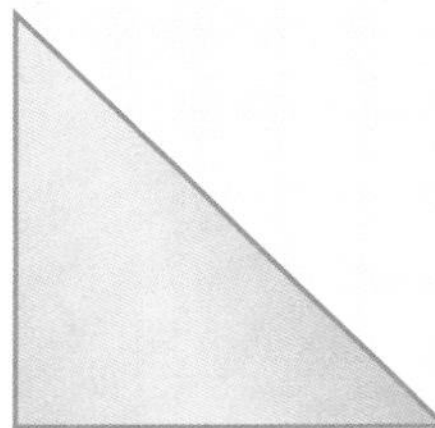

E

SP5a Core practical – Investigating refraction

Specification reference: P5.9

Aim

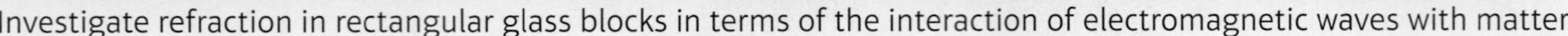

Investigate refraction in rectangular glass blocks in terms of the interaction of electromagnetic waves with matter.

A This photo was taken through the wall of an aquarium. Light reflected by the parts of the turtle under the water changes direction when it enters air, and makes it look as if the animal has been cut in half!

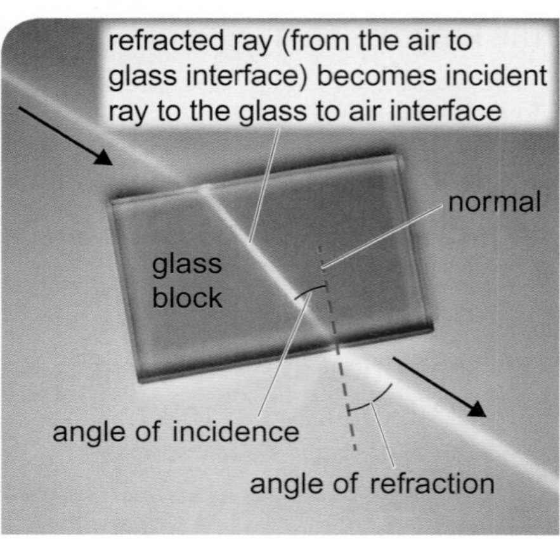

refracted ray (from the air to glass interface) becomes incident ray to the glass to air interface
normal
glass block
angle of incidence
angle of refraction

B

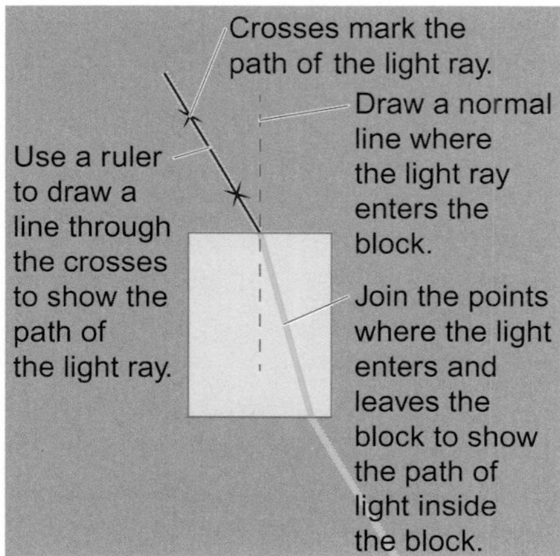

Crosses mark the path of the light ray.
Draw a normal line where the light ray enters the block.
Use a ruler to draw a line through the crosses to show the path of the light ray.
Join the points where the light enters and leaves the block to show the path of light inside the block.

C

Electromagnetic waves travel at different speeds in different materials. Light slows down when it goes from air into glass or water. If the light hits the interface at an angle, it changes direction. This is called refraction.

We can investigate refraction by measuring the angles between light rays and the normal (a line at right angles to the interface). The light ray approaching the interface is called the incident ray. The angle between this ray and the normal is called the angle of incidence (*i*). The angle between the normal and the light ray leaving the interface (the refracted ray) is called the angle of refraction (*r*).

Your task

Your task is to investigate how the direction of a ray of light changes as it enters and leaves a rectangular glass block.

Method

A Place a piece of plain paper on the desk. Set up the power supply, ray box and single slit so that you can shine a single ray of light across the paper on your desk. Take care, as ray boxes can become very hot.

B Place a rectangular glass block on the paper. Draw around the block.

C Shine a ray of light into your block. Use small crosses to mark where the rays of light go.

D Take the block off the paper. Use a ruler to join the crosses to show the path of the light, and extend the lines so they meet the outline of the block. Join the points where the light entered and left the block to show where it travelled inside the block.

E Measure the angles of incidence and refraction where the light entered the block, and measure the angles where it left the block.

F Repeat steps C to E with the ray entering the block at different angles.

G Move the ray box so that the light ray reaches the interface at right angles. Note what happens to the light as it enters and leaves the block.

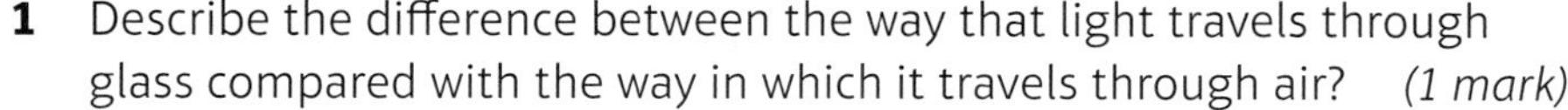

Exam-style questions

1 Describe the difference between the way that light travels through glass compared with the way in which it travels through air? *(1 mark)*

2 State what the following terms mean:

a normal *(1 mark)*

b angle of incidence *(1 mark)*

c angle of refraction. *(1 mark)*

3 Table D shows a student's results from this investigation.

a Draw a diagram to show the glass block and a light ray going into the glass at an angle of incidence of 30°. *(2 marks)*

b Draw in the refracted ray. *(1 mark)*

4 a Use the data in table D to plot a scatter graph to show the results for light going from air to glass. Put the angle of incidence on the horizontal axis, and join your points with a smooth curve of best fit. *(5 marks)*

b Use table D and your graph to write a conclusion for this part of the investigation. *(3 marks)*

c Use your graph to find the angle of refraction when the angle of incidence is 15°. *(1 mark)*

5 a Use the data in table D to plot a scatter graph to show the results for light going from glass to air. Put the angle of incidence on the horizontal axis, and join your points with a smooth curve of best fit. *(5 marks)*

b Use your graph to write a conclusion for this part of the investigation. *(3 marks)*

c Use your graph to find the angle of incidence when the angle of refraction is 45°. *(1 mark)*

6 If light passes through a glass block with parallel sides, the ray that comes out should be parallel with the ray that goes in. This means that the angle of incidence for air to glass should be the same as the angle of refraction from glass to air.

Look at table D. Suggest one source of random error that may have caused the differences in these angles. *(1 mark)*

Air to glass		Glass to air	
i	*r*	*i*	*r*
10°	6°	6°	9°
20°	13°	13°	20°
30°	20°	20°	31°
40°	25°	25°	40°
50°	30°	30°	50°
60°	34°	34°	58°
70°	38°	38°	69°
80°	40°	40°	78°

D

SP5b Colour

Specification reference: P5.2P; P5.3P

Progression questions

- What are specular and diffuse reflection?
- Why do surfaces have different colours?
- How do filters make coloured light?

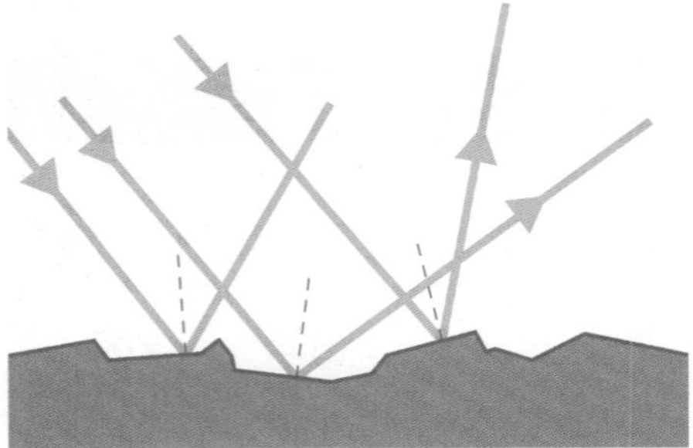

A diffuse reflection

You see **luminous** objects when light from them enters your eyes. You see non-luminous objects because they reflect light.

Most materials have rough surfaces if you examine them closely, so the reflected light is scattered in all directions. This is called **diffuse reflection**. Very smooth surfaces, such as mirrors, reflect the light evenly. This is called **specular reflection**.

 **1** 'The law of reflection applies to all surfaces'. Look at diagram A and explain why this statement is correct.

The light from the Sun or from lamps is called **white light**. White light is actually a mixture of different colours of light that our eyes see as white. White light can be split up into the colours of the **visible spectrum** using a prism.

When white light hits a coloured surface, some of the colours that make it up are absorbed and some are reflected. A yellow object looks yellow because it reflects yellow light and **absorbs** all the other colours. A white object looks white because it reflects all of the colours.

2 Look at photo B. Explain why some of the powders appear these colours.

 **a** yellow

 **b** blue

 **3** **a** Explain why a white shirt looks white.

 **b** Suggest why black objects look black.

B These powders are used in the Indian 'Festival of Colours'. Each powder reflects different colours of light.

Theatres use spotlights to produce effects on stage. Spotlight lamps produce white light but this can be made into coloured light using a **filter**. Filters are pieces of transparent material that absorb some of the colours in white light. For example a blue filter **transmits** (allows through) blue light and absorbs all the other colours.

4 Explain which colours in white light are transmitted and absorbed by:

a the kind of glass used to make house windows

b red glass in stained glass windows.

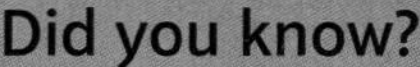

Did you know?

You see different colours because you have three types of cone cells in the part of your eyes that detects light (the retina). Each cone detects red, green or blue light which are the primary colours of light (not the same as the primary colours for paint). If red cones and green cones both detect light, you see the light as yellow. If all three sets of cones detect light, you see it as white.

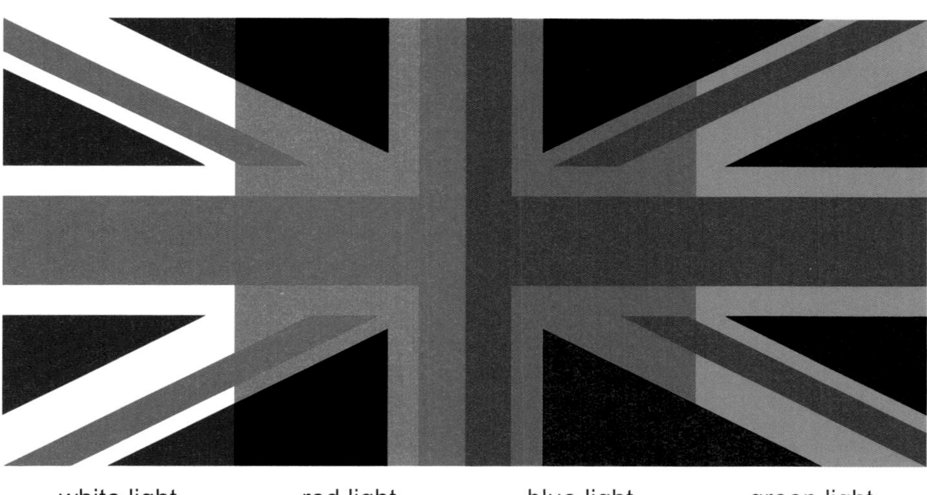

D Coloured objects look different when different coloured light shines on them.

5 Diagram D shows a union flag illuminated with different coloured lights. Explain why:

a the white parts look blue in blue light

b the blue parts look black in red light

c the red and blue parts look black in green light.

Exam-style question

Compare and contrast the way light is reflected by a mirror and by a sheet of paper. *(2 marks)*

C Stained glass windows are filters.

Checkpoint

How confidently can you answer the Progression questions?

Strengthen

S1 Describe what happens to light from the Sun when it hits a red flower.

S2 Describe what a filter is and what it does.

Extend

E1 Explain how you see a post box and why it looks red.

E2 Explain what colour the post box will appear if it is illuminated by a blue spotlight.

SP5c Lenses

Specification reference: P5.4P; P5.5P; P5.6P

Progression questions

- What factors affect the power of a lens?
- How do different shaped lenses refract light?
- How do lenses produce real and virtual images?

A Raindrops on this spider's web are acting as lenses.

A lens is a piece of transparent material shaped to refract light in particular ways. The **power** of a lens describes how much it bends light that passes through it. A more powerful lens is more curved and bends the light more.

1 What does the power of a lens describe?

2 State which is the most powerful: a ×10 magnifying glass or a ×20 one.

A **converging lens** is fatter in the middle than at the edges. It makes parallel rays of light converge (come together) at the **focal point**. The **focal length** is the distance between the focal point and the centre of the lens. A **diverging lens** is thinner in the middle than at the edges. The focal point is the point from which the rays seem to be coming after passing through the lens.

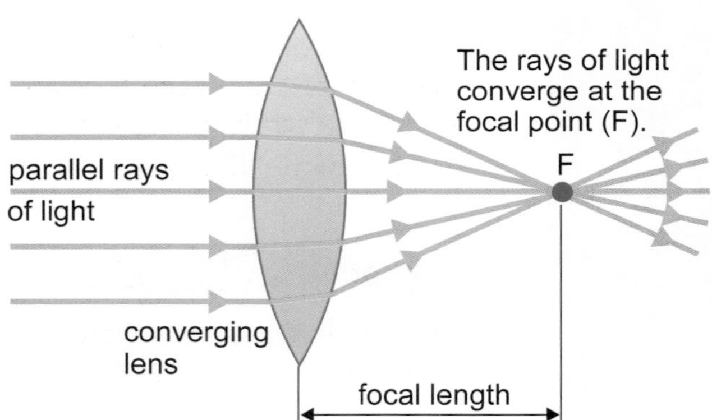

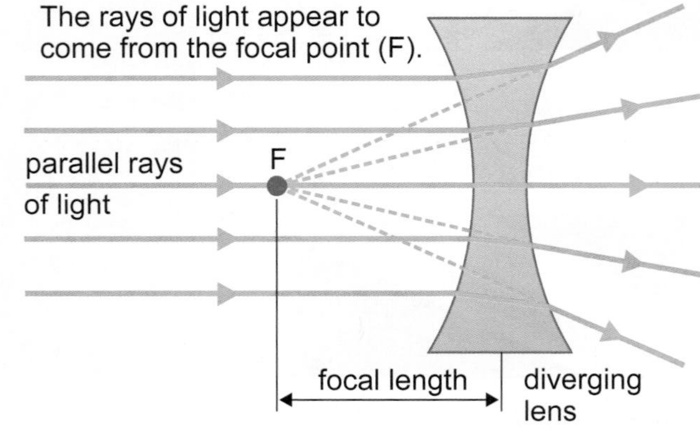

B

3 Look at the converging lens in diagram B.

a How would this lens look different if it were less powerful?

b How would its focal length be different if it were more powerful?

4 Look at the lenses in diagram B. Describe one similarity and one difference in the way they affect light.

The kind of image formed by a converging lens depends on where the **object** is. A converging lens can be used to focus rays of light onto a screen. An image that can be projected onto a screen in this way is called a **real image**. Real images can only be formed by light rays that come together.

Diagram C shows how a converging lens forms a real image. This image is also inverted (upside down) and smaller than the object.

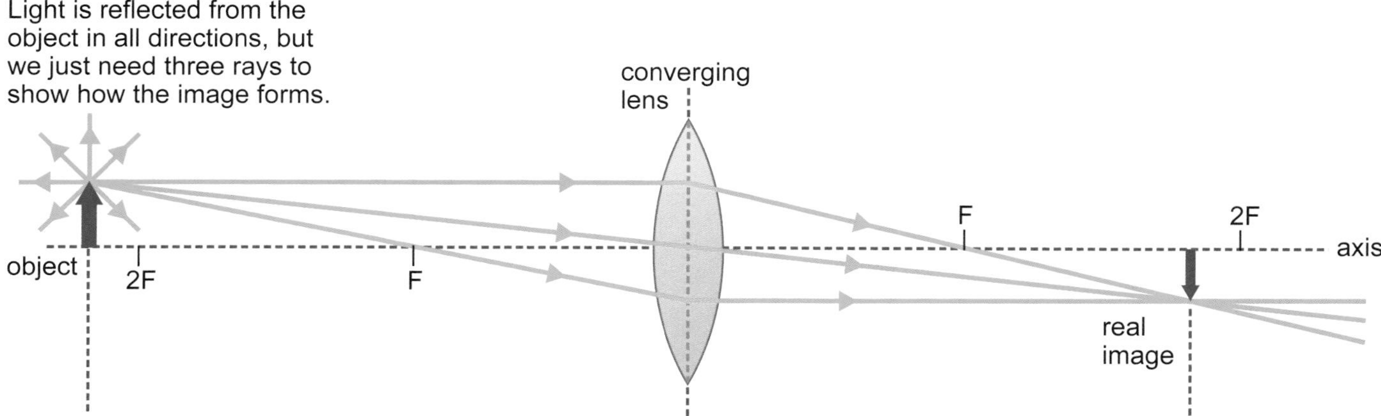

C A converging lens forms a real image of a distant object.

An object close to a converging lens will form a **virtual image**. It is called virtual because it cannot be projected onto a screen. The image appears to be on the same side of the lens as the object, and is upright and magnified. A magnifying glass is a converging lens.

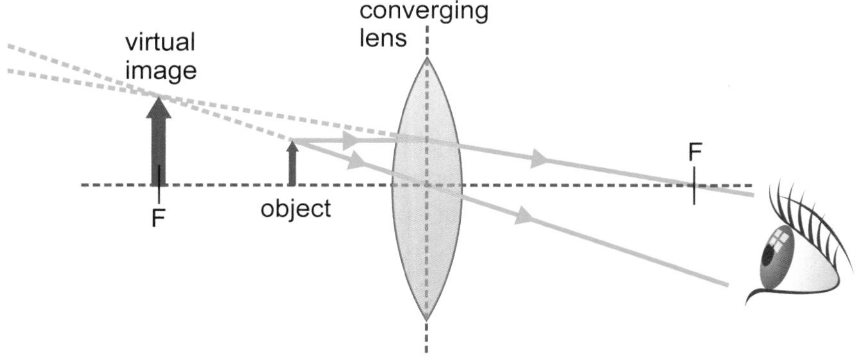

D a ray diagram showing a converging lens being used as a magnifying glass

Diverging lenses always produce virtual images that are the same way up, much smaller and closer to the lens than the object.

5 You can start a fire by using a lens to focus energy from the Sun onto paper. Explain which kind of lens you need to use.

6 Cinema projectors use lenses to project images on the film. Explain whether these projectors use converging or diverging lenses.

Exam-style question

Describe how a converging lens can be used to form a virtual image, and describe one use for a lens used in this way. *(2 marks)*

Did you know?

The largest lens in use today has a diameter of 102 cm and a focal length of 19.4 m. It is used in a telescope in the USA.

Checkpoint

How confidently can you answer the Progression questions?

Strengthen

S1 Explain how the power of a lens depends on its shape.

S2 A camera uses lenses to focus light. Explain whether cameras have converging or diverging lenses.

Extend

E1 'Converging lenses only produce real images'. Explain what real and virtual images are and why this statement is not correct.

SP5d Electromagnetic waves

Specification reference: P5.7; P5.8; P5.12

Progression questions

- What are some examples of electromagnetic waves?
- What do all electromagnetic waves have in common?
- Which electromagnetic waves can our eyes detect?

A A marsh marigold flower seen in visible light (left) and ultraviolet light (right).

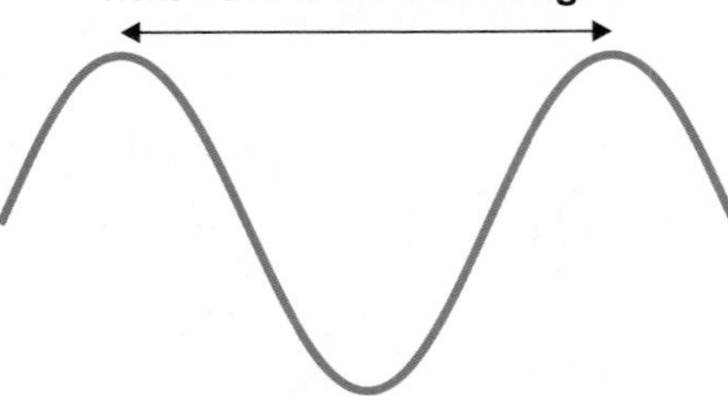

The distance from a point on one wave to a point in the same position on the next wave is the **wavelength**.
The number of waves passing a point each second is the frequency.

B Electromagnetic waves are transverse waves.

We see things when light travels from a source and is reflected by an object into our eyes. The light transfers energy from the source to our eyes. Light is a type of **electromagnetic wave**.

Our eyes can detect certain **frequencies** of light, and we refer to these frequencies as **visible light**. Different frequencies cause us to see different colours. Lower frequencies of visible light appear more red and higher frequencies appear more blue.

1 We see visible light as a range of colours from red to green to violet. Explain whether red or violet light has the higher frequency.

Some animals, such as birds, can also detect electromagnetic waves with frequencies that are higher than visible light. Electromagnetic waves with frequencies a little higher than visible light are called **ultraviolet** (**UV**).

2 Look at the photos in A. Describe what a marsh marigold flower would look like to a bird.

All electromagnetic waves are **transverse** waves. This means that the electromagnetic vibrations are at right angles to the direction in which the energy is being transferred by the wave. All electromagnetic waves travel at the same speed (3×10^8 m/s) in a **vacuum**. Like all waves, electromagnetic waves transfer energy from a source to an observer.

3 a What types of waves are electromagnetic waves?

b State two ways in which electromagnetic waves differ from one another.

Electromagnetic waves with frequencies slightly lower than visible light are called **infrared** (**IR**). All objects emit energy by infrared radiation. The hotter the object the more energy it emits. The photo on the opening page for this unit shows what penguins would look like if our eyes could detect infrared radiation. We can feel the effects of infrared radiation when energy is transferred from the Sun to our skin.

4 Write down two similarities and two differences between infrared radiation and ultraviolet radiation.

5 Look at the opening page for this unit. Which parts of the penguins are the hottest?

Discovering infrared

The first person to investigate infrared radiation was the British astronomer William Herschel (1738–1822). He put dark, coloured filters on his telescope to help him observe the Sun safely. He noticed that different coloured filters heated up his telescope to different extents and he wondered whether the different colours of light contained different 'amounts of heat'.

To test his idea he used a prism to split sunlight into a spectrum and then put a thermometer in each of the colours in turn. He also measured the temperature just beyond the red end of the spectrum, where there was no visible light.

6 Look at photo D.

a Which colour of visible light caused the greater temperature rise?

b Compare the energy transferred to the thermometer by infrared radiation and by visible light.

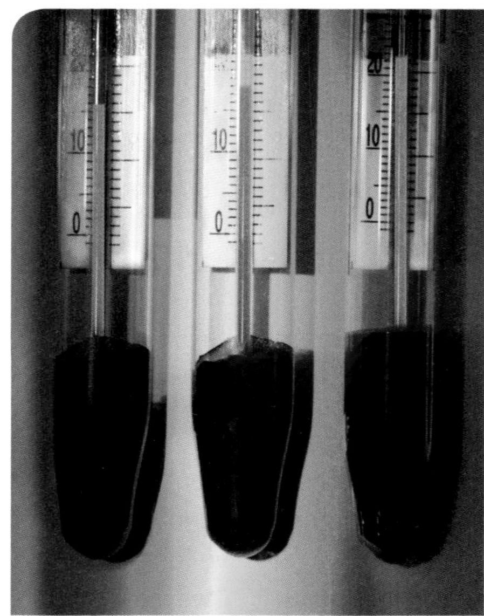

D A modern version of Herschel's experiment.

Exam-style question

State two characteristics that all electromagnetic waves have in common. *(2 marks)*

Did you know?

Some animals have special sense organs to detect infrared radiation. Many snakes, such as pit vipers, have these organs under their eyes, which help them to detect warm-blooded prey.

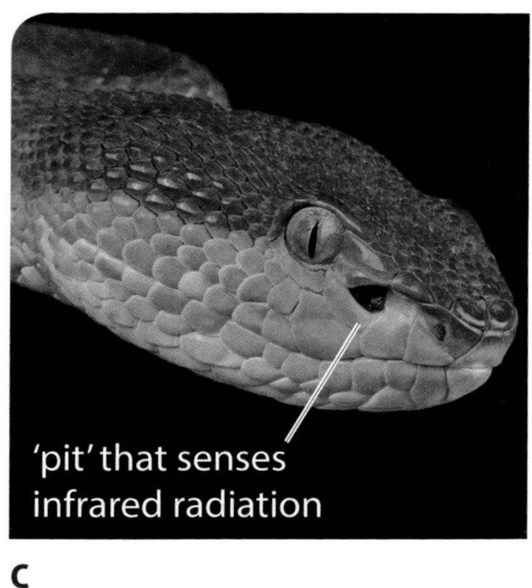

C

Checkpoint

How confidently can you answer the Progression questions?

Strengthen

S1 Describe how we can see a flower on a sunny day. Use the words 'energy' and 'transferred' in your answer.

S2 Explain why the image we see may not be the same as the image a bird sees.

Extend

E1 Compare and contrast infrared, visible light and ultraviolet radiation.

E2 Does photo D show that violet light transfers less energy than red light? Explain your answer.

SP5e The electromagnetic spectrum

Specification reference: P5.10; P5.11; H P5.13

Progression questions

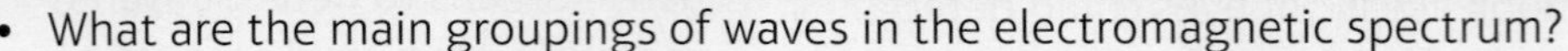

- What are the main groupings of waves in the electromagnetic spectrum?
- What characteristics of electromagnetic waves are used to group them?
- H What are some of the differences in the behaviour of waves in different parts of the electromagnetic spectrum?

A A rainbow shows the colours of the visible spectrum.

Visible light is part of a family of waves called electromagnetic waves. Our eyes can detect different colours in visible light. Scientists describe seven colours in the visible spectrum:

red, orange, yellow, green, blue, indigo, violet.

You can remember the order of the colours using a phrase such as ROY G BIV.

Did you know?

The colours in visible light were described by Sir Isaac Newton (1642–1727). He originally divided the spectrum into five colours, which were all that he could see. However, he thought there was a mystical connection between the colours, the days of the week and the number of known planets, so he ended up describing seven colours.

The colour of visible light depends on its frequency. If the frequency of an electromagnetic wave is lower than that of red light, human eyes cannot see it. Infrared, **microwaves** and **radio waves** have lower frequencies than red light.

1 Name three different types of electromagnetic waves.

Ultraviolet radiation has a higher frequency than visible light. Even higher frequencies and shorter wavelengths are present in **X-rays** and then **gamma rays**.

The full range of electromagnetic waves is called the **electromagnetic spectrum**. The spectrum is continuous, so all values of frequency are possible. Higher frequency waves have shorter wavelengths, and lower frequency waves have longer wavelengths. It is convenient to group the spectrum into seven wavelength groups, as shown in diagram B.

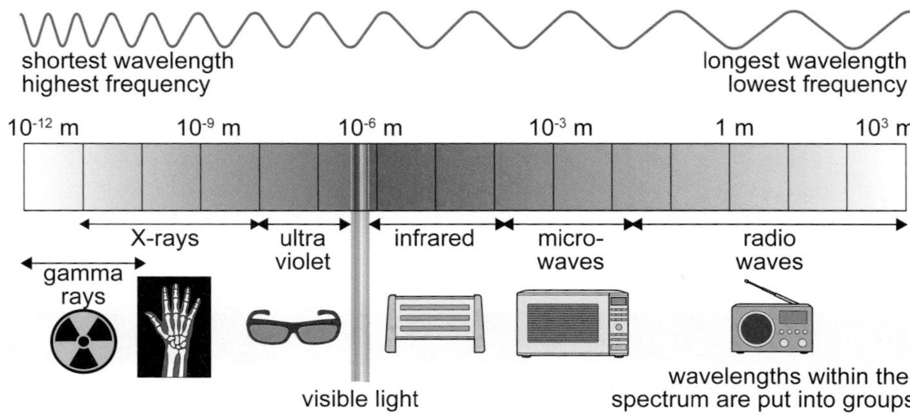

B the electromagnetic spectrum (not to scale)

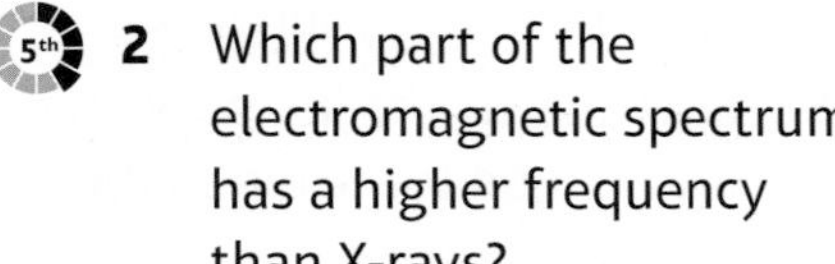

2 Which part of the electromagnetic spectrum has a higher frequency than X-rays?

3 What type of electromagnetic wave has a wavelength between those of visible light and X-rays?

4 How do we know that electromagnetic waves can travel through a vacuum, such as space?

H

Stars and other space objects can emit energy at all wavelengths. Astronomers use telescopes to study this radiation but they need to use different kinds of telescope to study different wavelengths. This is because different materials affect electromagnetic waves depending on the wavelength. For example, diagram C shows which wavelengths pass through the atmosphere and which are absorbed.

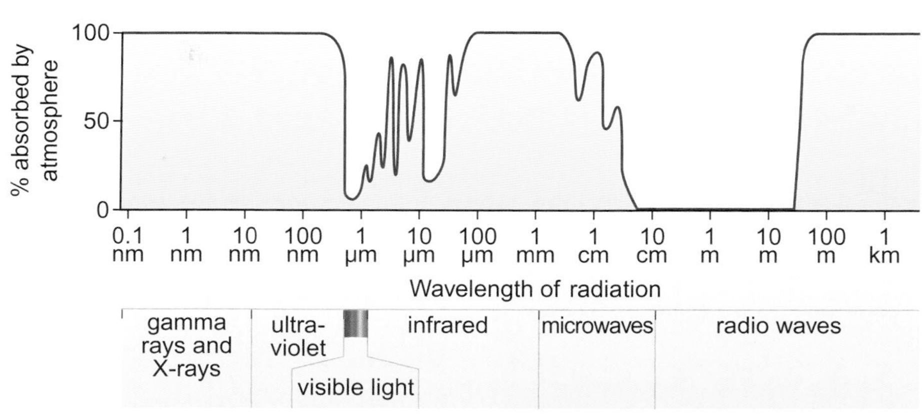

C Absorption of electromagnetic radiation by the atmosphere. You do not need to recall the details of this diagram.

Most telescopes use curved mirrors to focus the electromagnetic radiation onto a central sensor. The type of material used for the mirror and the size of the telescope depend on the wavelength of the radiation being studied.

5 Look at diagram C. Explain why telescopes that detect infrared radiation from objects in space are put into orbit around the Earth.

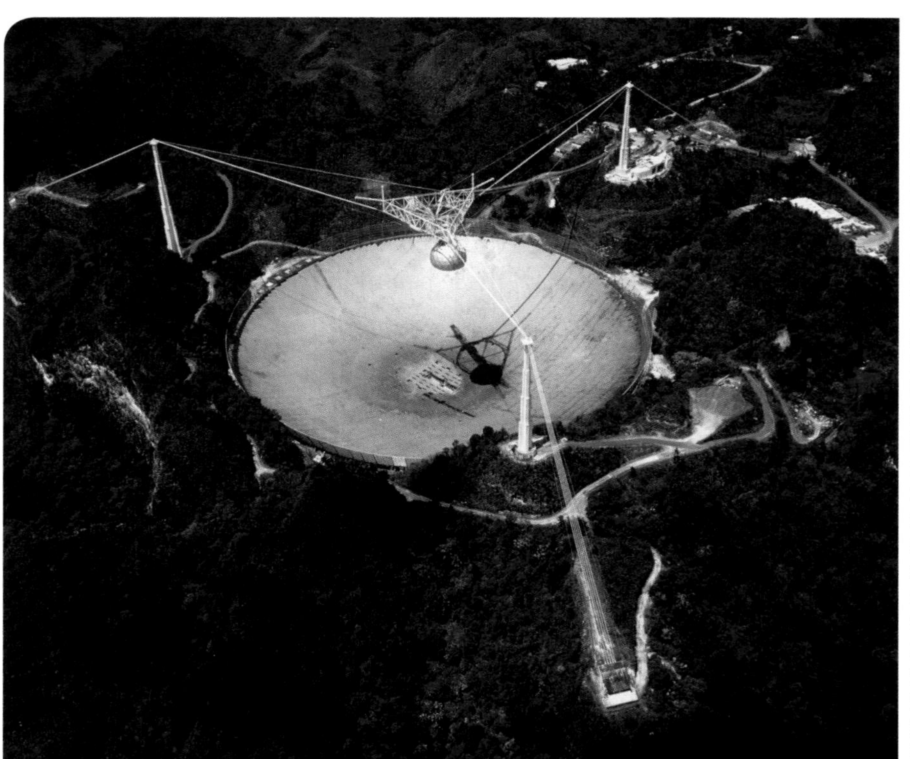

D The Arecibo telescope in Puerto Rico has a reflector dish that is over 300 m in diameter and contains nearly 40 000 aluminium panels.

Exam-style question

State one way in which X-rays are similar to visible light, and one way in which they are different. *(2 marks)*

Checkpoint

How confidently can you answer the Progression questions?

Strengthen

S1 List the seven parts of the electromagnetic spectrum in order and describe how the wavelength and frequency change from one end of the spectrum to the other.

Extend

E1 **H** Explain how the locations and types of instruments astronomers use depend on the wavelength of the electromagnetic waves that they study.

SP5f Using the long wavelengths

Specification reference: H P5.13; H P5.14; P5.22; H P5.23

Progression questions

- What are some uses of radio waves, microwaves and infrared?
- H How are radio waves produced and detected?
- H How do different substances affect radio waves, microwaves and infrared?

A This sculpture is made from optical fibres which act as 'light pipes'. Visible light and infrared can both be sent along optical fibres.

Did you know?

Microwave cooking was invented by Percy Spencer (1894–1970). The heating effects of radio waves had been known for years, but Spencer applied the idea to cooking after he noticed that waves from a radar apparatus he was working with had melted a chocolate bar in his pocket.

The uses for the waves in different parts of the electromagnetic spectrum depend on their wavelengths.

Visible light

Visible light is the part of the electromagnetic spectrum that our eyes detect. Light bulbs are designed to emit visible light, while cameras detect it and record images.

Infrared

Infrared radiation can be used for communication at short ranges, such as between computers in the same room or from a TV to its remote control unit. The information sent along optical fibres is also sent using infrared radiation.

A grill or toaster transfers energy to food by infrared radiation. The food absorbs the radiation and heats up. Thermal images show the amount of infrared radiation given off by different objects.

Security systems often have sensors that can detect infrared radiation emitted by intruders. Some buildings are fitted with systems of infrared beams and detectors – someone walking through one of these beams breaks it and sets off the alarm.

Microwaves

Microwaves are used for communications and satellite transmissions, including mobile phone signals. In a microwave oven, microwaves transfer energy to the food, heating it up.

B Pilots communicate with each other and with ground controllers using radio waves.

Radio waves

Radio waves are used for transmitting radio broadcasts and TV programmes as well as other communications. Some radio communications are sent via satellites. Controllers on the ground communicate with spacecraft using radio waves.

1 Which parts of our body can detect:

a visible light

b infrared radiation?

 **2** List three parts of the electromagnetic spectrum used for communication.

 **3** Describe how two different parts of the electromagnetic spectrum are used for cooking.

 **4** Suggest why security systems have sensors that detect infrared rather than other wavelengths of electromagnetic radiation.

H

Radio waves are produced by **oscillations** (variations in current and voltage) in electrical circuits. A metal rod or wire can be used as an aerial to receive radio waves. The radio waves are absorbed by the metal and cause oscillations in electric circuits connected to the aerial.

Waves travel in straight lines unless they are reflected or refracted. Refraction is the bending of the path of a wave due to a change in velocity. Some frequencies of radio waves can be refracted by a layer in the atmosphere called the ionosphere. If radio waves reach the ionosphere at a suitable angle, they may be refracted enough to send them back towards the Earth. Microwaves are not refracted in the Earth's atmosphere.

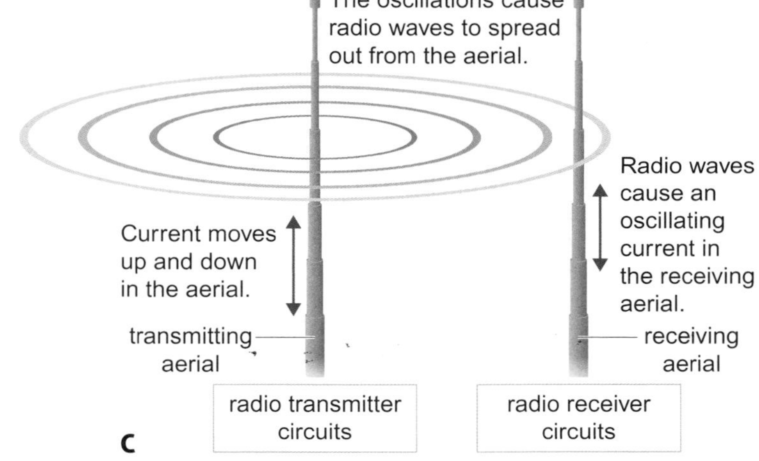

The oscillations cause radio waves to spread out from the aerial.
Radio waves cause an oscillating current in the receiving aerial.
Current moves up and down in the aerial.
transmitting aerial
receiving aerial
radio transmitter circuits
radio receiver circuits

C

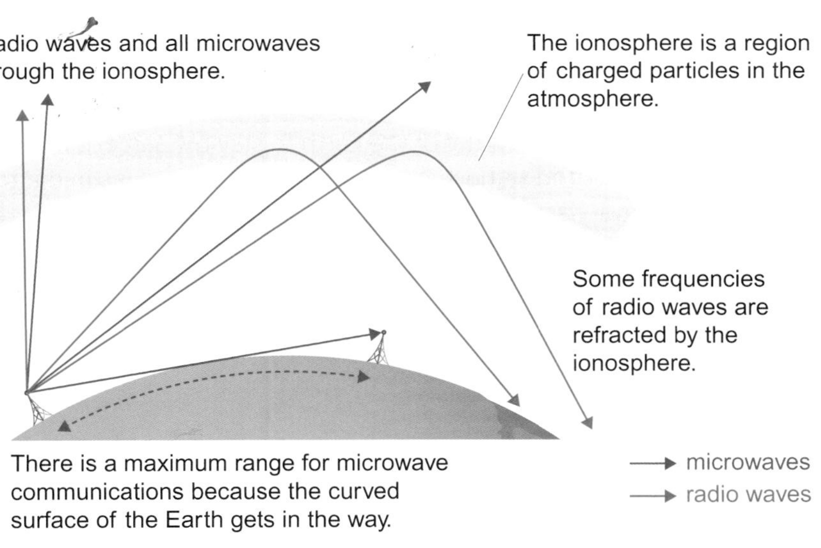

Some radio waves and all microwaves pass through the ionosphere.
The ionosphere is a region of charged particles in the atmosphere.
Some frequencies of radio waves are refracted by the ionosphere.
There is a maximum range for microwave communications because the curved surface of the Earth gets in the way.
microwaves
radio waves

D The maximum distance (range) of radio communication is much greater than for microwave communication.

5 Microwaves and radio waves are both used for communication between different places on the Earth. Explain why a satellite is needed to give microwaves a similar range to radio waves.

Checkpoint

How confidently can you answer the Progression questions?

Strengthen

S1 Draw a table or make a list of bullet points to show the uses for visible light, infrared, microwaves and radio waves.

Extend

E1 Compare the uses for visible light, infrared, microwaves and radio waves.

Exam-style question

Compare the ways in which infrared and microwaves are used in cooking. *(2 marks)*

SP5g Radiation and temperature

Specification reference: P5.15P; H P5.16P; H P5.17P; H P5.18P

Progression questions

- How does the radiation emitted by a body depend on its temperature?
- H How does the temperature of a body depend on the amount of power it absorbs and radiates?
- H How is the temperature of the Earth affected by different factors?

A a lava flow in Hawaii

The intensity (amount) of radiation emitted by an object increases as its temperature increases. The wavelengths of the radiation emitted also change with temperature – the higher the temperature the shorter the wavelengths.

1 Explain which emits more radiation, a cup of tea at 75 °C or a bowl of soup at 50 °C.

All of the lava in photo A is hot but only some of it is hot enough to emit radiation in visible wavelengths. The parts glowing yellow are hotter than the orange parts, which are hotter than the red parts.

2 Explain why astronomers think that blue stars are hotter than yellow stars.

H

Constant temperatures

The amount of energy transferred in a certain time is the **power**. It is measured in **watts** (**W**) (1 W = 1 J/s). For a system to stay at a constant temperature it must absorb the same amount of power as it radiates.

3 A butterfly house at 26 °C radiates 30 kW. What must happen for it to stay at 26 °C?

Earth's energy balance

The Earth's surface absorbs about half of the radiation that reaches it from the Sun. It re-radiates this energy as infrared radiation, which can warm up the atmosphere. For the temperature of the Earth to stay the same, it must radiate energy into space at the same average rate it is absorbed, as in diagram B.

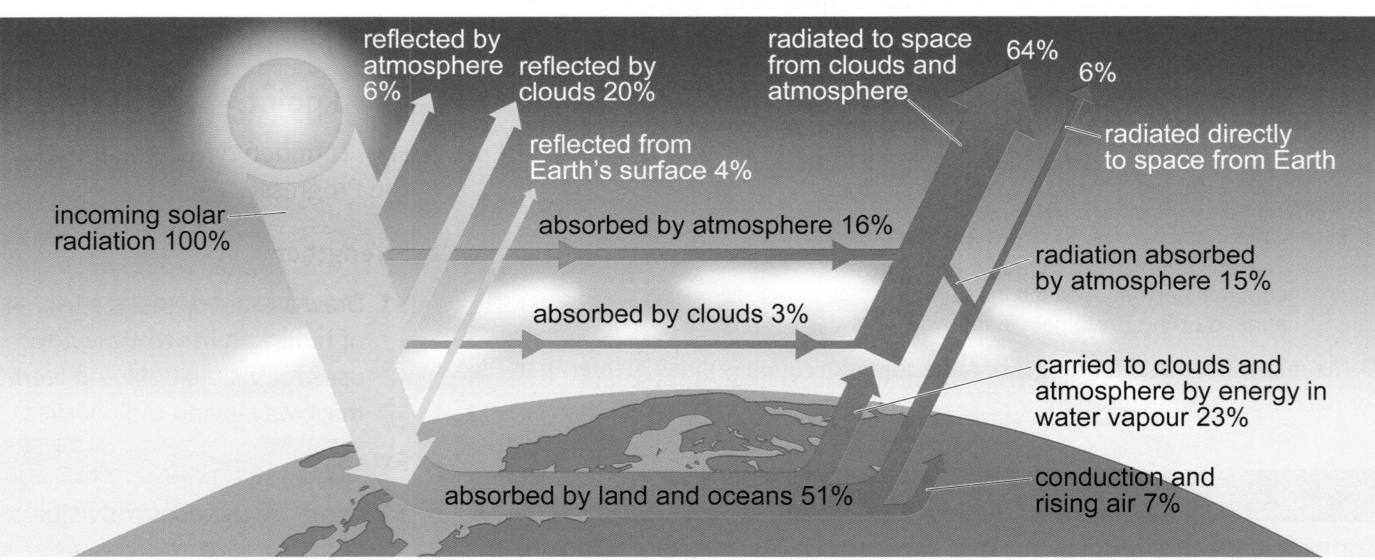

B The Earth's energy balance: the amount of energy leaving the atmosphere is the same as the amount coming in.

H

Some gases in our atmosphere (such as carbon dioxide) naturally absorb some energy, keeping the Earth at a higher temperature than if there were no atmosphere. This is the **greenhouse effect** and these gases are often called **greenhouse gases**. Many scientists think that humans have upset this natural balance and that the Earth is warming up because of an increase in greenhouse gases.

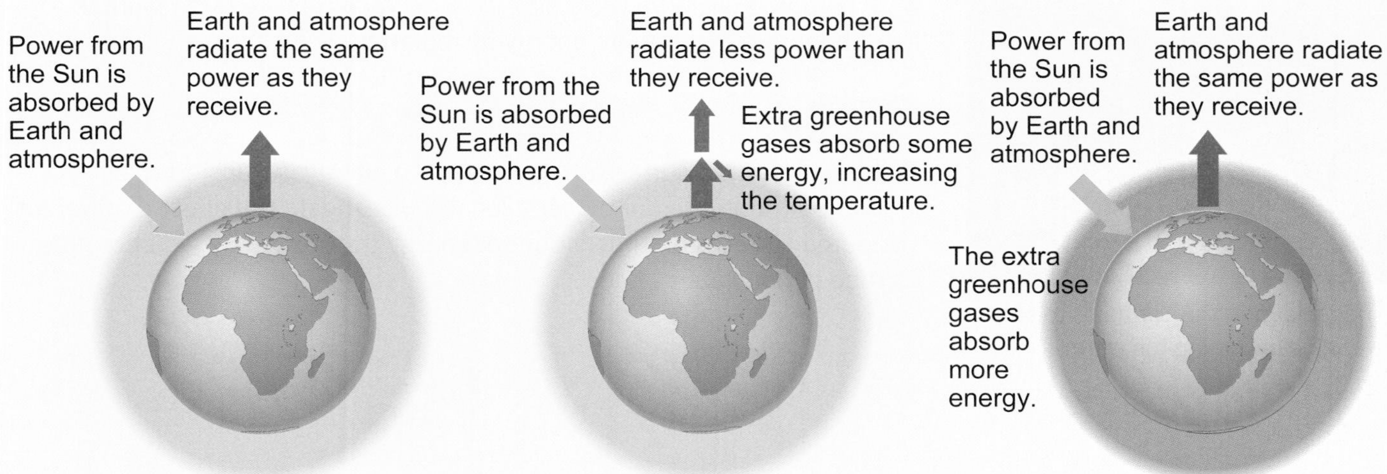

Power from the Sun is absorbed by Earth and atmosphere.
Earth and atmosphere radiate the same power as they receive.
The Earth and atmosphere are at a constant temperature.
Power from the Sun is absorbed by Earth and atmosphere.
Earth and atmosphere radiate less power than they receive.
Extra greenhouse gases absorb some energy, increasing the temperature.
The Earth and atmosphere are getting warmer.
Power from the Sun is absorbed by Earth and atmosphere.
Earth and atmosphere radiate the same power as they receive.
The extra greenhouse gases absorb more energy.
The Earth and atmosphere are at a new constant temperature.

C how changes in the atmosphere can warm the Earth

If some greenhouse gases were removed from the atmosphere, the atmosphere would be able to hold less energy and its temperature would decrease.

4 Without the atmosphere, each square metre of the Earth's surface could receive an average of 343 W of solar power. Use diagram B to answer the following questions.

a What is the power absorbed by each square metre of Earth on average?

b Calculate the power being re-radiated from each square metre that goes directly into space.

c Describe what would happen if less than this amount went into space.

5 If the Earth's average temperature rises to a new steady level, what can you say about the power absorbed and radiated by the Earth and atmosphere?

Did you know?

One idea to stop the Earth's temperature rising is to place a huge white screen, 2000 km by 2000 km, in space.

Exam-style question

Compare the radiation emitted by a stove at 100 °C and one at 150 °C.

(2 marks)

Checkpoint

How confidently can you answer the Progression questions?

Strengthen

S1 Blacksmiths heat iron before hammering it into a new shape. Explain how looking at the colour of the heated iron can tell them whether it is hot enough.

Extend

E1 Explain what effect giant white screens in space would have on the temperature of the Earth.

SP5g Core practical – Investigating radiation

Specification reference: P5.19P

Aim

Investigate how the nature of a surface affects the amount of thermal energy radiated or absorbed.

A The radiator in this car helps to stop the engine becoming too warm.

The radiator on the car in photo A is designed to transfer energy to the outside air, to stop the engine overheating. Radiators to cool car engines were patented over 100 years ago, in 1879. Radiators are often painted to help them to transfer more energy by radiation.

Your task

Different types of surface affect how much energy is transferred by radiation from different objects. You will investigate the effect of different coloured surfaces on the amount of energy transferred by radiation from a tube of hot water.

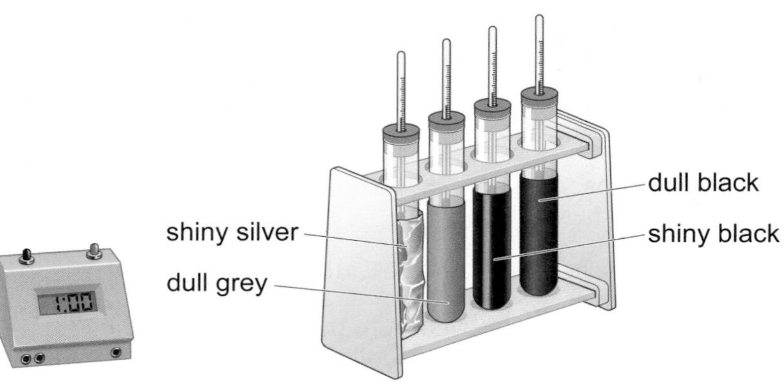

B

Method

A Cover four or more boiling tubes in different coloured materials (as shown in diagram B).

B Pour the same volume of hot water from a kettle into each tube.

C Insert a bung with a thermometer into each tube. Measure the temperature of the water in each tube and start a stop clock.

D Record the temperature of the water in each tube every 2 minutes for 20 minutes.

Exam-style questions

1. Which part of the electromagnetic spectrum transfers energy by heating? *(1 mark)*
2. Some very hot objects emit visible light. Explain why the water in the boiling tubes does not emit visible light. *(2 marks)*
3. Explain one safety precaution that should be taken while carrying out this investigation. *(2 marks)*

4 The table shows a set of results from Ali's radiation investigation. Ali recorded the temperatures every five minutes.

a Draw a suitable chart or graph to show these results, with all four sets of results on the same axes. Draw a smooth line through each set of points. *(4 marks)*

Time (min)	Temperature of water (°C)			
	shiny silver	dull black	shiny black	dull grey
0	80	80	80	80
5	70	66	68	69
10	63	57	59	60
15	58	51	54	55
20	53	47	49	46

b Identify any points you think may be errors. *(1 mark)*

c Describe what your graph shows. *(4 marks)*

5 Use your graph to write a conclusion for the investigation. *(2 marks)*

6 Explain what is the best colour for a car radiator. *(2 marks)*

7 **a** Calculate the overall temperature change for the water in each tube, using the results as given in the table. *(4 marks)*

b What is the advantage of plotting a graph to help you to draw a conclusion for this experiment? *(2 marks)*

8 Ellie says 'Silver surfaces emit less radiation than grey ones'. Explain why this is not necessarily a valid conclusion from Ali's investigation. *(3 marks)*

9 The investigation described above was looking at whether the colour of a surface affects the amount of radiation it emits. Describe an experiment to investigate whether the colour of a surface affects how much radiation it absorbs. Your description should include the apparatus needed, the method to be followed and how you will make sure your test is fair. *(5 marks)*

10 The world is getting warmer, which is causing many glaciers to melt faster than they used to. However, scientists have also discovered that deposits of soot and dust on the surface of ice in Greenland are causing the icecap there to melt even more quickly.

a Explain what this report suggests about the difference in the amount of infrared radiation absorbed by light and dark surfaces. *(2 marks)*

b Use your ideas about the way different coloured surfaces reflect visible light to suggest how the colour of a surface affects the amount of infrared radiation it absorbs. *(3 marks)*

C investigating ice melting in on a glacier

SP5h Using the short wavelengths

Specification reference: H P5.13; P5.22

Progression questions

- What are some uses of ultraviolet waves?
- What are some uses of X-rays and gamma rays?
- H How do different substances affect ultraviolet, X-rays and gamma rays?

Did you know?

Obtaining clean drinking water is a problem in many parts of the world. Microorganisms in water can be killed by putting the water in clear plastic bottles and leaving them in sunshine for several hours. Infrared energy from the sun heats up the water and the high temperature and the ultraviolet radiation both help to kill microorganisms.

A

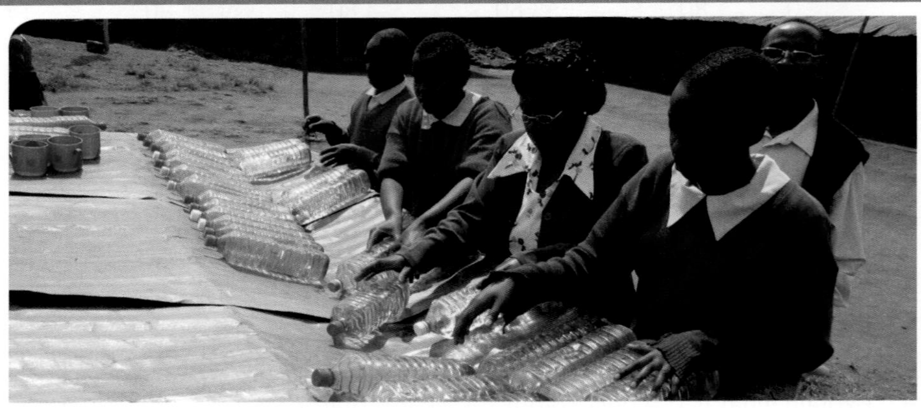

Ultraviolet

Ultraviolet radiation transfers more energy than visible light. It is absorbed by most of the same materials that absorb visible light, including our skin. The energy transferred can be used to disinfect water by killing microorganisms in it.

 **1** Why would there be UV lamps at a sewage works?

Some materials absorb ultraviolet radiation and re-emit it as visible light. This is called **fluorescence**. Fluorescent materials are often used in security markings – they are only visible when ultraviolet light shines on them.

Many low energy light bulbs are fluorescent lamps. A gas inside these lamps produces ultraviolet radiation when an electric current passes through it. A coating on the inside of the glass absorbs the ultraviolet and emits visible light.

 **2** Why might someone write their postcode on a TV or computer using a pen with fluorescent ink that is not visible in normal light?

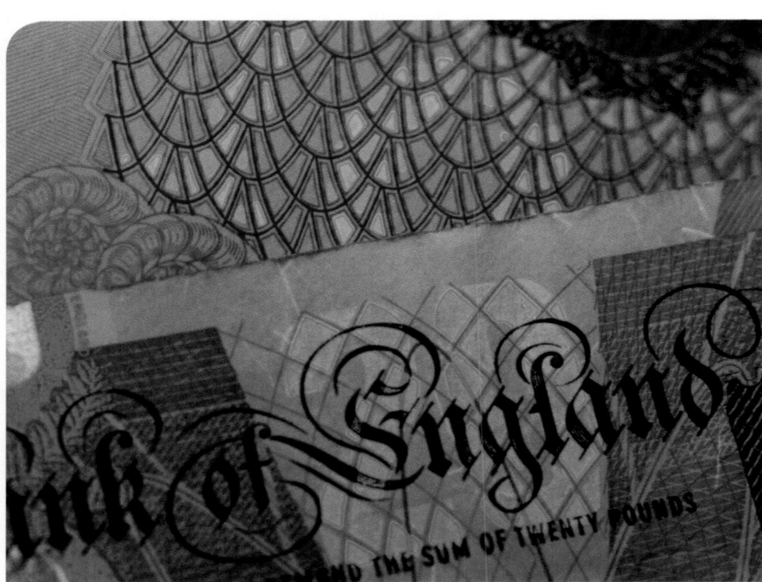

B Forged banknotes can be detected using UV light because they do not have markings that glow. These are real notes.

X-rays

X-rays can pass through many materials that visible light cannot. For example, they can pass through muscles and fat easily but bone absorbs some X-rays. This means X-rays can be used in medicine to make images of the inside of the body. X-rays can also be used to examine the insides of metal objects and to inspect luggage in airport security scanners.

3 Suggest two reasons why security staff at airports use X-ray scanners to check luggage instead of looking inside the luggage.

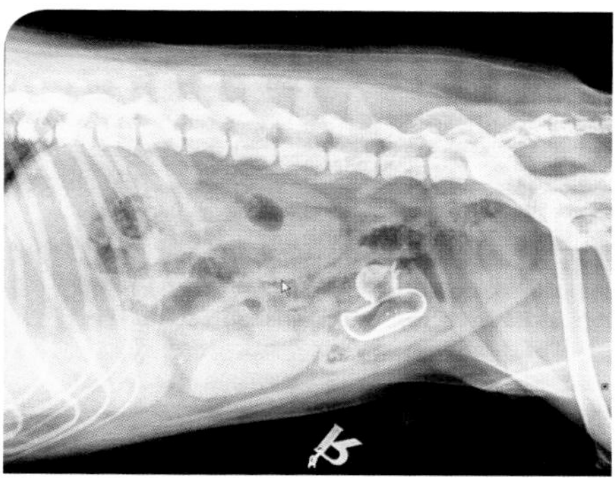

C This X-ray image shows that the dog has swallowed a toy duck.

Gamma rays

Gamma rays transfer a lot of energy, and can kill cells. For this reason, they are used to sterilise food and surgical instruments by killing potentially harmful microorganisms.

Gamma rays are used to kill cancer cells in **radiotherapy**. They can also be used to detect cancer. A chemical that emits gamma rays is injected into the blood. The chemical is designed to collect inside cancer cells. A scanner outside the body then locates the cancer by finding the source of the gamma rays. Gamma rays can pass through all the materials in the body.

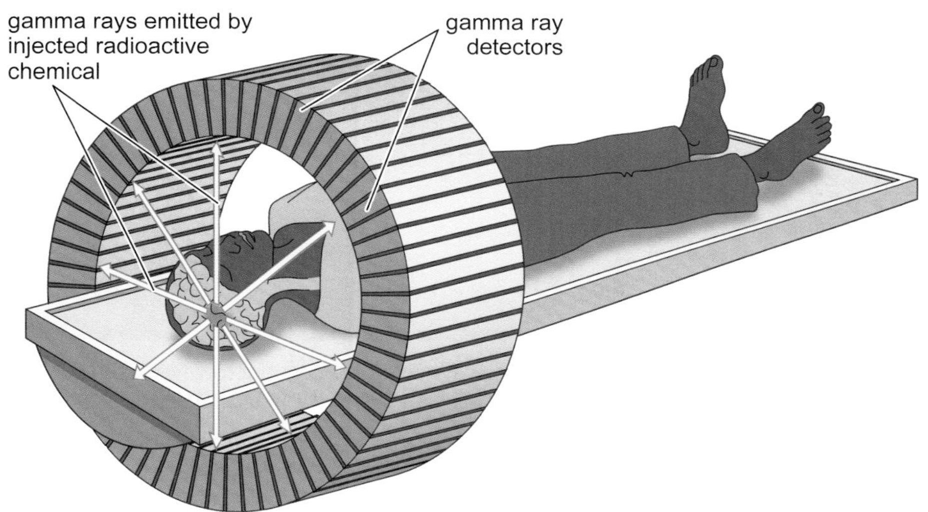

D a gamma ray medical scanner

4 Describe two ways in which gamma rays can be used for medical purposes.

5 H Describe the differences in the way muscle, fat and bone absorb or transmit X-rays and gamma rays.

Exam-style question

Describe three different ways in which electromagnetic radiation with frequencies greater than that of visible light can be used in medicine.

(3 marks)

Did you know?

Lenses, such as spectacle lenses, work because light travels more slowly in glass than in air and changes direction as it changes speed. The change of speed when X-rays enter different materials is very small, so X-rays can only be focused using several metal lenses together.

Checkpoint

How confidently can you answer the Progression questions?

Strengthen

S1 Describe three uses for:

- **a** ultraviolet radiation
- **b** X-rays
- **c** gamma rays.

Extend

E1 State two similarities and two differences between:

- **a** gamma rays and ultraviolet radiation
- **b** gamma rays and X-rays.

SP5i EM radiation dangers

Specification reference: P5.20; P5.21; P5.24

Progression questions

- What are the dangers of electromagnetic radiation?
- How is the danger associated with an electromagnetic wave linked to its frequency?
- How is electromagnetic radiation linked to changes in atoms and their nuclei?

A Mobile phone transmitters use different frequencies of microwaves compared with microwave ovens.

All waves transfer energy. A certain microwave frequency can heat water and this frequency is used in microwave ovens. This heating could be dangerous to people because our bodies are mostly water and so the microwaves could heat cells from the inside. Mobile phones use different microwave frequencies. Current scientific evidence tells us that, in normal use, mobile phone signals are not a health risk.

B The metal grid in the door of the microwave oven reflects microwaves but the holes allow visible light through.

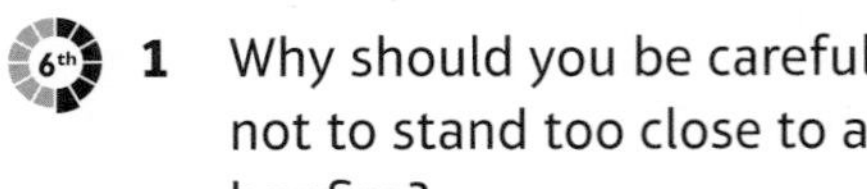

1 Why should you be careful not to stand too close to a bonfire?

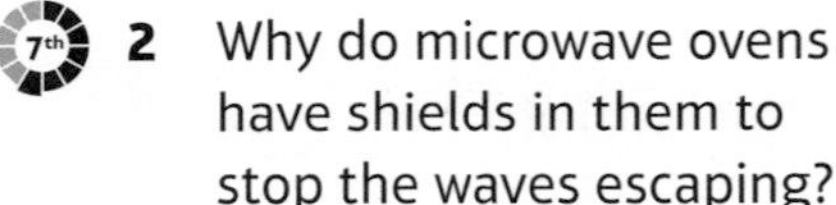

2 Why do microwave ovens have shields in them to stop the waves escaping?

Infrared radiation is used in grills and toasters to cook food. Our skin absorbs infrared, which we feel as heat. Too much infrared radiation can damage or destroy cells, causing burns to the skin.

Higher-frequency waves transfer more energy than low-frequency waves and so are potentially more dangerous. Sunlight contains high frequency ultraviolet radiation, which carries more energy than visible radiation. The energy transferred by ultraviolet radiation to our cells can cause sunburn and damage **DNA**. Too much exposure to ultraviolet radiation can lead to **skin cancer**. We can help to protect our skin by staying out of the strongest sunshine, covering up with clothing and hats, and using sun cream with a high SPF (sun protection factor).

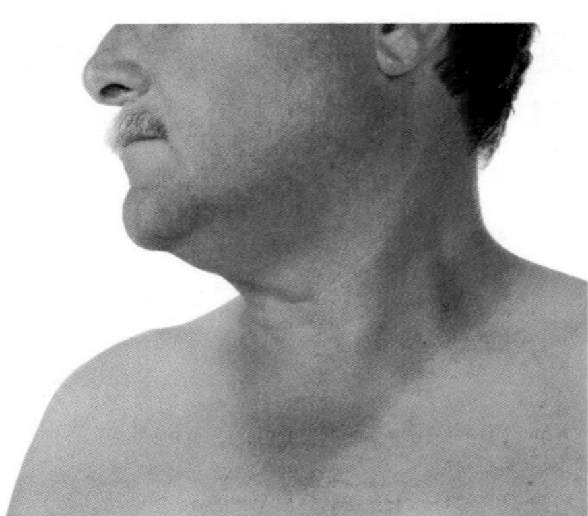

C Sunburn is caused by too much ultraviolet radiation being absorbed by the skin.

The ultraviolet radiation in sunlight can also damage our eyes. Skiers and mountaineers can suffer temporary 'snow blindness' because so much ultraviolet radiation is reflected from snow. We can protect our eyes using sunglasses.

D This photo was taken using ultraviolet light. The dark spots show parts of the skin that may have been damaged by exposure to lots of ultraviolet light from the Sun. Some of this damage could eventually turn into skin cancer.

 **3** State three ways to protect your body against damage by UV radiation when in bright sunlight.

X-rays and gamma rays are higher frequency than ultraviolet radiation and so transfer more energy. They also can penetrate the body. Excessive exposure to X-rays or gamma rays may cause **mutations** in DNA that can kill cells or cause cancer.

 **4** Why do people have hospital X-ray photographs taken if X-rays are so dangerous?

 **5** Two different electromagnetic waves have frequencies of 10 000 Hz and 100 000 Hz. Explain which wave is likely to cause the most harm if absorbed by your body.

Radiation and atoms

Electromagnetic radiation is produced by changes in the electrons or the nuclei in atoms. For example, when materials are heated, changes in the way the electrons are arranged can produce infrared radiation or visible light. Changes in the nuclei of atoms can produce gamma radiation.

Radiation can also cause changes in atoms, such as causing atoms to lose electrons to become ions. You will learn more about this in Unit *SP6 Radioactivity*.

 **6** Explain why gamma radiation produces positive ions.

Exam-style question

Look back at diagram C on *SP5e The electromagnetic spectrum*. Explain why changes in the composition of the atmosphere could cause an increase in skin cancer. *(2 marks)*

Did you know?

The world's oldest snow goggles are around 2000 years old and were made from leather, bone or wood. They protected the wearer from snow blindness by having only narrow slits to see through.

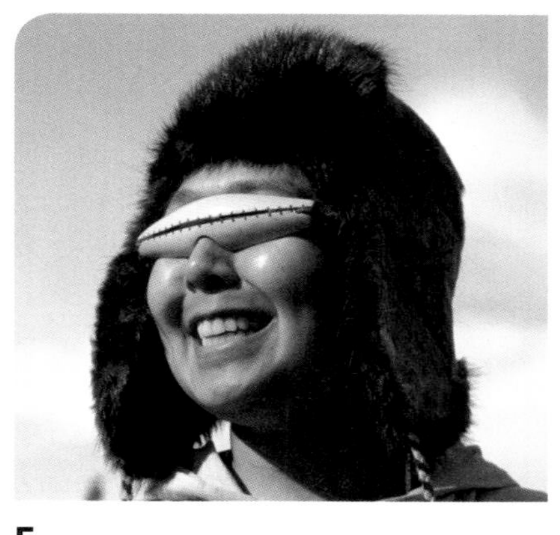

E

Checkpoint

How confidently can you answer the Progression questions?

Strengthen

S1 State one danger to your body of:

- **a** microwaves
- **b** infrared radiation
- **c** ultraviolet radiation
- **d** X-rays and gamma rays.

Extend

E1 Draw a table with a row for each part of the electromagnetic spectrum mentioned on these pages, with frequency increasing down the table. In the second column list any hazards that you know of for each part of the spectrum. Use your table to explain how the frequency relates to the potential danger.

Electromagnetic waves

Infrared and ultraviolet waves have different frequencies. Both types of wave can have harmful effects on humans. Compare and contrast the harmful effects of infrared and ultraviolet waves. **(6 marks)**

Student answer

Infrared waves and ultraviolet waves can both damage our skin [1]. *Infrared waves can cause skin burns, for example when we have been sunbathing* [2]. *Ultraviolet waves can damage our eyes and damage the cells in our skin which can lead to skin cancer* [3]. *Ultraviolet waves have higher frequencies than infrared waves, which is why they cause more harm* [4].

[1] This is a similarity between the harm caused by the two types of radiation. This question asks you to 'compare and contrast', so you need to include both similarities and differences.

[2] The student has mentioned a harmful effect of infrared.

[3] This mentions harmful effects of ultraviolet, and so is contrasting the harm with the harm caused by infrared.

[4] This last part explains why ultraviolet causes more harm by linking its frequency with the danger.

Verdict

This is a strong answer. It mentions the different types of harm that can be caused by each type of wave, and mentions similarities between the waves. The answer also links ideas together by pointing out that the higher frequency waves are more harmful.

Exam tip

Once you have written your answer, read the question again to make sure you have answered all parts of the question. In this case, don't forget to relate the harm caused by the different types of wave to their frequency.

Paper 1

SP6 Radioactivity

In the first half of the twentieth century, radium paint was used to make objects such as watch hands and numbers glow in the dark. The paint was radioactive. The women who painted the watch faces used to 'point' their brushes by putting them in their mouths. The women did not know about the harmful effects of this until they noticed that their teeth were beginning to fall out and their jaw bones were collapsing. In this unit you will find out more about atoms and their structure, and how atoms can produce radioactivity when they change.

The learning journey

Previously you will have learnt at KS3:

- about the particle model of matter
- that atoms contain smaller charged particles called electrons
- about nuclear fuel as a non-renewable energy resource.

In this unit you will learn:

- how the particles inside atoms are arranged
- how to represent atoms using symbols
- about the different types of radiation and how they affect atoms
- about the background radiation that is all around us
- about uses of radioactivity in the home and industry
- about the dangers of radiation and how we can protect ourselves
- how radioactive materials are used to diagnose and treat cancer
- about the advantages and disadvantages of nuclear power
- what fusion and fission nuclear reactions are.

SP6a Atomic models

Specification reference: P6.1; P6.2; P6.17

Progression questions

- What particles make up atoms?
- How big are atoms?
- How has our model of the atom changed over time?

1.2×10^{-10} m
(0.000 000 000 12 m, or 0.12 millionths of a millimetre)
a carbon dioxide molecule modelled using spheres
an oxygen atom modelled as a sphere

A Atoms and molecules are often modelled as spheres.

Particle theory (or **kinetic theory**) is a model that helps explain the properties of solids, liquids and gases. The particles are usually represented as spheres.

We can explain some of the properties of different **elements** by thinking about the particles that each contains. We call these particles **atoms**. Chemical reactions occur when the different atoms in substances become joined in different ways.

6th

1 How does particle theory explain why solids have a fixed shape?

These ideas helped scientists throughout the 1800s. However, in 1897 J.J. Thomson (1856–1940) carried out some experiments that showed that atoms contain much smaller **subatomic particles** called **electrons**. These had a negative **charge** and hardly any mass.

In the 1900s, Thomson supported using a new model for atoms that could explain this new evidence. This new model described the atom as a 'pudding' made of positively charged material, with negatively charged electrons (the 'plums') scattered through it.

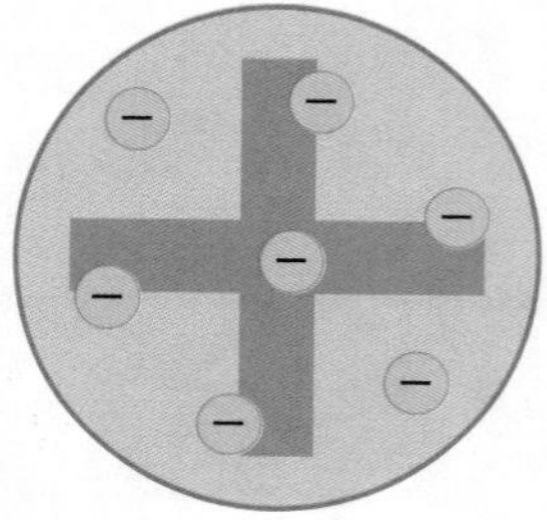

B the plum pudding model of the atom

8th

2 Positive and negative charges cancel each other out. How does the plum pudding model explain that atoms have no overall charge?

8th

3 Carbon atoms are roughly the same size as oxygen atoms. How long is the carbon dioxide molecule shown in diagram A?

Between 1909 and 1913, a team of scientists led by Ernest Rutherford (1871–1937) carried out a series of experiments that involved studying what happened when positively charged subatomic particles, called **alpha particles**, passed through various substances (such as gold foil).

In the experiment shown in diagram C, the scientists discovered that most of the alpha particles passed through the gold foil, but a few bounced back. The plum pudding model could not explain this result.

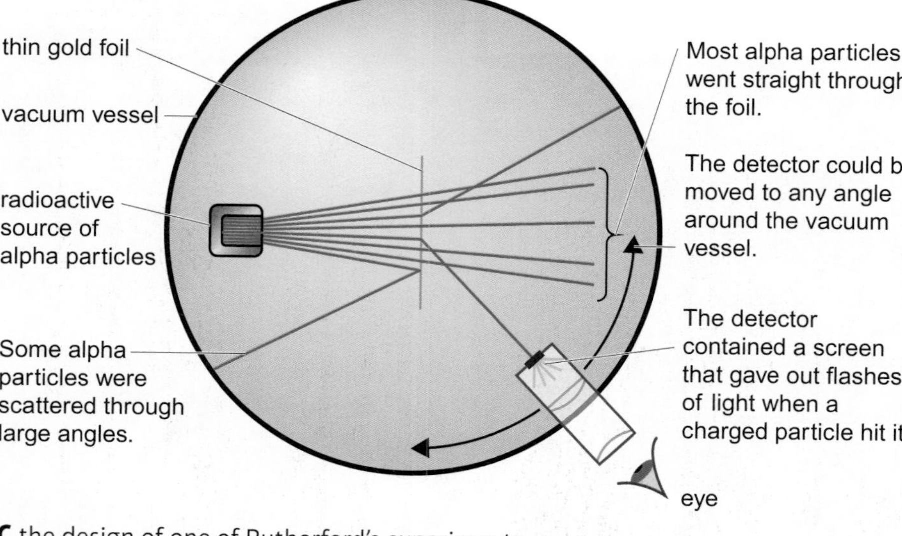

thin gold foil
vacuum vessel
radioactive source of alpha particles
Some alpha particles were scattered through large angles.
Most alpha particles went straight through the foil.
The detector could be moved to any angle around the vacuum vessel.
The detector contained a screen that gave out flashes of light when a charged particle hit it.
eye

C the design of one of Rutherford's experiments

Rutherford suggested that atoms were mostly empty space, with most of their mass in a tiny central **nucleus** with a positive charge and electrons moving around the nucleus. Figure D shows how this model of the atom explains the results.

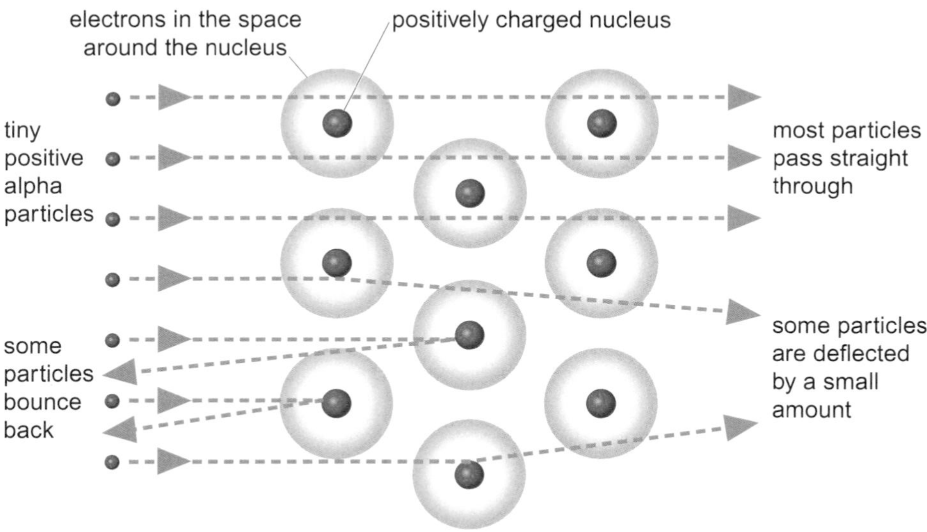

D The small nucleus in Rutherford's model explained why a small number of alpha particles were deflected by the gold foil.

Today, we know that the radius of a nucleus is about 1×10^{-15} m (0.000 000 000 000 001 m). The radius of an atom is about 1×10^{-10} m (0.000 000 000 1 m). So the atom itself is 100 000 times bigger than the nucleus inside it.

E If an atom had the same diameter as the dome of the O2, its nucleus would be the size of this dot ●.

 **5** A dot one millimetre in diameter represents the nucleus of an atom. To the same scale, how far across would the whole atom be?

Exam-style question

Compare and contrast Rutherford's model of the atom with the plum pudding model. *(4 marks)*

Did you know?

Rutherford did not expect the results the team found. He said 'It was almost as incredible as if you fired a 15-inch shell at a piece of tissue paper and it came back and hit you.'

 **4** Which part of Rutherford's atomic model is responsible for some alpha particles bouncing back?

Checkpoint

How confidently can you answer the Progression questions?

Strengthen

S1 Describe Rutherford's model of the atom. Include these words in your description: charge, electron, mass, nucleus.

Extend

E1 Draw a labelled diagram to show Rutherford's model of the atom and explain why it is not drawn to scale. (Note that neutrons had not been discovered when Rutherford was carrying out his experiments.)

E2 Describe two pieces of evidence that support this model of the atom.

SP6b Inside atoms

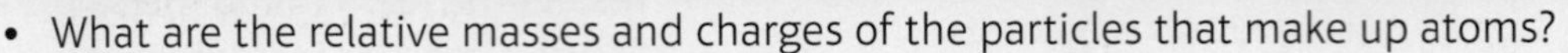

Specification reference: P6.3; P6.4; P6.5; P6.6

Progression questions

- What are the relative masses and charges of the particles that make up atoms?
- What are isotopes of an element?
- How can isotopes be represented using symbols?

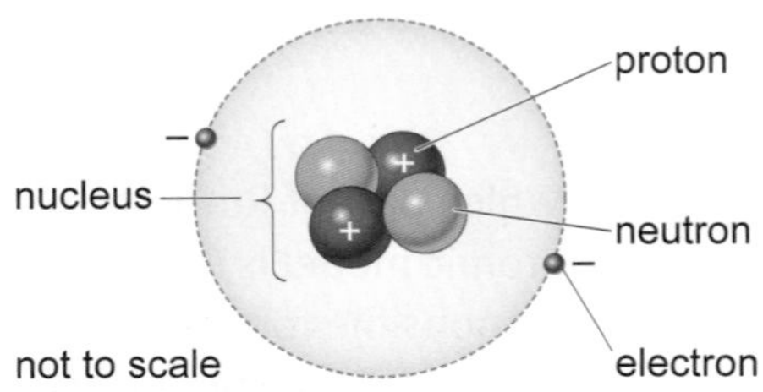

A the structure of an atom

The mass of an atom is concentrated in its nucleus. The nucleus itself is made up of smaller particles called **nucleons**. Nucleons can be **protons** or **neutrons**. All subatomic particles have very small masses so it is easier to describe their **relative masses**. We give the proton a mass of 1 and we compare the masses of the other subatomic particles relative to this. Table B summarises the subatomic particles within atoms.

Subatomic particle	Location in atom	Relative charge	Relative mass
proton	nucleus	+1 (positive)	1
neutron	nucleus	0	1
electron	around nucleus	−1 (negative)	$\frac{1}{1835}$ (negligible)

B The mass of an electron is so small that it is usually ignored when talking about the mass of an atom.

1 Which two subatomic particles:

 **a** are nucleons

 **b** have a charge

 **c** have a relative mass of 1?

2 What do these numbers tell you about the numbers of protons, neutrons and electrons in the atoms of an element?

 **a** the atomic number

 **b** the mass number

All atoms of a particular element have the same number of protons. This number is called the **atomic number** or **proton number** of the element. Atoms of different elements have different numbers of protons and so have different atomic numbers.

Neutrons have no charge and so it is the protons that give the nucleus its positive charge. Atoms have the same number of electrons as protons and so atoms are always electrically neutral (they have no overall charge).

The number of neutrons in an atom can vary. The **mass number** or **nucleon number** is the total number of protons and neutrons in the nucleus.

We can represent the atomic number and mass number of an element in symbol form, as shown in diagram C.

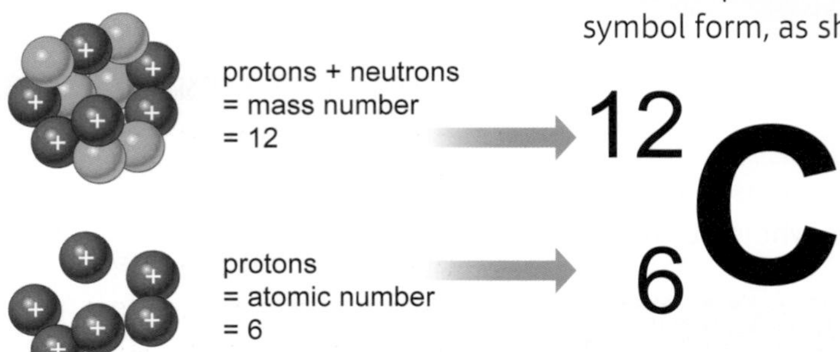

$^{12}_{6}C$

C In symbol form, the mass number is always above the atomic number in both value and position.

Worked example

An atom of nitrogen has 7 protons and 7 neutrons. Show this atom in symbol form, with its mass and atomic numbers.

mass number = 7 + 7 = 14

In symbols, this is written as $^{14}_{7}N$.

Isotopes

Two atoms of the same element will always have the same atomic number, but they can have different mass numbers if they contain different numbers of neutrons. Atoms of a single element that have different numbers of neutrons are called **isotopes**.

For example, carbon can occur naturally as carbon-12, carbon-13 or carbon-14. The number in the name is the mass number of the isotope. The atomic number of carbon is 6, so an atom of carbon-14 has 6 protons and 8 neutrons in its nucleus.

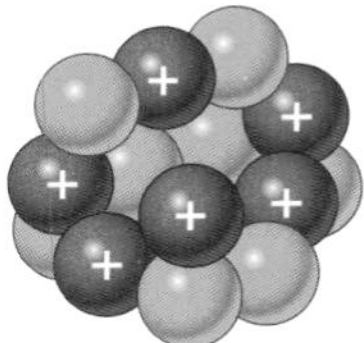

D a carbon-14 nucleus

Did you know?

The age of the remains of extinct animals such as this mammoth can be found by examining the different isotopes of carbon in them.

E

Exam-style question

Sam says that as two different atoms have the same mass number, they must be isotopes. Is Sam correct? Explain your answer. *(2 marks)*

3 A helium nucleus has 2 protons and 2 neutrons. Write down the symbol for this isotope of helium.

4 The symbol for an isotope of beryllium is ${}^{9}_{4}Be$. How many of the following particles does it contain?

a protons

b neutrons

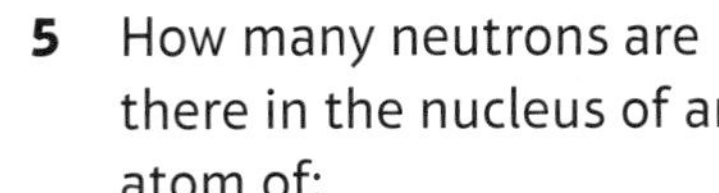

5 How many neutrons are there in the nucleus of an atom of:

a oxygen-18

b carbon-13?

Checkpoint

How confidently can you answer the Progression questions?

Strengthen

S1 Write glossary definitions for these terms: atomic number, electron, isotope, mass number, neutron, proton.

Extend

E1 Hydrogen has three isotopes, hydrogen-1, hydrogen-2 and hydrogen-3. The atomic number of hydrogen is 1. Explain what an isotope is and give the similarities and differences between these three isotopes.

SP6c Electrons and orbits

Specification reference: P6.7; P6.8; P6.9; P6.17

Progression questions

- How are electrons arranged in an atom?
- What happens to atoms when they absorb or emit electromagnetic radiation?
- How do atoms become ionised?

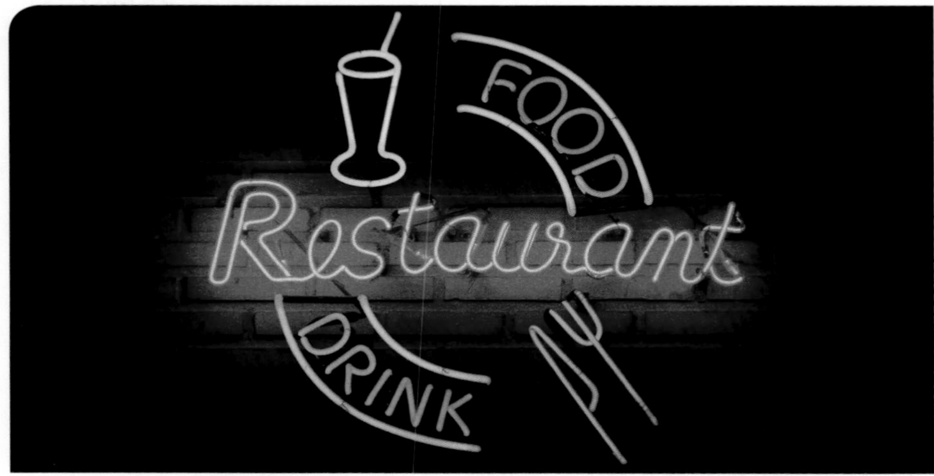

A The different colours in this sign are due to the different gases inside the tubes. The orange tubes contain neon gas.

These orbits (electron shells) are normally empty in neon atoms.

nucleus

If an atom absorbs energy, an electron can move to a 'higher' orbit.

When an electron returns to a lower orbit the atom emits energy as visible light of a particular wavelength.

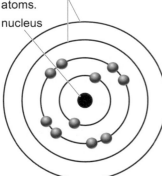

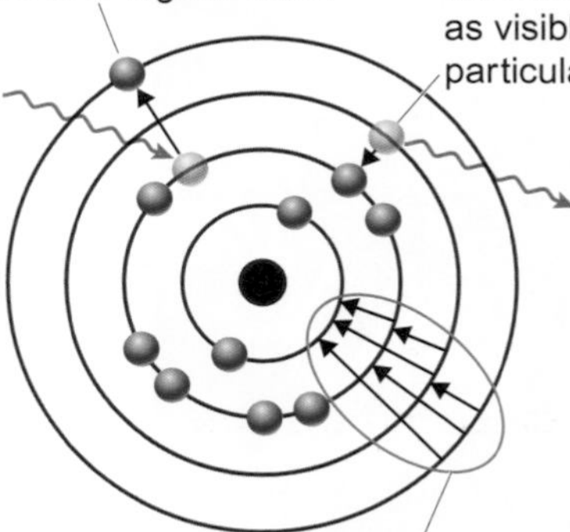

Electrons can make all of these different orbit changes. Each different change produces a different wavelength of light.

B electronic configuration and energy level changes for neon

emission

absorption

C emission and absorption spectra for neon

The tubes in photo A produce light when an electrical voltage makes electrons move within atoms of a gas.

The electrons in an atom can only exist in certain **orbits** around the nucleus, called **electron shells**. Each electron shell is at a different energy level. Diagram B shows the **electronic configuration** for neon.

In a neon tube, the neon atoms absorb energy transferred by the electricity because the electrons jump to higher shells. When the electrons fall back again they emit energy as **electromagnetic radiation** that we can see. The top part of photo C shows the colours of visible light emitted by neon. Each colour is a different **wavelength** of light. This is called an **emission spectrum**. The emission spectrum is different for each element.

1 Look at photo A. How do you know that several different gases are used in the sign?

2 Look at diagram B. Explain why neon can emit light at lots of different wavelengths.

3 Suggest an explanation for why neon tubes glow orange.

Gases can also absorb energy transferred by electromagnetic radiation, such as **visible light**. The bottom part of photo C shows the parts of the **visible spectrum** that neon gas absorbs when light passes through it. This is the **absorption spectrum** for neon. The wavelengths of light that neon gas absorbs are the same wavelengths that it emits.

Niels Bohr (1885–1962) amended Rutherford's model of the atom to explain observations like these by suggesting that electrons can only be in certain fixed orbits (electron shells) around the nucleus. They cannot be part-way between two orbits. This model could explain the lines in emission and absorption spectra.

Ionisation

Sometimes an atom gains so much energy that one or more of the electrons can escape from the atom altogether. An atom that has lost or gained electrons is called an **ion**. Radiation that causes electrons to escape is called **ionising radiation**.

An atom has the same number of protons and electrons, so overall it has no charge. If an atom loses an electron, it then has one more proton than it has electrons. It has an overall positive charge and is called a **positive ion**.

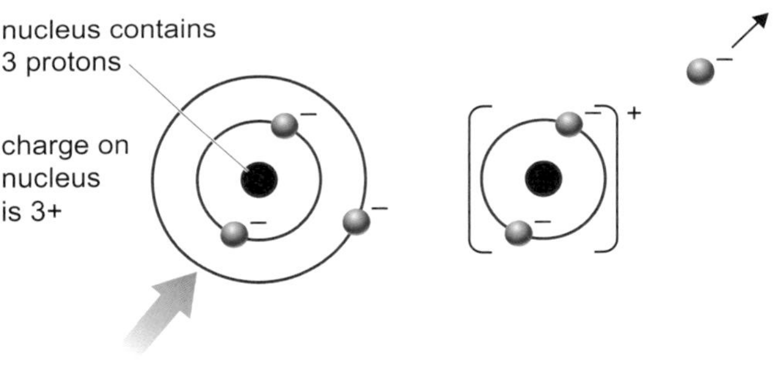

D ionisation of a lithium atom

 **5** Describe what happens if an atom absorbs less energy than the amount needed to ionise it.

6 Sodium has an atomic number of 11 and loses 1 electron when it forms an ion. How many protons and electrons are in:

 **a** a sodium atom

 **b** a sodium ion?

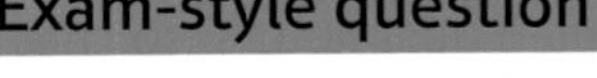

Exam-style question

Describe one piece of evidence that supports Bohr's idea of atoms having fixed orbits for electrons. *(2 marks)*

 **4** **a** What is the Rutherford model of the atom? (You may need to look back at *SP6a Atomic models*.)

 **b** How is Bohr's model different from Rutherford's model?

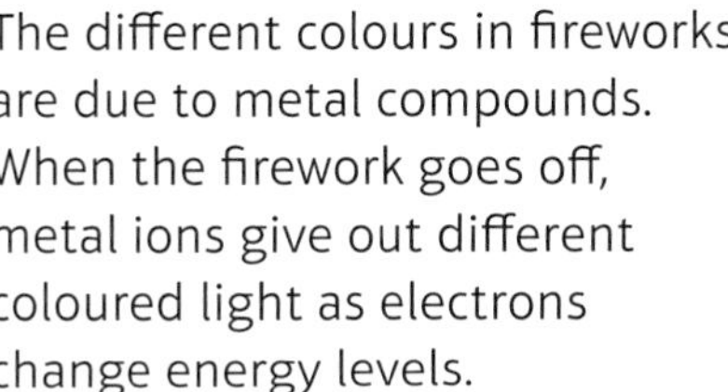

Did you know?

The different colours in fireworks are due to metal compounds. When the firework goes off, metal ions give out different coloured light as electrons change energy levels.

Checkpoint

How confidently can you answer the Progression questions?

Strengthen

S1 Draw a labelled diagram to describe the Bohr model of the atom.

S2 Describe how an atom becomes a positive ion.

Extend

E1 Describe the link between energy, the emission of light and the orbits of electrons in an atom.

SP6d Background radiation

Specification reference: P6.12; P6.13; P6.14

Progression questions

- What is meant by background radiation?
- What are the sources of background radiation?
- How is radioactivity detected and measured?

A This radon outlet pipe sucks air containing radon from beneath a solid concrete floor, stopping it from entering the house.

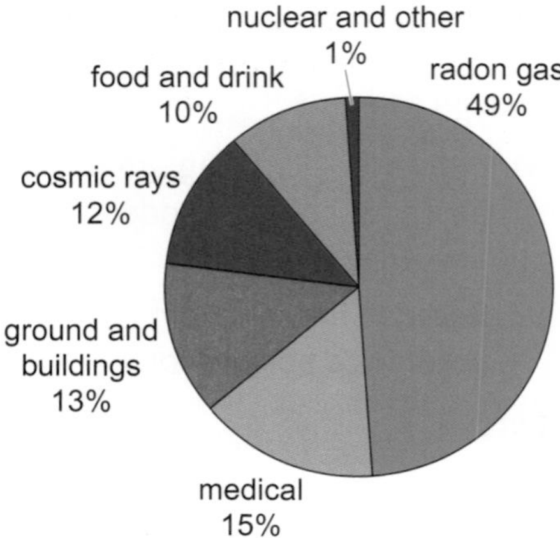

B sources of background radiation in the UK

We are constantly being exposed to ionising radiation at a low level, from space and from naturally radioactive substances in the environment. This is called **background radiation**.

Sources of background radiation

Chart B below shows the sources of background radiation averaged over the UK. The main source is radon gas. This radioactive gas is produced by rocks that contain small amounts of uranium. Radon diffuses into the air from rocks and soil and can build up in houses, especially where there is poor ventilation. The amount of radon in the air depends on the type of rock and its uranium content. Rock type and building stone vary around the country and so does the amount of radon.

 **1** Explain why background radiation varies in different parts of the UK.

Some foods contribute to your exposure to background radiation because they naturally contain small amounts of radioactive substances. Hospital treatments, such as X-rays, gamma-ray scans and cancer treatments, also contribute to people's exposure to background radiation.

High-energy, charged particles stream out of the Sun and other stars. They are known as **cosmic rays** and are a form of radiation. Many cosmic rays are stopped in the upper atmosphere but some still reach the Earth's surface.

 **2** Approximately how much background radiation in the UK comes from natural sources?

 **3** Suggest how your food and drink become naturally radioactive.

 **4** During a solar storm, the Sun's output of high-energy charged particles increases dramatically. Explain why some scientists suggest that aeroplanes should fly at lower altitudes during a solar storm.

Measuring radioactivity

Radioactivity can be detected using photographic film, which becomes darker and darker as more radiation reaches it. However, the film has to be developed in order to measure the amount of radiation (the **dose**). People who work with radiation often wear film badges (called dosimeters) to check how much radiation they have been exposed to. Newer dosimeters use materials that change colour without needing to be developed.

The radioactivity of a source can also be measured using a **Geiger-Müller (GM) tube**. Radiation passing through the tube ionises gas inside it and allows a short pulse of current to flow.

A GM tube can be connected to a counter, to count the pulses of current, or the GM tube may give a click each time radiation is detected. The **count rate** is the number of clicks per second or minute.

Did you know?

Henri Becquerel (1852–1908) discovered radioactivity when he put some substances he was investigating on top of a photographic plate in an envelope. When he developed the plate he found it had become dark and he suggested that some 'rays' emitted by the material had caused this effect.

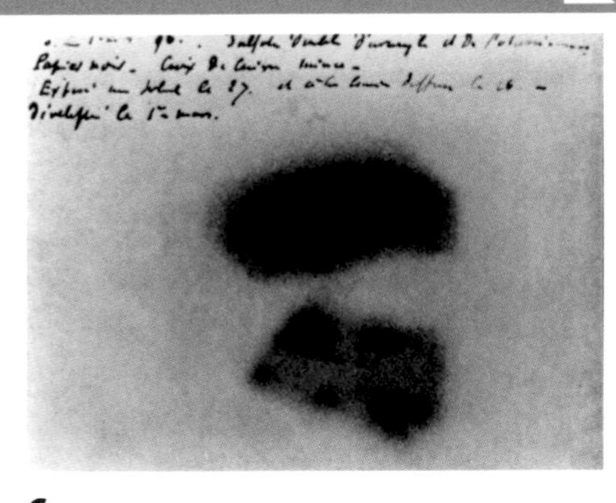

C

D using a GM tube and counter to measure radiation

When scientists measure the radioactivity of a source, they need to measure the background radiation first by taking several readings and finding the mean. This mean value is then subtracted from measurements.

 **5** Tom records background counts of 15, 22 and 17 counts per minute. He then records a count rate of 186 counts per minute from a sample of granite. What is the corrected count rate for the sample's activity?

Exam-style question

The following was found on a blog: 'Natural radiation won't hurt you but human-made radiation will.' Comment on this statement. *(4 marks)*

Checkpoint

How confidently can you answer the Progression questions?

Strengthen

S1 List three sources of background radiation.

S2 Describe two ways of detecting radiation.

Extend

E1 Explain why the measurements of the activity of a radioactive source must be corrected.

SP6e Types of radiation

Specification reference: P6.5; P6.10; P6.11; P6.15; P6.16

Progression questions

- What are alpha particles, beta particles and gamma radiation?
- How do the different kinds of radiation compare in their ability to penetrate materials?
- How do the different kinds of radiation compare in their ability to ionise atoms?

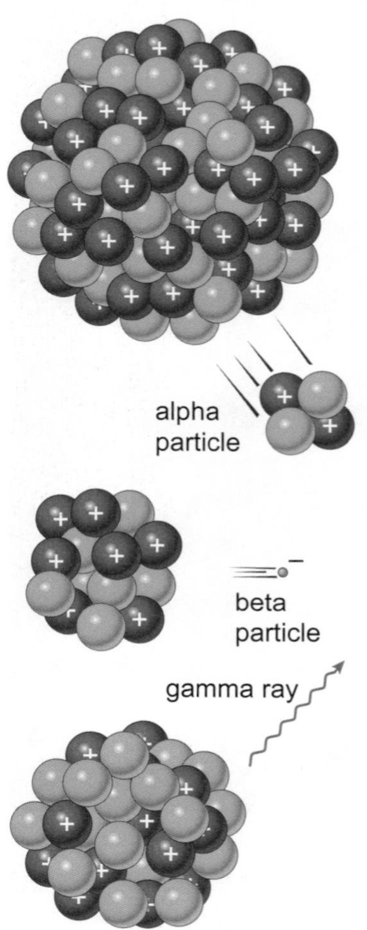

A Three types of radiation that can be emitted by unstable nuclei.

 **1** Draw a table to summarise the charges and relative masses of the five different types of radiation emitted from atomic nuclei.

 **2** Carbon-12 is a stable isotope and carbon-14 is an unstable isotope. Explain which of these isotopes will decay.

The nucleus of a radioactive substance is **unstable**, which means it can easily change or **decay**. When decay occurs, radiation is emitted which causes the nucleus to lose energy and become more stable. You cannot predict when a nucleus will decay – it is a **random** process.

Types of radiation

There are different sorts of radiation that a nucleus can emit when it decays.

Alpha particles contain two protons and two neutrons, just like the nucleus of a helium atom. They have a relative mass of 4. They have no electrons and so have a charge of +2. They can be written as $\boldsymbol{\alpha}$ or $^{4}_{2}\mathbf{He}$.

The electrons that are beta particles come from the *nuclei* of atoms when a neutron transforms into a proton. Beta particles do not ionise the atoms as they leave them.

Beta particles are high-energy, high-speed electrons. They have a relative mass of $\frac{1}{1835}$ and a charge of −1. They can be written as $\boldsymbol{\beta}^-$ or $^{\;0}_{-1}\mathbf{e}$.

This number represents charge.

Positrons are high-energy, high-speed particles with the same mass as electrons but a charge of +1. They can be written as $\boldsymbol{\beta}^+$ or $^{\;0}_{+1}\mathbf{e}$.

Gamma rays (γ) are high-frequency electromagnetic waves (they travel at the speed of light). They do not have an electric charge.

Neutrons can also be emitted from an unstable nucleus. They have a relative mass of 1 and no electric charge.

Did you know?

Electrons are normal matter particles and positrons are antimatter particles. If a positron meets an electron, the two particles will annihilate each other, releasing energy. The energy released is far more than can be obtained from normal chemical fuels and NASA are carrying out research to see if antimatter can be used to power spacecraft.

B

Ionising and penetrating radiation

The types of radiation described above are all examples of ionising radiation, and they can all **penetrate** (pass through) materials.

Alpha particles are emitted at high speeds. Due to this and their high relative mass, they transfer a lot of energy and so are good at ionising atoms they encounter. However, each time they ionise an atom they lose energy. Since they produce many ions in a short distance, they lose energy quickly and have a short penetration distance.

Beta particles are much less ionising than alpha particles and so can penetrate much further into matter than alpha particles can. Gamma-rays are about ten times less ionising than beta particles and can penetrate matter easily.

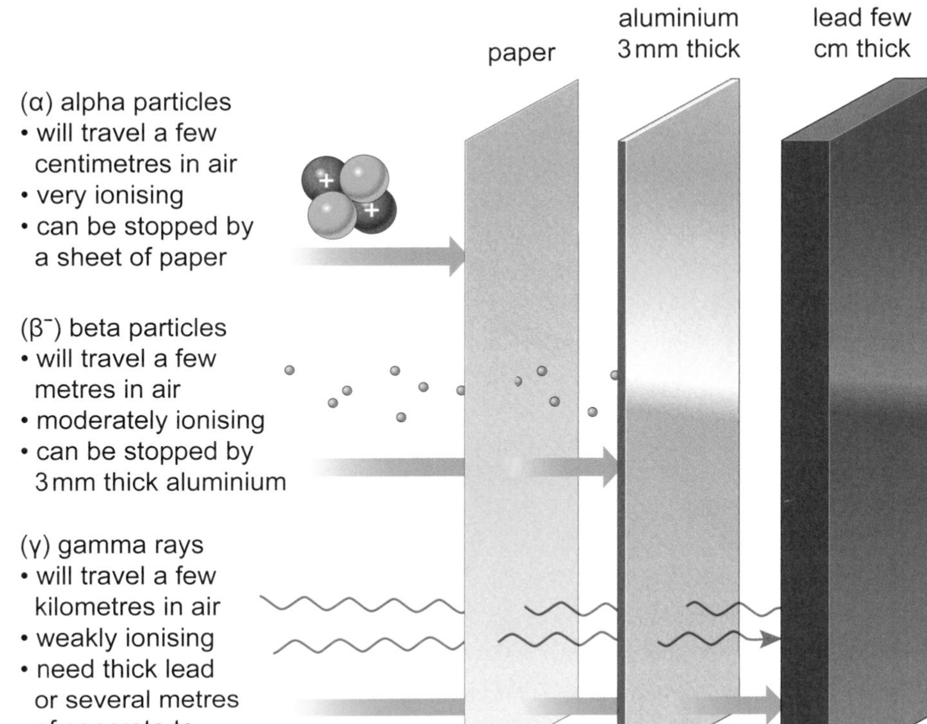

paper
aluminium 3 mm thick
lead few cm thick
(α) alpha particles
• will travel a few centimetres in air
• very ionising
• can be stopped by a sheet of paper
(β⁻) beta particles
• will travel a few metres in air
• moderately ionising
• can be stopped by 3 mm thick aluminium
(γ) gamma rays
• will travel a few kilometres in air
• weakly ionising
• need thick lead or several metres of concrete to stop them

C the penetrating properties of alpha, beta and gamma radiation

D Gamma-rays can be used to check the inside of lorries, to help prevent the movement of illegal goods.

8th
3 What materials will absorb and stop beta particles?

10th
4 The reactor in a nuclear power station is surrounded by large amounts of concrete. Why is this necessary?

10th
5 Look at photo D. Explain why gamma radiation is used for checking lorries.

10th
6 Explain how an oxygen molecule in the air might become an ion by being near a radioactive source.

Exam-style question

Compare and contrast the emission of an electron during the ionisation of an atom with the emission of an electron during β^- decay. *(4 marks)*

Checkpoint

How confidently can you answer the Progression questions?

Strengthen

S1 Describe alpha particles, beta particles and gamma radiation.

S2 Draw up a table to summarise the penetration and ionisation properties of the three types of radiation in question **S1**.

Extend

E1 Use the characteristics of alpha and beta particles to explain the differences in their abilities to ionise and penetrate.

SP6f Radioactive decay

Specification reference: P6.18; P6.19; P6.20; P6.21; P6.22

Progression questions

- How does beta decay occur?
- How are atomic and mass numbers affected by different kinds of decay?
- How can radioactive decays be represented in nuclear equations?

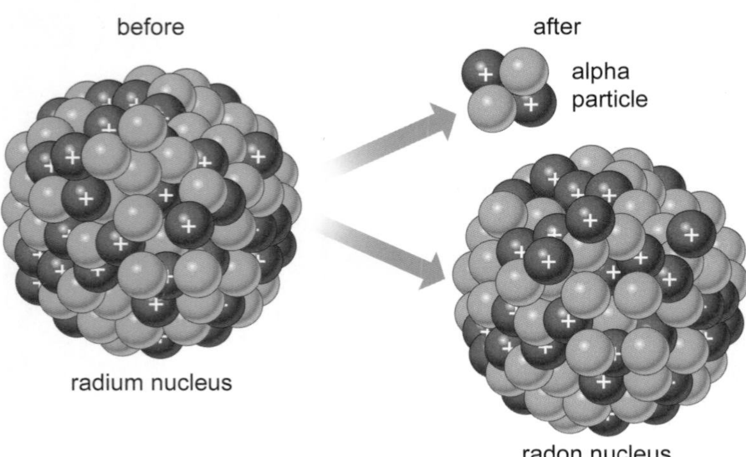

A Alpha decay turns a radium nucleus into a radon nucleus.

When an unstable nucleus changes and emits particles, the atomic number can change. If this happens, the atom becomes a different element. For example, when an alpha particle is emitted, the mass number of the nucleus goes down by 4 and the atomic number goes down by 2.

During β^- decay, a neutron changes into a proton and an electron. The electron is ejected from the atom. The atomic number increases by 1 but there is no change to the mass number.

In β^+ decay a proton becomes a neutron and a positron. The atomic number goes down by 1 but the mass number does not change.

1 Carbon-14 decays by emitting a beta particle. How is the beta particle formed?

If a neutron is ejected from a nucleus, the mass number goes down by 1 but the atomic number does not change.

Nuclei may also lose energy as gamma radiation when the subatomic particles in the nucleus are rearranged. This helps to make them more stable.

Did you know?

Long-lasting lights can be made using tritium, a radioactive isotope of hydrogen (hydrogen-3). The tritium is placed in a sealed glass tube coated on the inside with a material that glows when it is hit by the particles produced as the tritium decays.

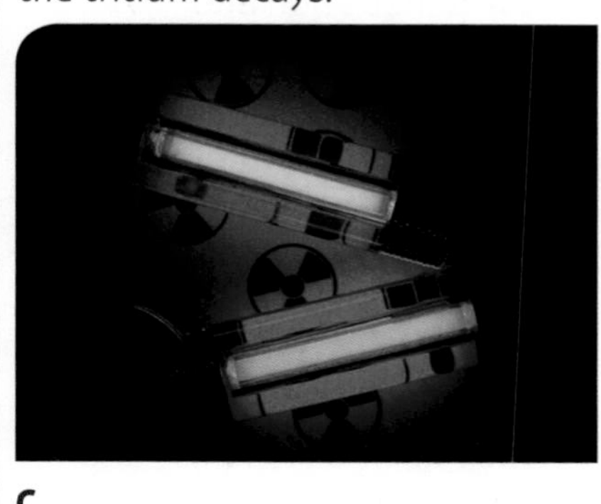

C

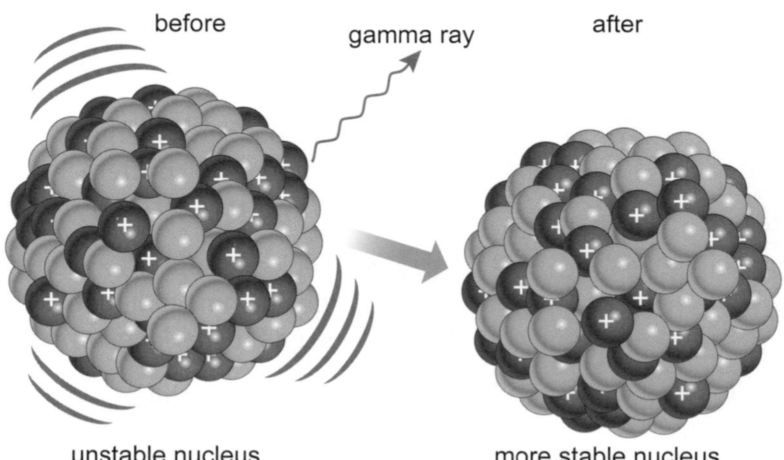

B gamma radiation

2 Explain why the numbers of nucleons do not change if a nucleus emits only gamma radiation.

3 When a neutron is ejected, explain why:

a the mass number drops **b** the atomic number stays the same.

Nuclear equations

A **nuclear equation** shows what happens during radioactive decay. The equation must be balanced – the total mass number must be the same on each side and the total charges must be the same.

Table D shows the symbols used to represent the various particles. The atomic (proton) number for a nucleus or an alpha particle also represents the amount of positive charge. We give beta particles and positrons a –1 or +1 to indicate their charge.

Particle	Symbol	
alpha	α	${}^{4}_{2}He$
beta	β^-	${}^{0}_{-1}e$
positron	β^+	${}^{0}_{+1}e$
neutron		n

D symbols used in nuclear equations

Worked example W1

Radium-226 emits an alpha particle. What is the other product?

$${}^{226}_{88}Ra \rightarrow {}^{4}_{2}He + ?$$

On the right of the arrow the nucleus has an atomic number of 88 – 2 = 86. This is radon (Rn). (The atomic numbers also represent the positive charges.)

Mass numbers must also balance. The radon nucleus has a mass number of 226 – 4 = 222.

$${}^{226}_{88}Ra \rightarrow {}^{4}_{2}He + {}^{222}_{86}Rn$$

Worked example W2

Iodine-131 undergoes β^- decay. What is the other product?

$${}^{131}_{53}I \rightarrow {}^{0}_{-1}e + ?$$

The mass number stays the same. The atomic number goes up by 1 to 54. This is xenon (Xe). The atomic numbers represent positive charges and the –1 on the beta particle represents a negative charge.

$${}^{131}_{53}I \rightarrow {}^{0}_{-1}e + {}^{131}_{54}Xe$$

4 Write balanced nuclear equations to show the following decays. You will need to use a periodic table.

a Polonium-208 (${}^{208}_{84}Po$) undergoes α decay.

b Technetium-99 (${}^{99}_{43}Tc$) undergoes β^- decay.

c Potassium-37 (${}^{37}_{19}K$) undergoes β^+ decay.

5 Explain the difference between radioactive decay and a chemical reaction in the way new substances are formed.

Exam-style question

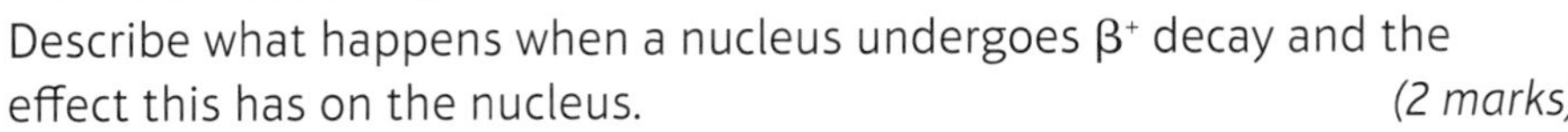
Describe what happens when a nucleus undergoes β^+ decay and the effect this has on the nucleus. *(2 marks)*

Checkpoint

How confidently can you answer the Progression questions?

Strengthen

S1 Draw up a table to summarise the different types of radioactive decay, and what effect each one has on the atomic number and mass number of the nucleus.

S2 Carbon-14 (${}^{14}_{6}C$) undergoes β^- decay. Write a balanced nuclear equation to show this.

Extend

E1 Polonium-216 (${}^{216}_{84}Po$) undergoes alpha decay and the product then undergoes β^- decay. Write two nuclear equations to show this sequence of decay.

SP6g Half-life

Specification reference: P6.23; P6.24; P6.25; P6.26; P6.27

Progression questions

- How does the activity of a substance change over time?
- What does the half-life of a radioactive substance describe?
- How can the half-life be used to work out how much of a substance decays?

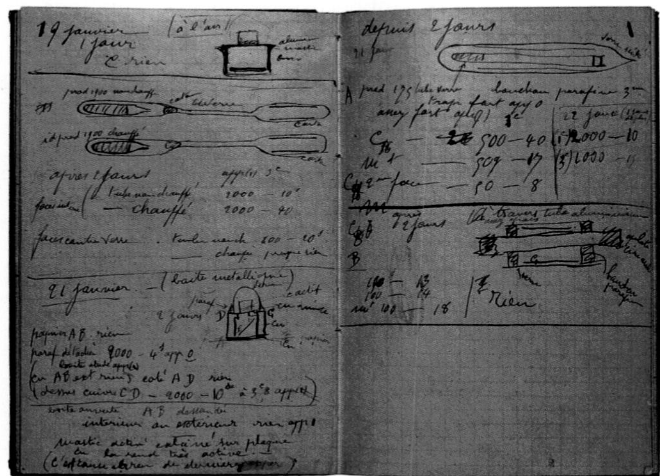

A one of Marie Curie's notebooks

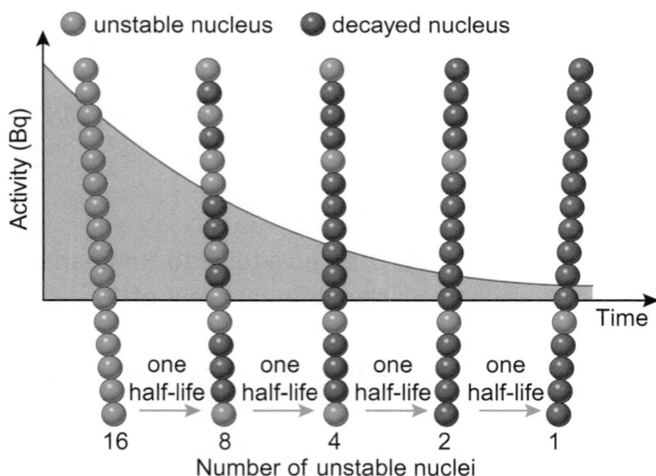

B After each half-life the number of unstable nuclei halves.

2 What does the half-life of an isotope describe?

3 A 10 kg sample of caesium-137 has a half-life of 30 years. What is the half-life of a 5 kg sample?

Marie Curie (1867–1934) made many important discoveries connected with radioactivity, before its dangers were known. Her laboratory notebooks are still radioactive and will be dangerous for at least another 1500 years.

When an unstable nucleus undergoes radioactive decay its nucleus changes to become more stable. The **activity** of any radioactive substance is the number of nuclear decays per second and is measured in **becquerels** (**Bq**). One becquerel is one nuclear decay each second.

1 A sample of uranium-235 has a decay rate of 2 Bq. How many nuclei decay each second?

Radioactive decay is a random process – we cannot predict when it will happen. When you throw a die, sometimes you get a six and sometimes you don't. The **probability** of getting a six is $\frac{1}{6}$.

In a similar way, at any given moment there is a certain probability that a particular unstable nucleus will decay.

The **half-life** is the time taken for half the unstable nuclei in a sample of a radioactive isotope to decay. This is shown in diagram B. We cannot predict the decay of an individual nucleus because it is a random process. However, the half-life does allow us to predict the activity of a large number of nuclei. The half-life is the same for any mass of a particular isotope. Some half-lives are shown in table C.

Isotope	Half-life
uranium-235	700 million years
carbon-14	5730 years
caesium-137	30 years
radon-222	3.8 days

C half-lives of some isotopes

After decaying, a nucleus may become stable. The more stable nuclei a sample of a substance contains, the lower its activity. The half-life of an isotope is therefore also a measure of how long it takes for the activity to halve. It can be found by recording the activity of a sample over a period of time.

Worked example

In figure D, the activity at 3 minutes is 800 counts per second. After one half-life the count rate will have decreased to 400 counts per second.

This occurs at 9.5 minutes, so the half-life is 9.5 – 3 = 6.5 minutes.

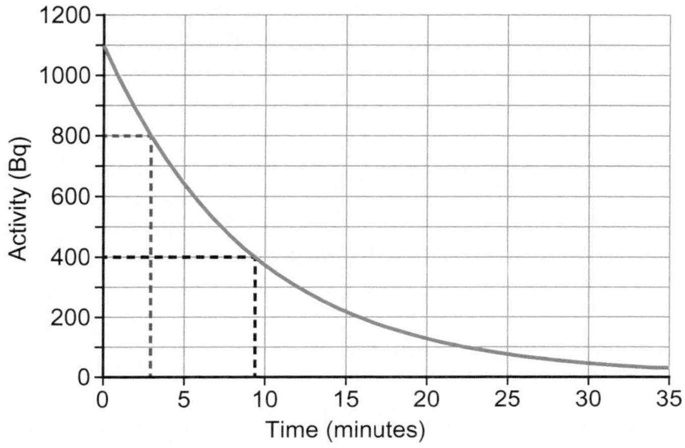

D graph of activity against time for a radioactive substance

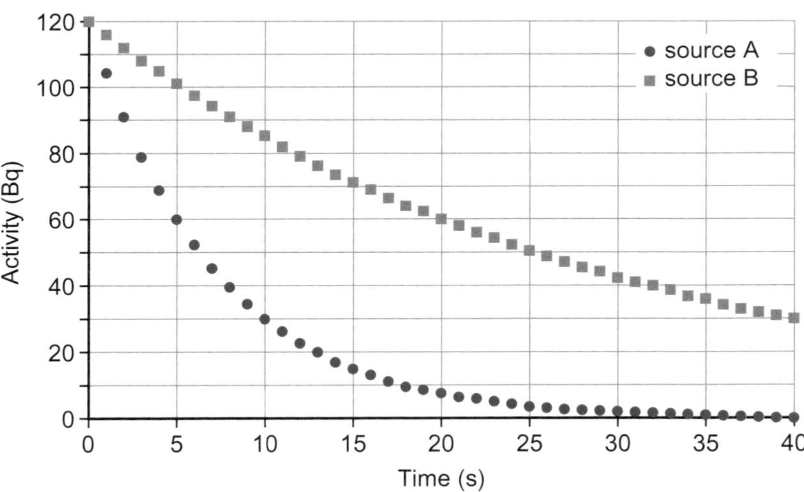

E graph showing the decay of two different radioactive substances

 **6** Work out the half-lives of the two sources shown in Figure E.

Exam-style question

Describe how you could find the half-life of a newly discovered radioactive substance. *(2 marks)*

4 Strontium-90 has a half-life of 29 years. How many strontium-90 half-lives is:

 **a** 29 years **b** 58 years

 **c** 116 years **d** 14.5 years?

5 There are 10 million atoms in a sample of radon-222. How many undecayed nuclei are left after:

 **a** 3.8 days **b** 7.6 days

 **c** 11.4 days **d** 1.9 days?

Did you know?

When the Earth was formed around 4.5 billion years ago, most of it was molten. The Earth has been cooling down ever since but part of the core is still molten, partly because of the energy released by radioactive isotopes in the Earth.

Checkpoint

How confidently can you answer the Progression questions?

Strengthen

S1 A sample of caesium-137 has an activity of 100 Bq. What will its activity be in 90 years time?

Extend

E1 Explain what the half-life of a radioactive sample tells you about how its activity and the number of unstable nuclei change over time.

E2 A sample containing carbon-14 has an activity of 1.5 Bq. How long ago would it have had an activity of 24 Bq?

SP6h Using radioactivity

Specification reference: P6.28P

Progression questions

- How can radioactivity be used to preserve food?
- How is radioactivity used in industry?
- How is radioactivity used in smoke alarms?

A All these strawberries are several days old. The ones on the left were irradiated with gamma rays.

Killing microorganisms

All foods contain microorganisms that eventually cause them to decompose. Some bacteria also cause food poisoning. Food can be **irradiated** with gamma rays to kill bacteria. This makes it safer to eat and also means that it can be stored for longer before going off. It does not make the food more radioactive, although some foods are naturally radioactive.

Surgical instruments need to be **sterilised** to kill microorganisms. The usual method is to heat them. Some instruments such as plastic syringes cannot be sterilised using heat, so they are sealed into bags and irradiated with gamma rays, which can penetrate the bags and the equipment.

1 What happens when food is irradiated?

2 Suggest why surgical equipment is sealed into bags before irradiation.

Radioactive detecting

Radioactive isotopes can be used as **tracers**. For example, a gamma source added to water is used to detect leaks in water pipes buried underground. Where there is a leak, water flows into the surrounding earth. A Geiger-Müller tube following the path of the pipe will detect higher levels of radiation where there is a leak.

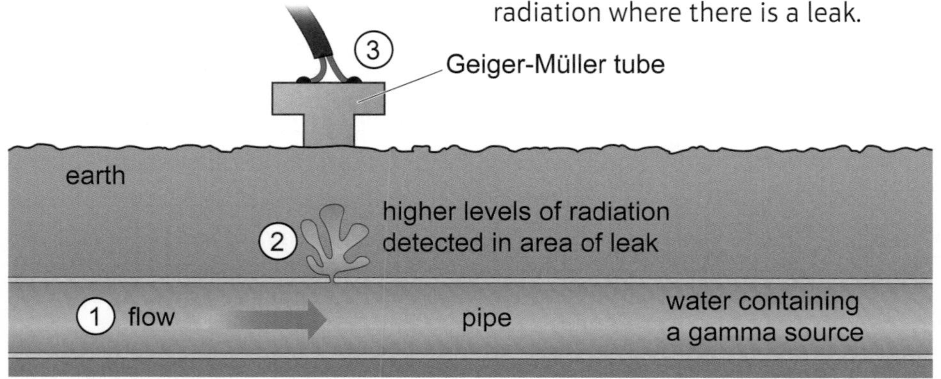

B using a gamma source to detect a leak

Cancer

Radioactivity can be used to help diagnose cancer using tracers in the body. It can also be used to treat cancer. You will learn more about this in *SP6j Radioactivity in medicine*.

3 Explain why gamma sources are used as tracers rather than beta sources.

Checking thickness

Paper is made by squeezing wood pulp between rollers. Paper can be made in different thicknesses and the rollers need to squeeze the wood pulp with a force that produces the correct thickness of paper. The detector in diagram C counts the rate at which beta particles get through the paper from a source on one side.

When the paper is too thin, more beta particles penetrate the paper and the detector records a higher count rate. A computer senses that the count rate has risen and reduces the force applied to the rollers to make the paper thicker. When the paper is too thick, the opposite happens.

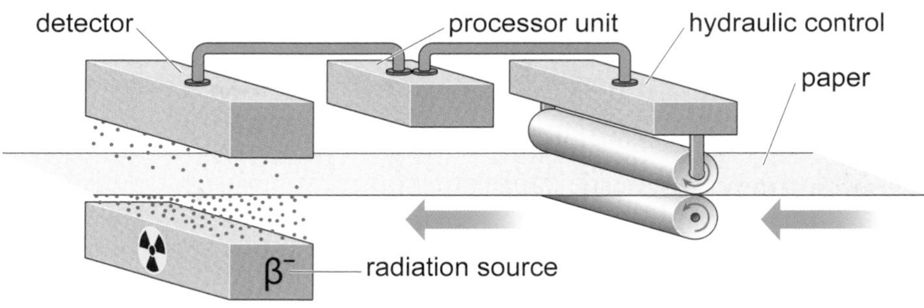

C A beta particle detector is used to control the thickness of paper during its manufacture.

 **4** Look at diagram C. Explain what happens when the paper is too thick.

 **5** Why would you not use an alpha source to monitor paper thickness?

Smoke alarms

A smoke alarm contains a source of alpha particles, usually a radioactive isotope called americium-241. The detector has an electrical circuit with an air gap between two electrically charged plates. The americium-241 source releases alpha particles, which ionise molecules in the air. These ions are attracted to plates with an opposite charge and so allow a small electrical current to flow.

As long as this current is flowing, the alarm will not sound. When smoke gets into the air gap the smoke particles slow down the ions. This means that the current flowing across the gap decreases. The alarm sounds when the current drops below a certain level.

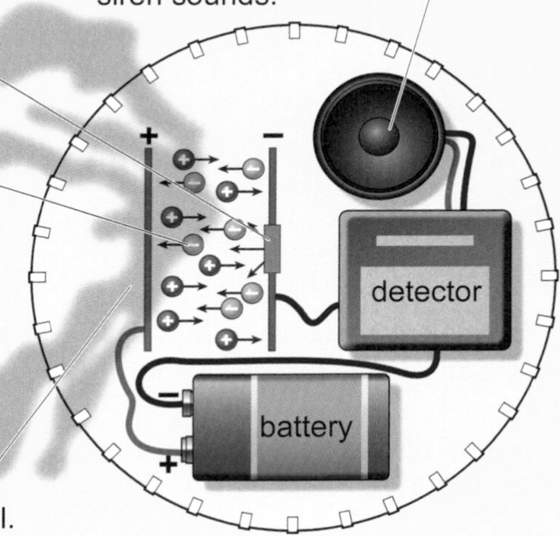

An americium-241 source gives off a constant stream of alpha particles.
Alpha particles ionise air molecules and these ions then move across the gap, forming a current.
Smoke in the device will slow down the ions and so make the current fall.
The detector senses the amount of current. If the current falls the siren sounds.
detector
battery

D how a smoke alarm works

 **6** What type of radiation is produced by americium-241?

 **7** The half-life of americium-241 is 432 years. Why does this make it a suitable source of alpha particles in a smoke alarm?

Exam-style question

Describe how you could use radioactivity to monitor the thickness of aluminium foil as it is being made. *(4 marks)*

Checkpoint

How confidently can you answer the Progression questions?

Strengthen

S1 Describe five uses for radioactivity.

Extend

E1 Explain the characteristics needed in radioactive sources used for each of the following: sterilising surgical equipment, as tracers, for controlling paper thickness and in smoke alarms.

SP6i Dangers of radioactivity

Specification reference: P6.29; P6.30P; P6.31; P6.32

Progression questions

- What are the dangers of ionising radiation?
- What precautions should be taken to protect people using radiation?
- What is the difference between contamination and irradiation effects?

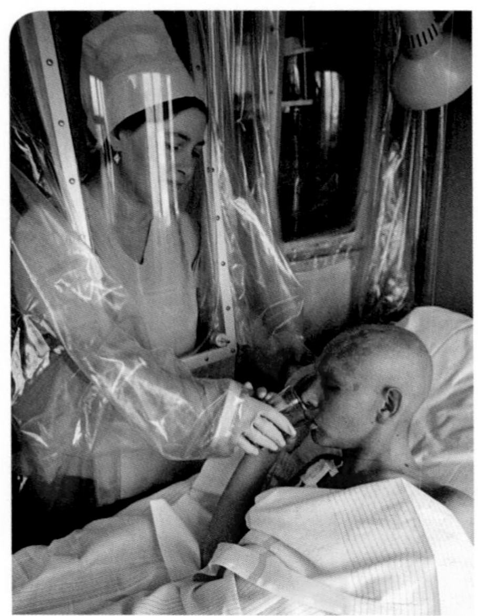

A This patient has radiation burns. He was among the first emergency personnel on the scene after the Chernobyl nuclear power plant disaster on 26 April 1986.

B Radioactive sources are handled with tongs.

A large amount of ionising radiation can cause tissue damage such as reddened skin (radiation burns) and also other effects that cannot be seen.

Small amounts of ionising radiation over long periods of time can damage the DNA inside a cell. This damage is called a **mutation**. DNA contains the instructions controlling a cell, so some mutations can cause the cell to malfunction and may cause cancer. Gene mutations that occur in gametes can be passed on to the next generation. However, not all mutations are harmful and cells are often capable of repairing the damage if the radiation **dose** is low.

1 Describe two effects of ionising radiation on the human body.

Radiation is a hazard, because it can cause harm. We are exposed to background radiation all the time, but we are only exposed to small amounts so the risk of harm is low. However, people who work with radioactive materials could be exposed to more radiation and so must take precautions to minimise the risks from radiation.

Handling radioactive sources

The intensity of radiation decreases with distance from the source, so sources are always handled with tongs. The risk can also be reduced by not pointing sources at people and storing them in lead-lined containers.

2 Explain why radioactive sources are handled with tongs and stored in lead-lined containers.

Radiation in hospital

Medical staff working with radioactive sources have their exposure limited in a number of ways, including increasing their distance from the source, shielding the source and minimising the time they spend in the presence of sources. Their exposure is also closely monitored using dosimeter badges (see *SP6d Background radiation*).

Some patients may be exposed to a dose of radiation for medical diagnosis or treatment (such as detecting and treating cancer). This is only done when the benefits are greater than the possible harm the radiation could cause, and the minimum possible dose is used, and sources with short half-lives are used to minimise the time for which the patient is exposed.

Nuclear accidents

Occasionally there is an accident in a nuclear power station, allowing radioactive materials to escape into the environment. Accidents such as this cause a hazard, as they may lead to people being irradiated or contaminated.

Someone is irradiated when they are exposed to alpha, beta or gamma radiation from nearby radioactive materials. Once the person moves away the irradiation stops.

Someone becomes **contaminated** if they get particles of radioactive material on their skin or inside their body. They will be exposed to radiation as the unstable isotopes in the material decay, and this will continue until the material has all decayed or until the source of contamination is removed (which is not always possible). Water and soils can also be contaminated, so contamination can spread into the food chain. Contamination with radioactive materials with long half-lives poses a greater hazard as the effects will last longer than for materials with shorter half-lives.

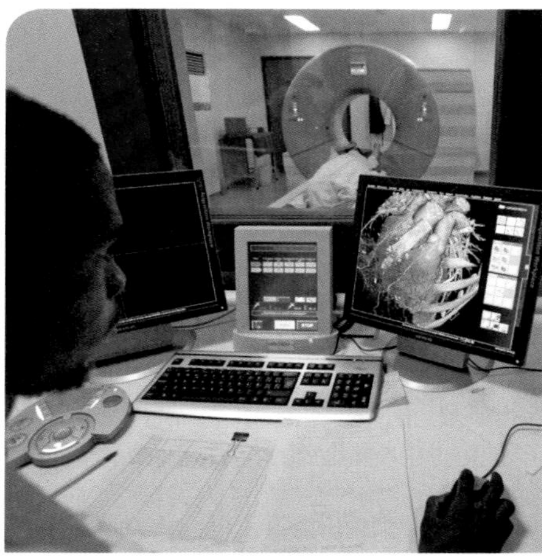

C behind a radiation shield in a hospital

Did you know?

When radioactivity was first discovered, people did not know of the harm it could cause. Radioactive substances such as thorium and radium were put into face creams and other products because people thought they might have health benefits.

D These workers are cleaning up after an accident at a nuclear reactor. The overalls stop their clothing becoming contaminated.

It is important to understand that the dangers of radiation from medicine, industry and power generation are small compared with many other aspects of our modern lives. However, many people are concerned that accidents may happen in nuclear power stations.

3 Look at photo D. Suggest why the workers are wearing masks.

4 Explain whether an alpha or beta source is the most harmful if both sources are:

a outside the body

b inside the body (e.g. if breathed in or swallowed).

5 Explain how the half-life of a radioactive source affects the potential hazard it poses.

Exam-style question

Describe the difference between contamination and irradiation. *(2 marks)*

Checkpoint

How confidently can you answer the Progression questions?

Strengthen

S1 What are the hazards posed by radiation?

S2 Describe three ways of minimising the risk from radiation.

Extend

E1 Explain two precautions that should be taken by people:

a working with radioactive sources in industry

b using radioactive sources in hospitals

c cleaning up after a nuclear accident.

SP6j Radioactivity in medicine

Specification reference: P6.33P; P6.34P; P6.35P

Progression questions

- What are some of the uses of radioactive substances in diagnosis?
- Why do isotopes used in PET scanners have to be produced nearby?
- How is radiation used to treat tumours?

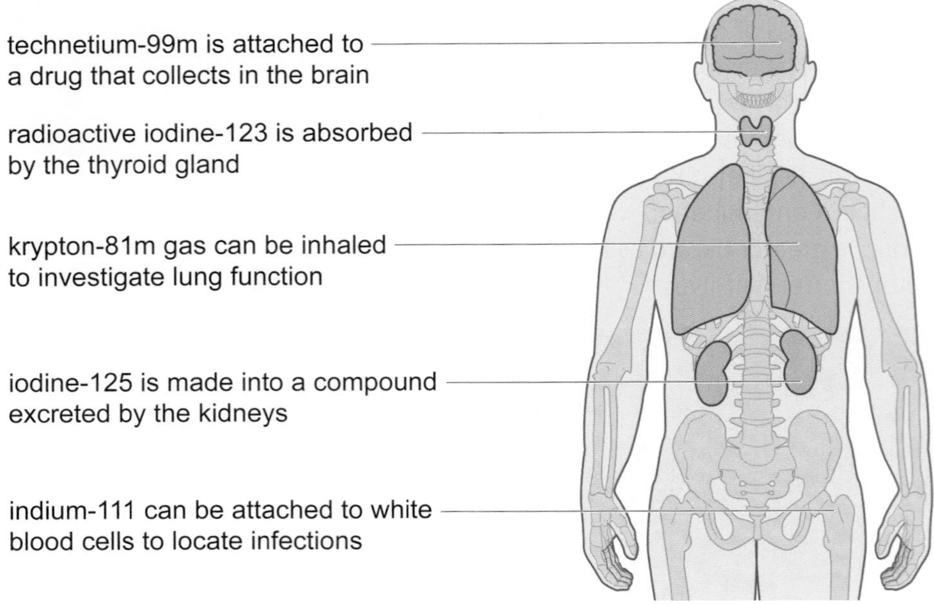
technetium-99m is attached to a drug that collects in the brain
radioactive iodine-123 is absorbed by the thyroid gland
krypton-81m gas can be inhaled to investigate lung function
iodine-125 is made into a compound excreted by the kidneys
indium-111 can be attached to white blood cells to locate infections

A Different tracers are absorbed by different parts of the body.

Diagnosing with gamma rays

Radioactive materials are used to diagnose medical conditions without having to cut into a patient's body. A radioactive tracer, which emits gamma rays, is put into the patient.

Tracers often contain a radioactive isotope attached to molecules that will be taken up by particular organs in the body. The tracer is usually injected into the bloodstream, but it may be swallowed, inhaled or injected directly into an organ. The location of the tracer in the body is detected using one or more **gamma cameras**.

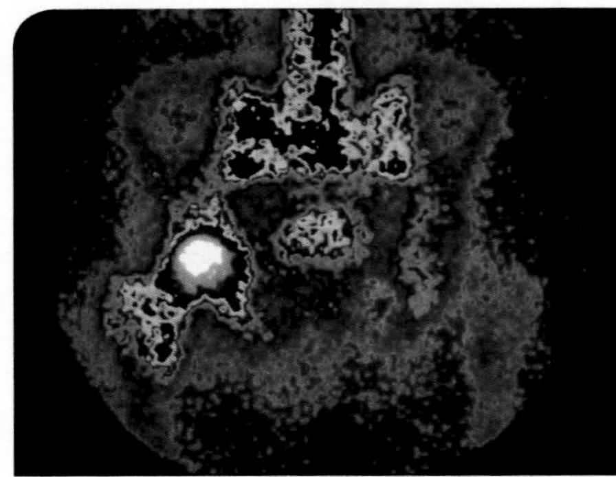

B This gamma camera scan shows a bone tumour. The brighter the colour, the more radiation has been detected.

Tracers can be injected into the blood to find sources of internal bleeding. Gamma cameras detect the area of highest gamma radiation, which is where the bleeding is occurring.

Gamma cameras are also used to detect **tumours**. The tracer is made using radioactive glucose molecules because very active cells, such as cancer cells, take up glucose more quickly than other cells.

6th

1 Name two medical conditions that can be investigated using gamma ray tracers.

8th

2 Look at photo B. Where is the main cancer in this person?

8th

3 Why are alpha and beta sources not used in medical radioactive tracers?

Diagnosing with positrons

Tracers that emit positrons can also be used to detect medical problems. The tracer emits a positron. When this meets an electron, both it and the electron are destroyed and two gamma rays are emitted in opposite directions. The detector in a **PET scanner** moves around the patient, building up a set of images showing where different amounts of gamma radiation are coming from.

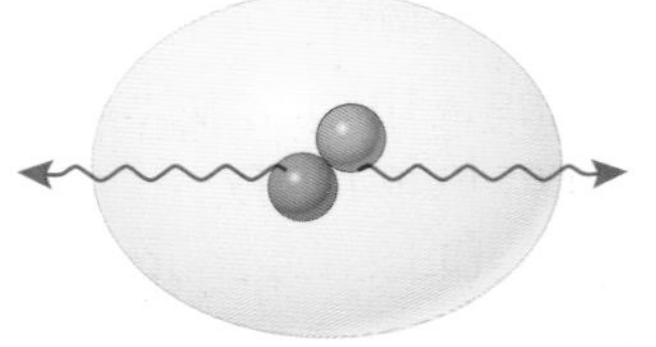

C The electron-positron annihilation causes two gamma rays to be emitted in opposite directions.

The radioactive isotopes used in all medical tracers need to have a short half-life so that other parts of the body are affected as little as possible. This means that they lose their radioactivity quickly and so must be made close to the hospital. They are often used within hours or even minutes of production.

4 Radioactive isotopes are produced in cyclotrons. Suggest why cyclotrons are located around the UK, sometimes within hospitals.

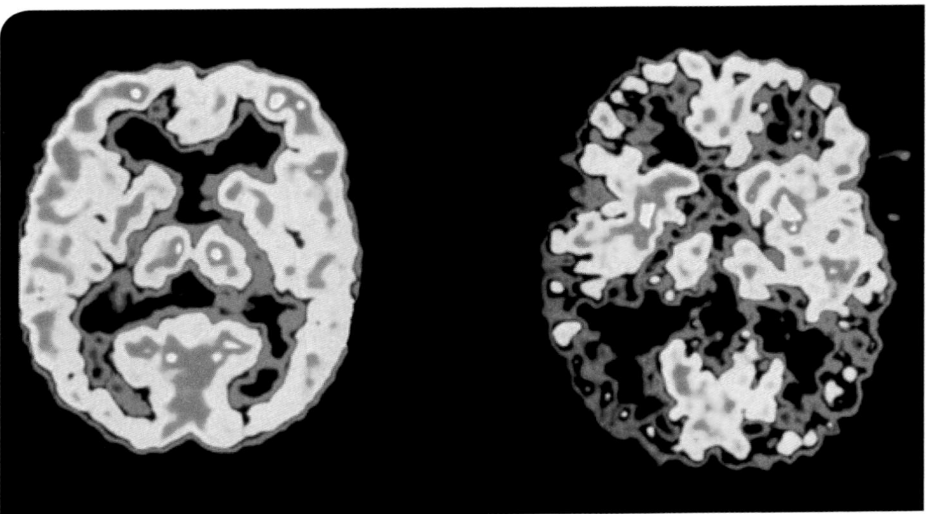

D PET scans showing the activity in a healthy brain (left) and one with Alzheimer's disease (right). The radioactive isotope is attached to a substance that is used by active brain cells.

Treating cancer

Cancer cells divide more rapidly than most other cells in the body and so are more susceptible to being killed by radiation.

Internal radiotherapy uses a beta emitter such as iodine-131 placed inside the body, within or very close to a tumour. This does not always require surgery – the patient stays in a room alone while the source is in place.

5 Suggest why patients are kept away from other people while being treated with internal radiotherapy.

Most radiotherapy is **external radiotherapy**, which uses beams of gamma rays, X-rays or protons directed at the tumour from outside the body. Several lower strength beams may be directed at the tumour from different directions so that only the tumour absorbs a lot of the energy and the surrounding tissues are harmed as little as possible.

6 Explain why beta emitters are usually used for internal but not for external radiotherapy.

Exam-style question

Explain why both the half-life and the type of radiation produced are taken into account when choosing tracers for use in medical diagnosis.

(4 marks)

Checkpoint

How confidently can you answer the Progression questions?

Strengthen

S1 Describe two ways in which radiation is used to diagnose medical conditions.

S2 Write down one similarity and one difference between internal and external radiotherapy.

Extend

E1 Explain how radioactive tracers are used to diagnose medical conditions.

E2 Compare and contrast the treatment of tumours using internal and external radiotherapy.

SP6k Nuclear energy

Specification reference: P6.36P; P6.37P; P6.42P

Progression questions

- What different types of nuclear reactions are there?
- What are the advantages of using nuclear power to generate electricity?
- What are the disadvantages of nuclear power?

A The nuclear submarine USS *Skate* was the first submarine to surface at the North Pole, in 1959. Nuclear fuel allows submarines to stay submerged for long periods.

In radioactive decay the radiation emitted by the unstable nuclei transfers energy. There are two other types of nuclear reaction that are used as a source of energy on a large scale.

- In **nuclear fission** large nuclei (such as uranium-235) break up to form smaller nuclei and release energy. Fission reactions are used in nuclear power stations.
- In **nuclear fusion** two small nuclei join together (fuse) to form a larger nucleus. Fusion reactions release energy inside the Sun.

1 Name a nuclear fuel.

Nuclear fuels store a lot more energy per kilogram than any other type of fuel. This makes them useful for naval ships and submarines. Nuclear fuels do not burn, so they do not need air to allow them to release energy and they do not produce carbon dioxide.

2 Give two reasons why nuclear fuels are useful in a submarine.

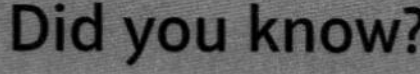

Did you know?

One kilogram of uranium-235 stores just over 80 million MJ of energy (8×10^{13} J). One kilogram of petrol stores around 44 MJ (4.4×10^{7} J).

Most nuclear energy is used in power stations to generate electricity. Uranium is a **non-renewable** fuel, but it is estimated that supplies could last for over 200 years at their current rate of use. This is much longer than other non-renewables, such as oil, will last.

B Although Heysham nuclear power station is built on the coast, around 100 000 people live within 10 km of it. They would all be directly affected if there were an accident at the power station.

Conventional power stations that burn **fossil fuels** produce carbon dioxide. The increasing amount of carbon dioxide in the atmosphere is contributing to **climate change**. Conventional power stations can also produce other forms of pollution, such as soot and acidic gases.

Although nuclear power stations do not emit gases they do produce waste that will stay radioactive for millions of years. This waste is expensive to treat, as it needs to be sealed into concrete or glass and buried safely.

Parts of a nuclear power station become radioactive as it is used, and this makes it very expensive to **decommission** (dismantle safely) a power station at the end of its life.

 **3** Explain one advantage and one disadvantage of nuclear power stations compared to fossil fuel power stations.

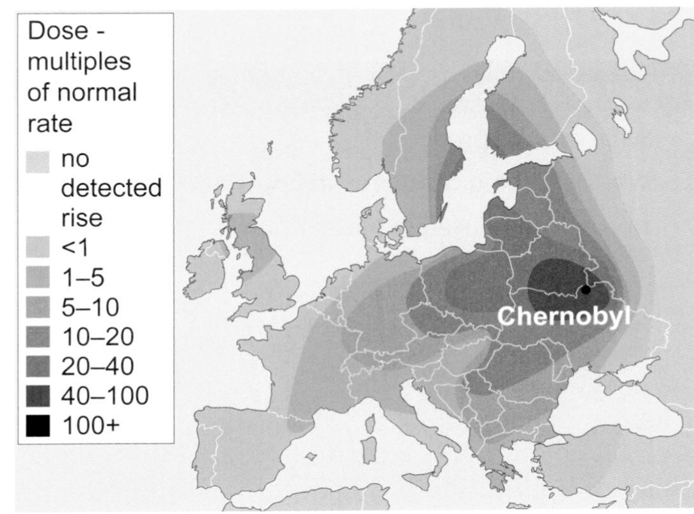

C A very serious nuclear accident occurred at the Chernobyl power station in Ukraine in 1986. The map shows the area affected by radioactive particles from the accident. There were many thousands of cases of cancer in the area and radiation in some soils in the UK meant that for 26 years sheep had to be tested for radiation before they could be sold for food.

Nuclear power stations are designed for safety. However accidents do sometimes occur, including small leaks of radioactive materials and reactor explosions. Major accidents can have very serious consequences for many people.

Many people do not think that the benefits of nuclear energy are worth the risks. At least 12 countries in Europe have voted to ban nuclear power stations.

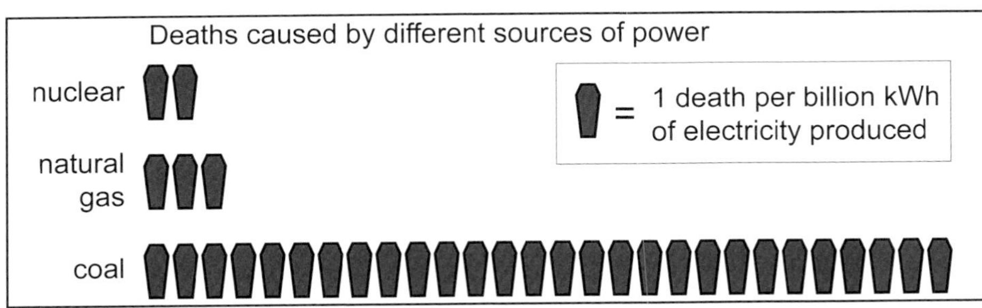

D the number of deaths caused by different fuels used to produce electricity, including mining and power station accidents and the effects of air pollution

 **4** Explain why people living near nuclear power stations worry about radioactive materials leaking from them.

 **5** Look at diagram D. Suggest why many people think that nuclear power has higher risks than using fossil fuels.

Exam-style question

Compare nuclear and fossil fuelled power stations in terms of any pollution they cause. *(3 marks)*

Checkpoint

How confidently can you answer the Progression questions?

Strengthen

S1 Describe three different types of nuclear reaction.

S2 Draw up a table to show the advantages and disadvantages of using nuclear energy to generate electricity.

Extend

E1 Explain the advantages and disadvantages of using nuclear power for generating electricity, in terms of the environment, safety and public opinion.

SP6l Nuclear fission

Specification reference: P6.38P; P6.39P; P6.40P; P6.41P

Progression questions

- What are the products of the fission of uranium-235?
- What is a chain reaction and how can it be controlled?
- How is fission used in nuclear power stations?

When a uranium-235 nucleus absorbs a neutron it immediately splits into two smaller **daughter nuclei**, which are also radioactive. Two or more neutrons are also released. Both daughter nuclei and the neutrons store a lot of kinetic energy because they are moving at high speeds. Energy is also transferred from the fission by heating.

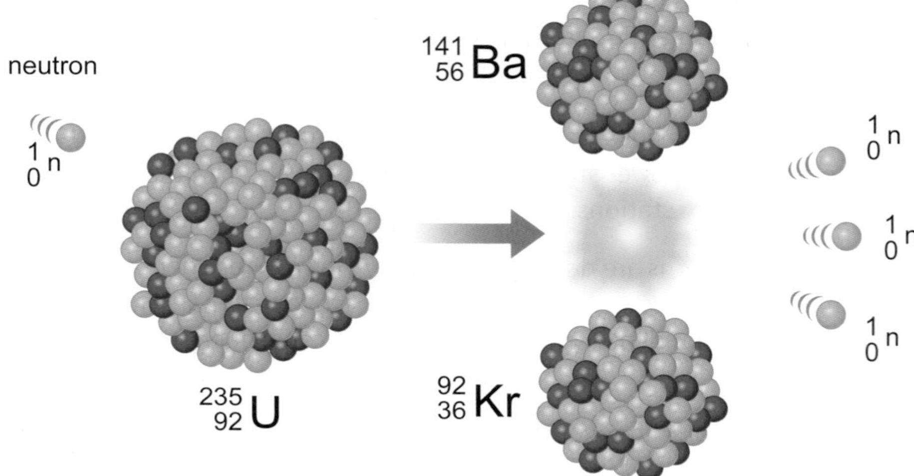

A one example of the fission of uranium-235 (other daughter nuclei can be formed)

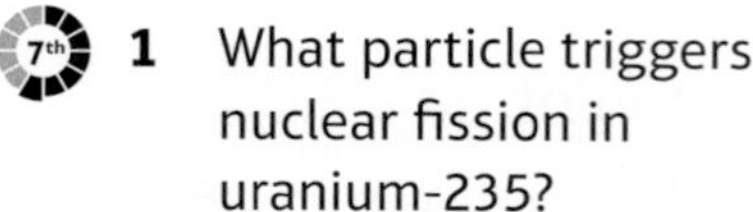

1 What particle triggers nuclear fission in uranium-235?

2 What are the products of the fission of uranium-235 given in the example in diagram A?

If the neutrons released are absorbed by other uranium-235 nuclei, these nuclei will become unstable and release more neutrons when their nuclei split. These neutrons can then be absorbed by yet more uranium nuclei, which in turn split up, releasing more neutrons. This is an uncontrolled nuclear **chain reaction**, such as occurs in an atomic bomb. The chain reaction can be controlled if other materials absorb some of the neutrons.

3 Uranium-235 can split into xenon-140 and strontium-94.

a Explain how many neutrons are produced in this fission.

b Suggest how the number of neutrons released by a fission reaction can affect the chain reaction.

Did you know?

The fission of uranium was discovered by Lise Meitner (1878–1968) and Otto Hahn (1879–1968). Meitner was the first woman to become a full professor of physics in Germany.

Nuclear reactors

In a nuclear reactor the fuel is made into **fuel rods**. As fission reactions occur, neutrons leave the fuel rods at high speed. They are slowed down to increase the chance they will be absorbed by another uranium-235 nucleus. Inside a reactor **core**, fuel rods are inserted into holes in a material called a **moderator**, which slows down the neutrons.

The chain reaction is controlled using **control rods**, which contain elements that absorb neutrons. These rods are placed between the fuel rods in the reactor core. If the rate of fission needs to be increased, the control rods are moved out of the core so that fewer neutrons are absorbed, and vice versa. When the control rods are fully lowered into the core, the chain reaction stops and the reactor shuts down.

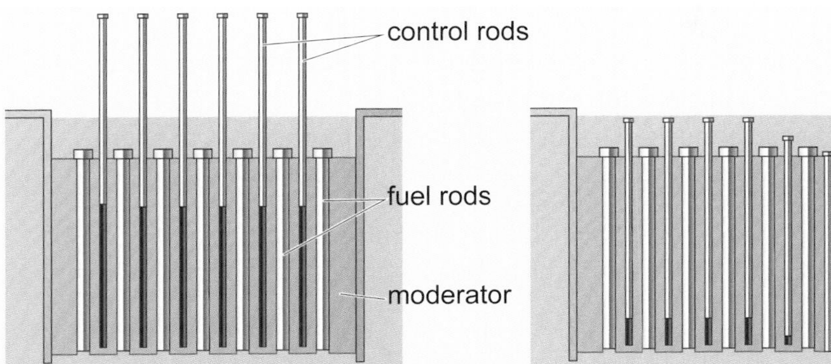

Not many of the neutrons from fission reactions are absorbed by the control rods so the chain reaction can proceed quickly and a lot of energy is released.

Most of the neutrons from fission reactions are absorbed by the control rods so the chain reaction will slow down.

B controlling the chain reaction in a reactor core

Generating electricity

Energy released from the core is transferred to a coolant, which is pumped through the reactor. The coolant can be water, a gas or a liquid metal. The hot coolant is pumped to a heat exchanger where it is used to make steam. The steam drives a turbine, which turns a generator to produce electricity.

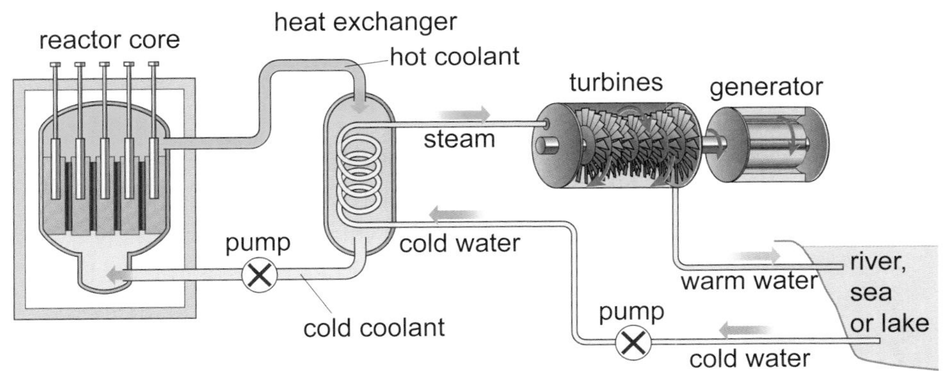

D Radioactive fuels are used to generate electricity in nuclear power stations.

6 Describe what is done inside a reactor core if the demand for electricity increases.

Exam-style question

Compare the functions of the moderator and the control rods in a nuclear reactor. *(4 marks)*

4 Why must the chain reaction in a nuclear reactor be controlled?

5 In a nuclear reactor, what are the functions of:

a the control rods

b the moderator?

C fuel rods being replaced in a nuclear reactor

Checkpoint

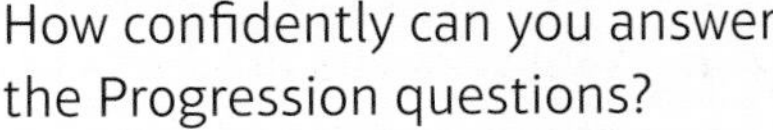

How confidently can you answer the Progression questions?

Strengthen

S1 Write glossary definitions for the words in bold on these pages.

S2 Draw a flow diagram to show how nuclear energy is used to generate electricity.

Extend

E1 Write a paragraph to explain how the energy released by fission reactions is controlled and used to generate electricity.

SP6m Nuclear fusion

Specification reference: P6.43P; P6.44P; P6.45P; P6.46P

Progression questions

- How is nuclear fusion different to nuclear fission?
- What are the conditions needed for nuclear fusion?
- Why haven't practical fusion power stations been developed yet?

Nuclear fusion occurs when small nuclei combine to form larger ones. The mass of the new nucleus formed is slightly less than the total of the masses of the two smaller nuclei. The lost mass has been converted to energy.

Fusion reactions in which hydrogen nuclei combine to form helium are the main energy source for stars, including our Sun. Diagram A shows part of a sequence of fusion reactions, which releases an enormous amount of energy.

1 What is the source of energy in the Sun?

2 Which element is produced by the fusion of hydrogen?

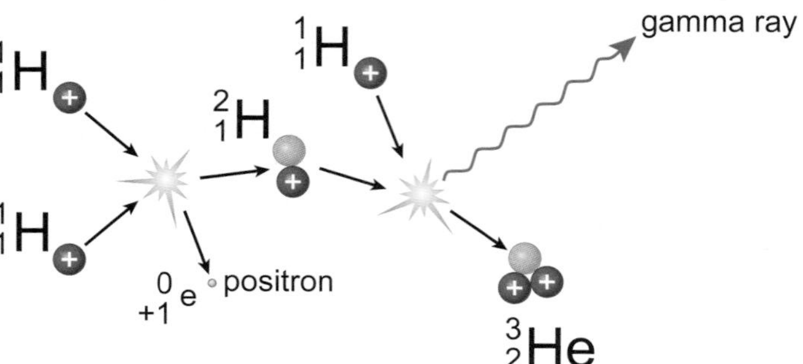

A part of the sequence of fusion reactions that release energy in the Sun

B Elements up to iron in the periodic table are formed inside stars. Elements heavier than iron are only formed when large stars explode in a supernova at the end of their lives.

Scientists are investigating using the fusion reactions of isotopes of hydrogen to generate electricity. Nuclei of hydrogen-2 (deuterium) and hydrogen-3 (tritium) have to be forced together. The protons in the nuclei are positively charged and like charges repel. This is called **electrostatic repulsion**.

For the nuclei to fuse, they need to get extremely close to each other (about 10^{-15} m). The Sun has a very strong gravitational field, which creates extremely high pressures at its centre. This forces all the nuclei to be very close to each other, so they are more likely to collide. Nuclei are also more likely to collide at higher temperatures, when they are moving faster. If nuclei are close enough or travelling fast enough, some can overcome their electrostatic repulsion and fuse.

A useful fusion reactor needs fusion to happen faster than it does in the Sun. It is also difficult to produce very high pressures on Earth, so the temperature inside a fusion reactor must be very high – hotter than the temperature inside the Sun.

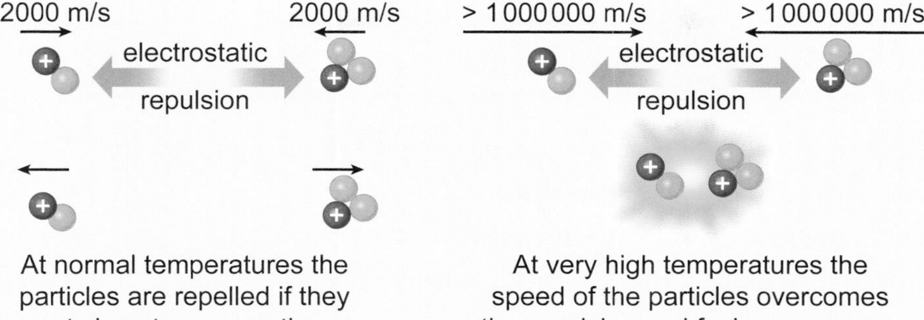
2000 m/s
2000 m/s
electrostatic
repulsion
At normal temperatures the particles are repelled if they get close to one another.
> 1 000 000 m/s
> 1 000 000 m/s
electrostatic
repulsion
At very high temperatures the speed of the particles overcomes the repulsion and fusion can occur.

C Electrostatic repulsion can be overcome if the nuclei are moving very fast.

Fusion reactors could theoretically produce a lot more energy than fission reactors. The helium produced in nuclear fusion is not radioactive but any materials used to contain fusion reactions do become radioactive. However, there are far fewer problems with safely disposing of radioactive waste materials from fusion reactors than from fission reactors.

It is very difficult to sustain the extreme temperatures and pressures required for fusion. So far none of the experimental reactors have produced more energy than has been put in, so fusion power is a long way off.

D The inside of the Joint European Torus fusion reactor. Magnetic fields are used to contain the very hot gases, as they would melt any materials they came into contact with.

 **5** Describe the difficulties of building a commercial fusion power station.

Exam-style question

Explain how electrostatic repulsion is overcome in fusion reactors.

(4 marks)

 **3** Why does fusion not happen at room temperature?

4 Why does fusion happen at a lower temperature in the Sun than in a fusion reactor on Earth?

Did you know?

The temperature in the centre of the Sun, where most of the fusion reactions take place, is nearly 16 million degrees Celsius (1.6×10^7 °C). Temperatures up to 100 million degrees Celsius (1×10^8 °C) can be produced inside fusion reactors.

Checkpoint

How confidently can you answer the Progression questions?

Strengthen

S1 Describe the differences between nuclear fusion and nuclear fission. You may need to look back at *SP6l Nuclear fission.*

Extend

E1 Explain the differences between nuclear fusion and nuclear fission, including their use as energy resources for generating electricity.

Background radiation

Some scientists carry out an experiment to measure the radioactivity from a source to be used in a factory.

They measure the background radiation before and after their experiment.

They take the background count at the same place they do their experiment.

Explain how this procedure helps to make sure that the results of the experiment are valid. **(6 marks)**

Student answer

Background radiation is around us all the time and comes from natural and human sources. Up to half of it can come from radon gas and a lot of it also comes from the ground and from hospitals [1]. *It is not the same in every place, so the background count needs to be measured at the place where you are doing an experiment* [2]. *If they don't measure the background count, their radiation measurements of the source will be wrong* [3].

[1] The question did not ask for information about different sources of background radiation. Putting in information that is not asked for does not gain any extra credit and wastes time.

[2] This explains why the background count has to be measured in the same place as the experiment.

[3] This does not actually explain how the measurements of background radiation are used to make the results valid. Valid results would only be obtained if the values for radioactivity only came from the source – so the background count needs to be measured and subtracted from the counts in the experiment.

Verdict

This is an acceptable answer. The student has used scientific knowledge to explain that background radiation varies from place to place, and so measurements should be taken in the place where the experiment is carried out.

The answer could be improved by linking facts with scientific reasons. For example, explaining that the results are only valid if they are only measuring radioactivity levels from the source. This is why the background count needs to be subtracted from the radioactivity of the source. The answer also needs to link the fact that the background count varies with time to the idea that the background count needs to be taken both before and after the experiment is carried out.

Exam tip

You should be able to apply your knowledge of practical work to other contexts. In this example you should be able to explain what has to be measured and why, in order to make the data as useful as possible.

Paper 1

SP7 Astronomy

This is an artist's impression of the *Cassini* orbiter flying past the surface of Enceladus, one of Saturn's moons. Enceladus has a surface of water ice and ejects plumes of tiny ice crystals. This indicates that there is an ocean of liquid water beneath the surface. Data from *Cassini* indicate that there are also simple organic molecules in the plumes, so it is possible that simple life forms may exist there.

In this unit you will learn about the Solar System and how gravity affects orbits. You will learn about the life cycles of stars and evidence for different theories of the origin of the Universe.

The learning journey

Previously you will have learnt at KS3:

- about the Solar System and how we find out about it
- about the Earth's gravitational field and what causes weight
- that there are stars and galaxies beyond the Solar System.

You will also have learnt in *SP2 Motion and Forces*:

- more about mass and weight.

In this unit you will learn:

- about the bodies in our Solar System and how ideas about the Solar System have changed over time
- how methods of observing the Universe have changed over time
- why gravity is different on different bodies and how this affects orbits
- what redshift is and what it shows
- about different theories on the origin of the Universe
- about the life cycles of stars.

SP7a The Solar System

Specification reference: P7.2P; P7.3P; P7.4P; P7.19P

Progression questions

- What objects make up the Solar System and how are they arranged?
- How have ideas about the Solar System changed with time?
- How have methods of observing the Universe changed with time?

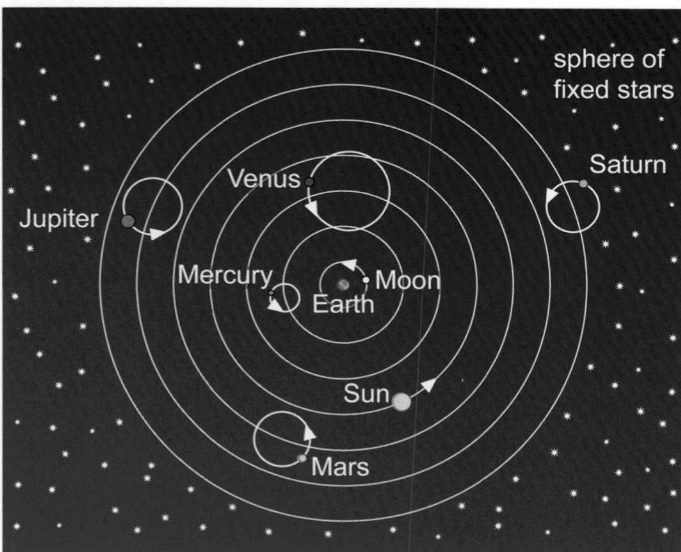

A In Ptolemy's geocentric model, planets moved in small circles as they orbited the Earth.

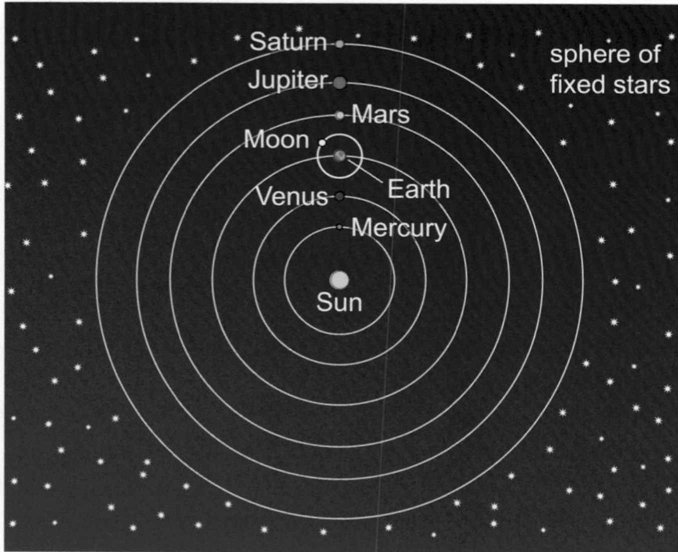

B Copernicus' heliocentric model of the Solar System

 **2** **a** Describe two differences between Ptolemy's and Copernicus' models.

 **b** Describe two similarities between the two models.

If you just observe the **planets** with your eyes, it seems that the Sun (our **star**) and the planets are all **orbiting** the Earth. Many ancient civilisations made detailed measurements of the movements of objects in the sky, and tried to explain them. One of the best known early models was made by the Greek astronomer Ptolemy (c100–170). His idea put the Earth in the centre of everything with the planets and the Sun orbiting around it – a **geocentric** model.

 **1** What is a geocentric model?

The Polish astronomer Nicolaus Copernicus (1473–1543) thought that Ptolemy's measurements fitted a different model – a **heliocentric** model with the Sun at the centre of the Solar System.

The invention of the **telescope** at the end of the 16th century allowed scientists to see objects in space in much greater detail and to find new objects. Using a telescope, the Italian astronomer Galileo Galilei (1564–1642) discovered four of Jupiter's moons. By plotting their movements, he showed that not everything orbited the Earth. This and other observations led him to support Copernicus's idea.

As telescopes improved, more discoveries were made, including the planets Uranus and Neptune and the **dwarf planet** Pluto. Smaller rocky bodies called **asteroids** were also discovered, most of which are found between the orbits of Mars and Jupiter. **Comets** are mostly made of ice, and some can be seen with the naked eye but many more have been found using telescopes.

Our current model of the Solar System includes eight planets, five dwarf planets, thousands of comets and millions of asteroids. These all move in **elliptical** (squashed circle) orbits around the Sun. Many of the planets also have **natural satellites** (**moons**) orbiting around them.

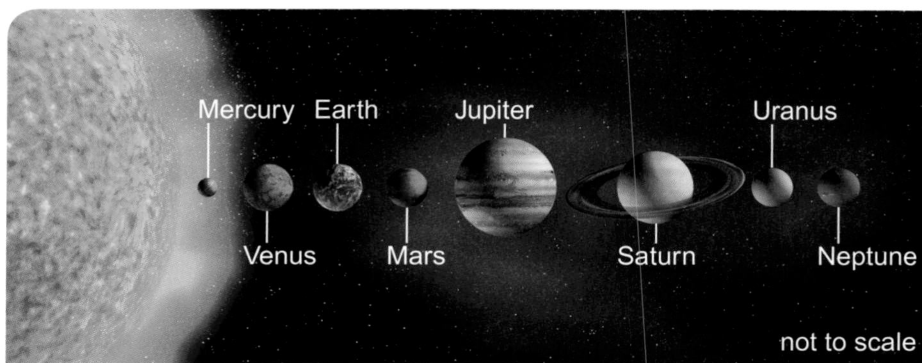

C the planets of the Solar System

Astronomy today

The invention of photography allowed astronomers to make more detailed observations and measurements than was possible by making drawings. Computers have further increased the speed and detail with which information from telescopes can be analysed. Today, photography enables astronomers to make detailed observations, and computers are used for analysis. Telescopes in orbit around the Earth give much clearer images than ground-based telescopes (since clouds and dust in the air do not interfere with the image). We also investigate the Solar System using space probes.

Many objects in space emit radio waves and infrared radiation. Different types of telescope are used to detect different types of electromagnetic waves. Some of these telescopes must be placed in orbit because the atmosphere absorbs some of the radiation they are designed to detect.

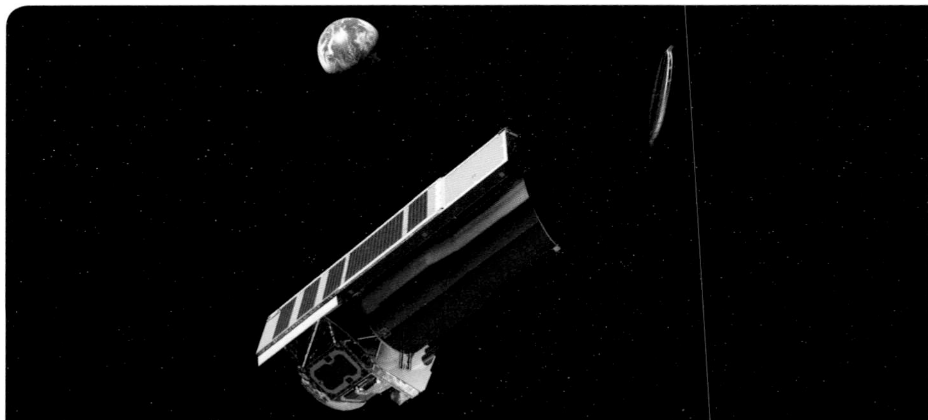

D The Spitzer Space Telescope detects infrared radiation from objects in space.

 **6** Name three types of electromagnetic waves detected by telescopes.

 **7** Suggest two reasons why modern telescopes are put into space.

Exam-style question

Compare and contrast Ptolemy's and Copernicus' models of the Solar System. *(4 marks)*

Did you know?

Until 2006 the Solar System had nine planets. In 2006 astronomers decided on a new category of Solar System object – the dwarf planet. Pluto is now classified as a dwarf planet. An asteroid called Ceres was also reclassified as a dwarf planet.

 **3** List the planets in the Solar System in order of their distance from the Sun.

 **4** Suggest one way in which our current model of the Solar System is different from Copernicus'.

 **5** How do Galileo's observations of Jupiter's moons support Copernicus' theory?

Checkpoint

How confidently can you answer the Progression questions?

Strengthen

S1 Describe the six different types of body in the Solar System mentioned on these pages.

S2 Give two ways in which observing the Universe is different for astronomers today compared with 500 years ago.

Extend

E1 Describe the ways in which changes to technology have increased our knowledge of the Universe.

SP7b Gravity and orbits

Specification reference: P7.1P; P7.5P; P7.6P; P7.7P

Progression questions

- Why is gravity different on different bodies in the Solar System?
- What kinds of different orbits are there?
- Why does the speed of a satellite affect the radius of its orbit?

A an artist's impression of *Philae* about to touch down

Body	Mass (kg)	Radius (m)	g (N/kg)
Earth	5.97×10^{24}	6 371 000	9.81
Moon	7.34×10^{22}	1 731 000	1.62
Mars	6.42×10^{23}	3 389 000	3.71
Ceres (a dwarf planet)	9.39×10^{20}	473 000	0.28

B gravitational field strength on different bodies in the Solar System

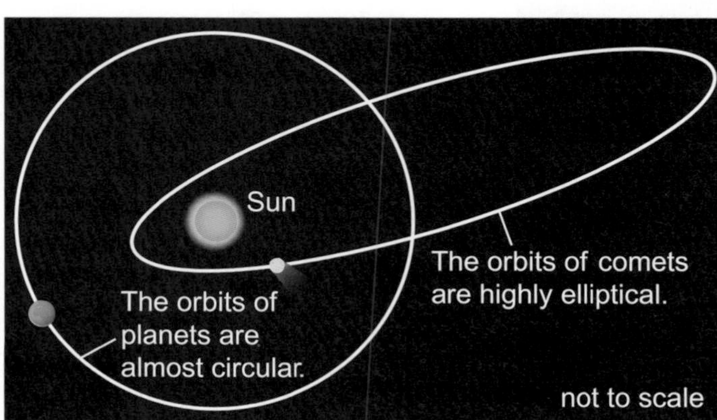

Sun
The orbits of comets are highly elliptical.
The orbits of planets are almost circular.
not to scale

C some different shapes of orbit

In 2014 the first 'soft' landing was made on a comet. The strength of gravity on the comet is thousands of times weaker than the gravity on Earth. The *Philae* lander was fitted with screws in the landing legs, harpoons and a small thruster, all designed to keep it on the surface. Unfortunately, these didn't work properly and the lander bounced over 1 km away from the surface before eventually landing.

Weight and gravity

Your **weight** is the force of gravity acting on you. Your weight depends on your mass and the **gravitational field strength** (***g***) of the Earth. On Earth, $g = 9.81$ N/kg, so the weight of a 1 kg mass is 9.81 N.

The gravitational field strength on the surface of a body (such as a planet or moon) depends on the mass of the body and the distance from its centre to its surface (its radius). The greater its mass and the smaller its radius, the greater its surface gravity.

 **1** **State two factors that affect the weight of an object on a planet.**

 **2** **Give one reason why *g* on Mars is greater than *g* on the Moon.**

Orbits

Most bodies in the Solar System are in elliptical orbits. **Artificial satellites** are used for communications and to observe the Earth and space. The type of orbit of an artificial satellite depends on what it is used for (as shown in diagram D).

Did you know?

The first artificial satellite was called *Sputnik*. It was launched by the Soviet Union in 1957, and went into a low Earth orbit.

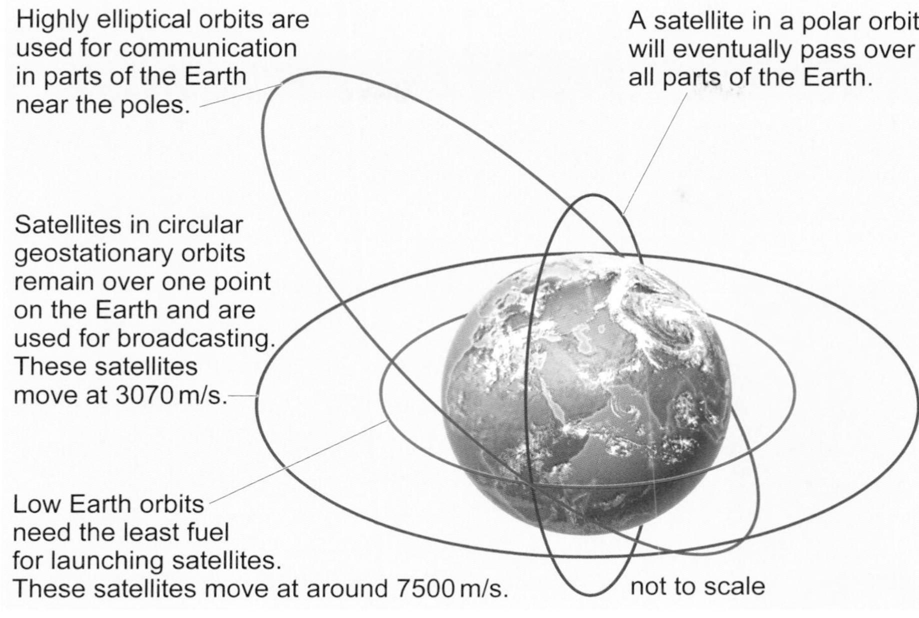

Highly elliptical orbits are used for communication in parts of the Earth near the poles.
A satellite in a polar orbit will eventually pass over all parts of the Earth.
Satellites in circular geostationary orbits remain over one point on the Earth and are used for broadcasting. These satellites move at 3070 m/s.
Low Earth orbits need the least fuel for launching satellites. These satellites move at around 7500 m/s.
not to scale

D Artificial satellites can be put into different types of orbit.

 **3** Describe three different types of orbit that artificial satellites can be put into.

Changing orbits

A satellite, planet or moon in a circular orbit has a constant speed as it travels. However, its direction is constantly changing. As **velocity** is a **vector quantity**, an orbiting body has a constantly changing velocity.

A moving object will continue to move in a straight line unless there is a force acting on it to make it change speed or direction. For a satellite in orbit, the gravitational force between the Earth and the satellite is at right angles to the direction of movement, so the force changes its direction but not its speed.

The gravitational force on a satellite in a low orbit is greater than that on a satellite in a high orbit. The satellite in the low orbit has to be moving much faster to stay in its orbit. If it slows down it will fall towards the Earth. It gains speed as it falls, until it is moving fast enough to stay in a new, lower orbit. If it goes low enough to encounter the top of the atmosphere, contact with the air will slow it down and it will eventually fall to Earth.

 **4** Explain why the velocity of a satellite in orbit is continually changing.

 **5** A satellite fires small rockets and it speeds up. Explain how this will affect the orbit it is in.

Exam-style question

Explain why the weight of an object changes if it is taken from the Earth to the Moon. *(3 marks)*

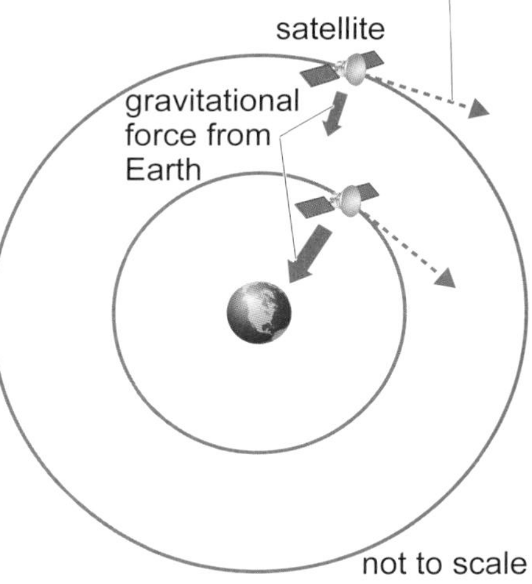

E forces acting on two satellites

Checkpoint

How confidently can you answer the Progression questions?

Strengthen

S1 Why is the weight of an object different on different planets?

S2 How is the orbit of a planet in our Solar System different to the orbit of a comet?

Extend

E1 Describe the factors that affect the surface gravity on a planet.

E2 Explain what happens to a satellite if its speed changes.

SP7c Life cycles of stars

Specification reference: P7.16P; P7.17P; P7.18P

Progression questions

- How do stars with masses similar to the Sun change over time?
- How do stars with much larger masses than the Sun change over time?
- How does the balance between thermal expansion and gravity affect stars?

A a possible view from the Earth in the future

The Sun provides enough energy to keep Earth at a temperature that supports life. However, one day it will become a red giant star. When this happens it will expand and may even swallow up the Earth. Long before that the extra heat Earth experiences will kill all life.

Star formation

A **nebula** is a cloud of dust and gases (mainly hydrogen). These materials can be pulled together by their own gravity. As the cloud contracts it becomes denser. The hydrogen becomes hotter as it spirals inwards and may start to glow. As more mass is attracted, the cloud's gravitational pull gets stronger and heats the material even more. This is a **protostar**.

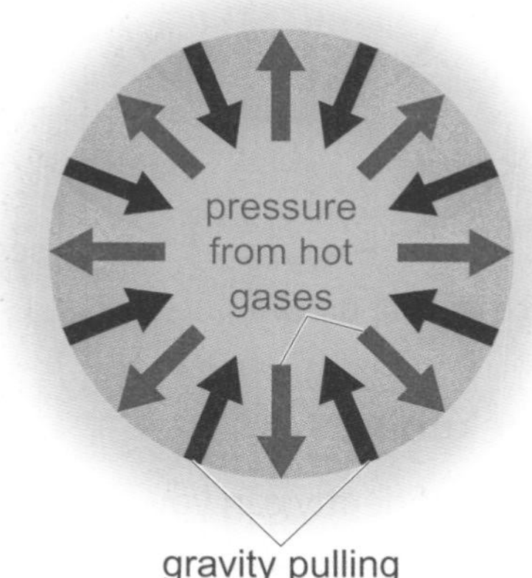

B Outward pressure from hot gases in the core of the Sun balances the gravitational pull.

Eventually the temperatures and pressures in the centre of the protostar become high enough to force hydrogen nuclei to fuse together and form helium. **Fusion reactions** like this release a lot of energy as **electromagnetic radiation**. The outward pressure from the hot gases just balances the compression due to gravity. The star is now in the **main sequence** part of its life-cycle. Our Sun is in this stage of its life cycle.

 **1** What 'fuel' is the Sun using to release energy?

 **2** Describe how a star forms.

Life cycles of stars like our Sun

Stars of similar sizes to our Sun remain stable for about 10 billion years. When they have fused most of their hydrogen into helium, the core is not hot enough to withstand gravity and it collapses. The outer layers expand to form a **red giant** star, much larger than the original star.

Other fusion reactions happen inside red giants, such as combining helium nuclei to form heavier elements. The star remains as a red giant for about a billion years before throwing off a shell of gas. The rest of the star is pulled together by gravity and collapses to form a **white dwarf** star. No fusion reactions happen inside a white dwarf and it gradually cools over about a billion years to become a black dwarf.

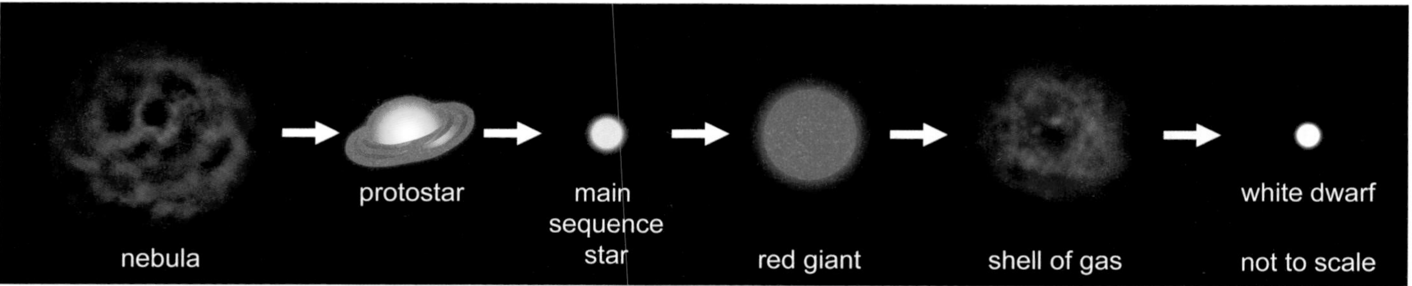

C the life cycle of a star like our Sun

Life cycles of massive stars

Stars with considerably more mass than the Sun are hotter and brighter. They fuse hydrogen into helium faster, and then become **red supergiants**. At the end of the red supergiant period the star rapidly collapses and then explodes in a **supernova**. The outer layers of the supergiant are cast off and expand outwards.

If what is left is four or more times the mass of the Sun, gravity pulls the remains together to form a **black hole**. The gravitational pull of a black hole is so strong that not even light can escape it. If the remains are not massive enough to form a black hole, gravity pulls them together to form a small, very dense star called a **neutron star**.

 **3** **a** What is a red giant?

 **b** What is a white dwarf?

 **4** Which stage of the Sun's life cycle does the artist's impression at the top of the previous page show?

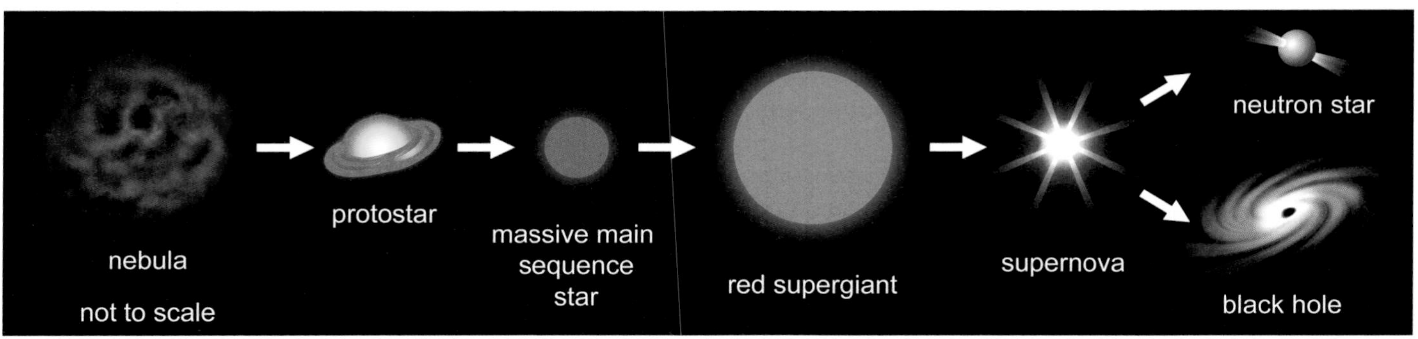

D the life cycle of a massive star

Did you know?

We cannot see black holes directly because light cannot escape from them. However they pull in gases from the space around them, and these become very hot as they fall into the black hole. The hot gases emit radiation that we can detect.

 **5** What is a supernova?

 **6** Why won't the Sun form a black hole?

Exam-style question

Describe two ways in which gravity has a part in the life cycle of a star.

(2 marks)

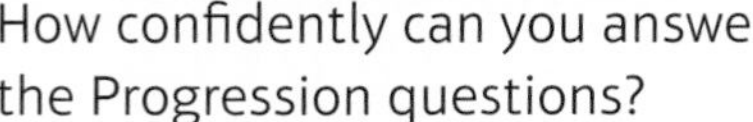

Checkpoint

How confidently can you answer the Progression questions?

Strengthen

S1 Draw two flow charts to describe the life cycles of stars:

a of similar masses to the Sun

b with masses much larger than the Sun.

Extend

E1 Compare and contrast the life cycles of stars of different masses.

SP7d Red-shift

Specification reference: P7.11P; P7.12P; P7.13P

Progression questions

- What is red-shift?
- How does the red-shift of stars vary with their distance from Earth?
- How does red-shift provide evidence for the expansion of the Universe?

As an emergency vehicle's siren travels away from you, its **pitch** gets lower. This is the **Doppler effect**. The pitch of a sound depends on the frequency of the sound wave. The sound waves behind a moving sound source become 'stretched', which makes their wavelength longer. This in turn lowers their frequency and so we hear the sound as a lower pitch. The opposite happens in front of the sound source. This only happens if the source of the sound is moving relative to the observer. If you are travelling in a car with the same velocity as the vehicle with the siren, its sound would not appear to change.

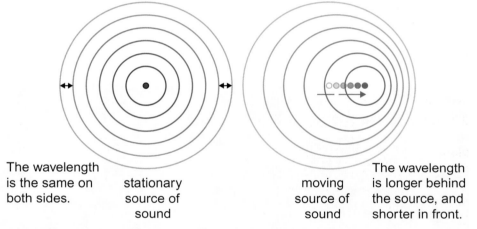

The wavelength is the same on both sides.

stationary source of sound

moving source of sound

The wavelength is longer behind the source, and shorter in front.

A how the movement of a source affects the waves detected (viewed from above)

1 a What is the difference in the frequency of the sound waves from an ambulance siren, between when it is approaching you and when it is heading away from you?

b How does the wavelength change?

A similar thing happens with light waves. The visible spectrum of light from stars contains patterns of dark lines. If these are **red-shifted** (moved towards the red end of the spectrum), the star is moving away from us. The further the lines are shifted, the faster the star is moving relative to us. The red-shift is a measure of how far along the spectrum the lines have moved.

If a star is moving towards us, the wavelength and frequency of the light waves become shorter and so the pattern of lines moves towards the blue end of the spectrum.

Did you know?

Speed cameras like the one in the photo use the Doppler effect. A beam of microwaves is sent towards the car. The car reflects the beam back to the detector. Because the car is moving, the microwaves arriving at the detector will be Doppler shifted – the faster the car is travelling the greater the Doppler shift. The Doppler shift is used to calculate the speed of the car.

B

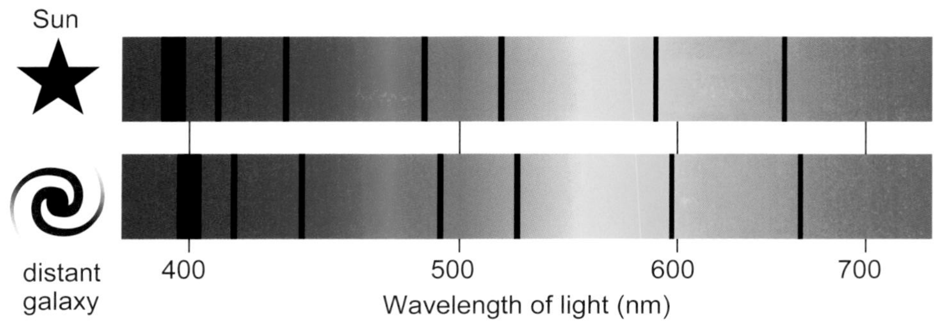
Sun
distant galaxy
400
500
600
700
Wavelength of light (nm)
Dark lines can be seen in light from the Sun.
The spectrum from a distant galaxy has the same pattern of lines, but shifted towards the red (longer wavelength) end of the spectrum.

C The light from the distant galaxy shows red-shift. This is explained if the galaxy is moving away from Earth. The movement of the source of light 'stretches' the wavelengths and so the pattern of lines has shifted towards the red end of the visible spectrum.

In the 1920s, Edwin Hubble (1889–1953) investigated how far the pattern of lines was shifted for around fifty galaxies in comparison to the Sun. He discovered that they were almost all red-shifted and concluded that these galaxies were moving away from us. He found that the further away a galaxy is, the greater its red-shift and so the faster it is moving away from us. We interpret this relationship to mean that the **Universe** is expanding.

 **2** What is red-shift?

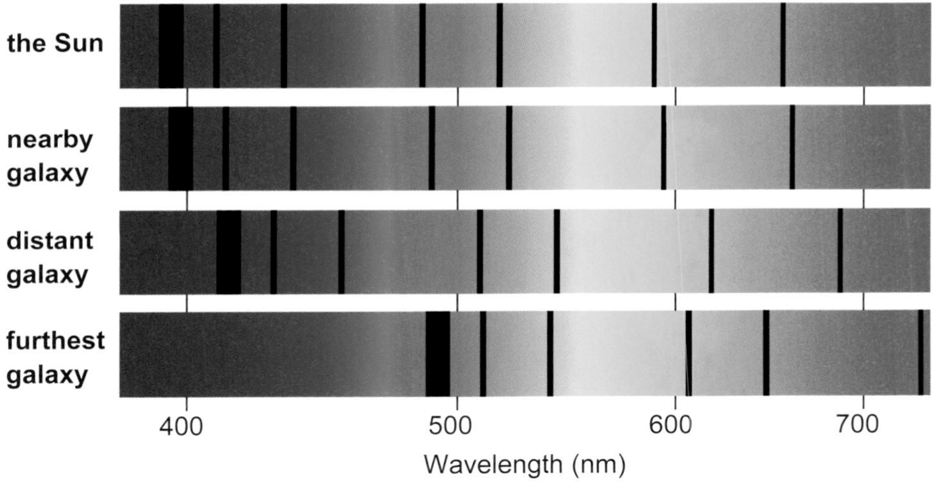
the Sun
nearby galaxy
distant galaxy
furthest galaxy
400
500
600
700
Wavelength (nm)

D The effect of red-shift on spectral lines for galaxies at different distances.

 **3** Draw a diagram to explain why an expanding Universe causes red-shift in the light from a distant galaxy.

 **4** Galaxy A is 4×10^{25} km away and Galaxy B is 3×10^{20} km away. Explain which galaxy is moving away from us more quickly.

 **5** The Andromeda galaxy has a blue-shift (the lines in the spectrum of a star are shifted towards the shorter wavelength (blue) end of the spectrum). Explain what this tells us.

Exam-style question

Explain how astronomers work out the red-shift of a galaxy. *(2 marks)*

Checkpoint

How confidently can you answer the Progression questions?

Strengthen

S1 What does red-shift in the light from a galaxy tell us?

S2 How does the red-shift of galaxies suggest that the Universe is expanding?

Extend

E1 Explain how the relationship between the red-shift of a galaxy and its distance away from us can be used as a way of estimating the distance to newly discovered galaxies.

SP7e Origin of the Universe

Specification reference: P7.8P; P7.9P; P7.10P; P7.14P; P7.15P

Progression questions

- What are the Steady State and Big Bang theories?
- What evidence supports the Big Bang theory?
- Why is the Big Bang theory the currently accepted model?

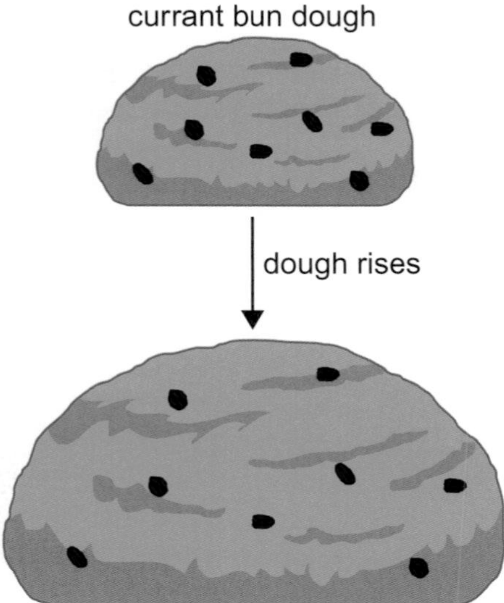

A A currant bun can be used as a model of the Universe. When the dough rises all the currants get further away from the other currants.

1 What does red-shift tell us about very distant galaxies?

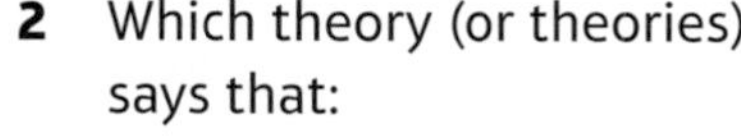

2 Which theory (or theories) says that:

a new matter is being created all the time

b the Universe began about 13.5 billion years ago?

3 Explain why the movement of galaxies supports both the Steady State and Big Bang theories.

By measuring the red-shift of galaxies and other distant objects, we know that more distant objects are moving away from us faster than closer ones. This can be explained if the Universe is expanding.

Astronomers use this information and other data to work out theories that explain the origin and present state of the Universe. One of these is known as the **Big Bang theory**, first suggested in the 1920s. This says that the whole Universe and all the matter in it started out as a tiny point of concentrated energy about 13.5 billion years ago. The Universe expanded from this point and is still expanding. As the Universe expanded, gravity caused matter to clump together to form stars.

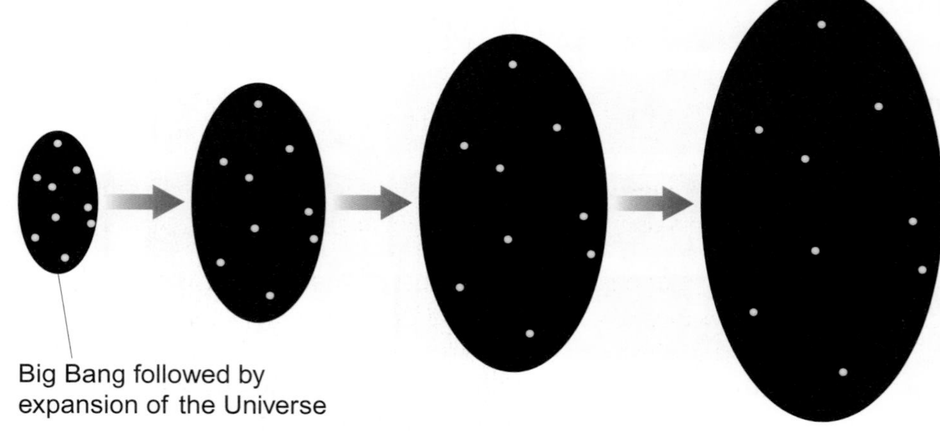

B the Big Bang theory of the Universe

An alternative theory, the **Steady State theory**, was suggested in 1948. This theory says that the Universe has always existed and is expanding. New matter is continuously created within the Universe as it expands.

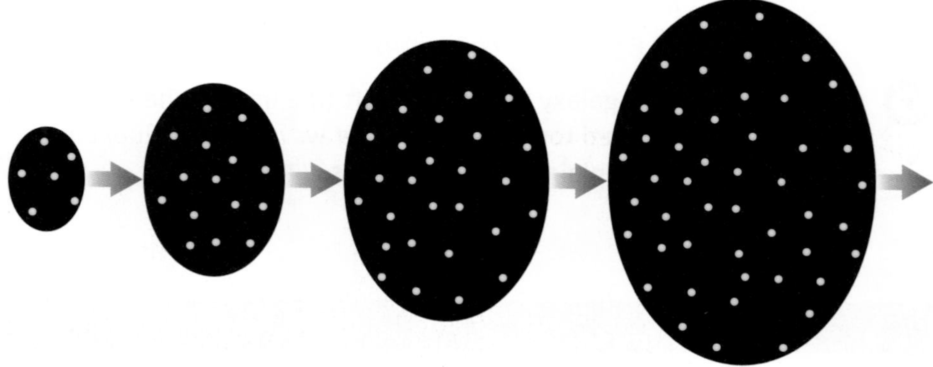

C the Steady State theory of the Universe

In 1964 two radio astronomers who were building a radio telescope detected microwave signals coming from all over the sky. At first they thought these were caused by a fault in their equipment, but eventually they realised that the signals were real. The astronomers realised that this was the radiation predicted by the Big Bang theory.

The Big Bang theory says that huge amounts of radiation were released at the beginning of the Universe. Because the Universe is expanding, the wavelength of this radiation has increased and so now it is only detectable as microwave radiation. It is called **cosmic microwave background** (**CMB**) **radiation**.

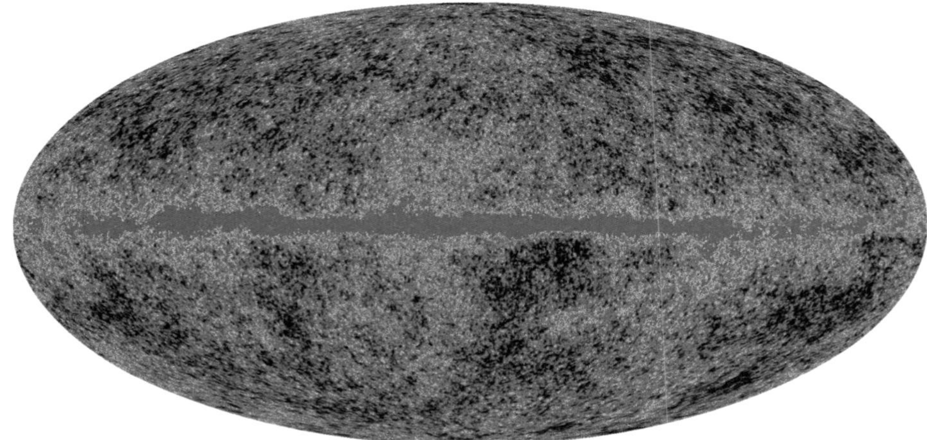

D This map of CMB radiation was made by the WMAP satellite. The colours show tiny variations from the average microwave wavelength of around 1 mm.

Both the Steady State and Big Bang theories say that the Universe is expanding. Observations of red-shift in the light from other galaxies can be used as support for both theories. However, CMB radiation provides supporting evidence for the Big Bang theory only. The Steady State theory cannot explain the CMB radiation. As there is more supporting evidence for it, the Big Bang theory is accepted by most astronomers today.

 **4** **a** What does CMB stand for?

 **b** Describe how astronomers explain the CMB radiation.

5 Which theory (or theories) is supported by the following evidence?

 **a** red-shift

 **b** the cosmic microwave background radiation

Exam-style question

Compare and contrast the Big Bang and Steady State theories. *(3 marks)*

Did you know?

Some theories of the Universe suggest that the Universe will eventually stop expanding and will collapse back in on itself in a Big Crunch.

Checkpoint

How confidently can you answer the Progression questions?

Strengthen

S1 Describe the two theories of origin of the Universe, and the evidence for them.

S2 Explain which theory is accepted by most scientists.

Extend

E1 Describe how and why ideas about the beginning of the Universe have changed since 1900.

The expanding Universe

Scientists believe that the Universe is expanding.

Describe how careful observation of electromagnetic radiation from distant galaxies as well as from the whole of space gave evidence supporting the Big Bang theory. **(6 marks)**

Student answer

The Big Bang theory says that the Universe started from a single point billions of years ago [1]. *Scientists looking at distant galaxies noticed that the light from them was red-shifted. This means the galaxies are moving away and means space is expanding* [2]. *Radiation coming from the whole of space is the cosmic microwave background radiation, which comes from all of the sky* [3]. *This also supports the Big Bang theory.*

[1] This description of the Big Bang theory is correct but will gain no marks as it was not asked for in the question.

[2] This partly describes the conclusion that scientists draw from the red-shifted light from galaxies, but it has not mentioned the link between the distance of the galaxy and the amount of red-shift.

[3] The student has correctly named the CMB radiation. They will get no extra credit for saying that it comes from all over the sky as the question tells them this.

Verdict

This is an acceptable answer because it mentions the two pieces of evidence that support the Big Bang theory.

The answer could be improved by explaining that the light from more distant galaxies has a greater red-shift and link this to the idea of the Universe expanding. More detail on the CMB radiation is also needed, linking its long wavelength to the idea that it is radiation from the Big Bang with its wavelength increased by the expansion of the Universe.

Exam tip

Don't waste time giving information not asked for in the question.

Paper 2

SP8 Energy – Forces Doing Work / SP9 Forces and their Effects

This base jumper is wearing a wing suit. As he falls, gravitational potential energy stored in his body will be transferred to a store of kinetic energy because of the force produced by the Earth's gravitational field. The wing suit will help him to fly away from the cliff.

In this unit you will learn more about how forces can transfer energy. You will also learn about force fields, and how to use vector diagrams to work out what happens when several different forces act on an object at the same time.

The learning journey

Previously you will have learnt at KS3:

- the different ways in which energy can be stored and transferred
- about resultant forces and the effects of balanced and unbalanced forces
- about moments as the turning effects of forces.

You will also have learnt in *SP1 Motion* and *SP3 Conservation of Energy*:

- the difference between vector and scalar quantities
- how to calculate changes in GPE and KE (also examined in this paper)
- about energy transfer diagrams and how to work out the efficiency of a transfer (also examined in this paper).

In this unit you will learn:

- how the energy in a system can be changed
- how to calculate power and work done
- how objects interact with each other, through force fields and contact forces
- about rotational forces, calculating moments and how levers and gears work
- **H** how to use vector diagrams to work out the effects of forces on an object.

SP8a Work and power

Specification reference: P8.1; P8.4; P8.5; P8.6; P8.7; P8.12; P8.13; P8.14

Progression questions

- How can the energy of a system be changed?
- What is work done and how can it be measured and calculated?
- What is power and how is it calculated?

A Animals are used instead of machines in many parts of the world.

Energy is transferred whenever things happen. Electricity transfers energy to an electric light bulb, which then transfers the energy to the surroundings by light and heating.

Energy can also be transferred when a force makes something move. The energy transferred by a force is called **work done**, and this amount of energy depends on the size of the force and on how far the force moved. We can calculate the work done using this equation:

work done (J) = force (N) × distance moved in the direction of the force (m)

This can be written as:

$E = F \times d$

where E represents work done

F represents force

d represents distance.

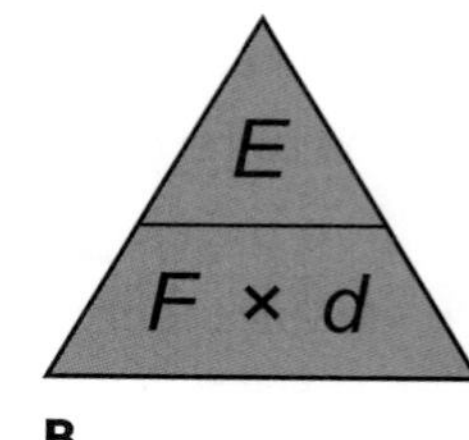

B

1 Give the scientific meaning for work done.

2 The log in photo A weighs 1800 N and the elephant lifts it up 1.5 metres. How much work does the elephant do?

Worked example

Danny is moving a box weighing 300 N. He pulls it 3 m along a sloping ramp using a force of 200 N. Calculate the work Danny does.

$E = F \times d$

$= 200\text{ N} \times 3\text{ m}$ — The force must be in the direction of the movement.

$= 600\text{ J}$

C An osprey does work when it lifts a fish.

3 Calculate the work done when:

a a weightlifter lifts a 280 N barbell weight 1.5 m straight up

b a boy cycles 760 m against friction forces of 140 N.

4 The total weight of the osprey and fish in photo C is 550 N. The osprey does 2200 J of work. Calculate how high it lifts the fish.

Power

Power is the rate at which energy is transferred. When energy is being transferred by forces, then power is also the rate of doing work. Power is measured in **watts** (**W**). 1 watt means 1 joule of work done per second.

If two ospreys lift the same sized fish 2 metres in the air, they will both have done the same amount of work. But if one lifts its fish in a shorter time, that bird has produced a greater power.

Power can be calculated using this equation:

$$\text{power (W)} = \frac{\text{work done (J)}}{\text{time taken (s)}}$$

This can be written as:

$$P = \frac{E}{t}$$

where E represents work done
P represents power
t represents time.

Did you know?

Some adverts still quote the power of vehicles in horsepower. This term was first used in the 1700s to compare the power of steam engines with the power of a horse.
1 horsepower = 746 watts.

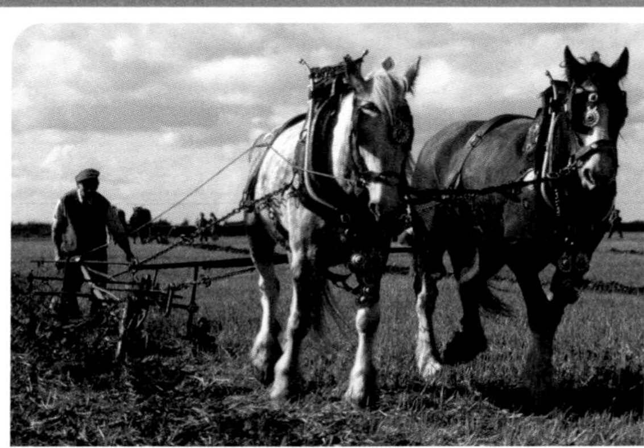

D Horses are still used for ploughing in some parts of the world.

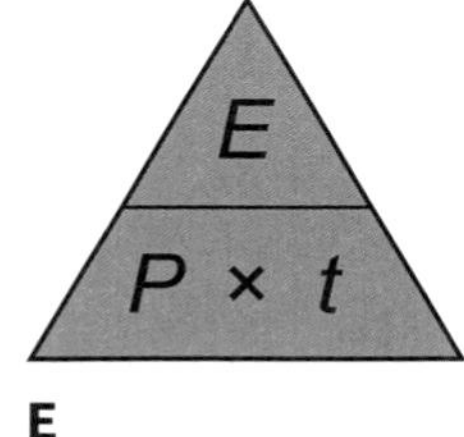

E

5 Iesha and Fran both weigh 500 N. Iesha runs up a 3 m high flight of stairs in 10 seconds. Fran takes 12 seconds.

 **a** Calculate the work done by Iesha in climbing the stairs.

 **b** Calculate the power of each girl.

6th **6** Explain how an electric motor can transfer energy to a system by heating. Include the word friction in your answer.

Exam-style question

Two children with the same mass run up the same flight of stairs. One child takes twice as long as the other. Compare the work done and the power of the two children. *(4 marks)*

Checkpoint

How confidently can you answer the Progression questions?

Strengthen

S1 Explain what you would need to measure to find out how much work the horses in photo D do when they plough a field.

S2 A car has a weight of 10 000 N. A man takes 10 seconds to push it 5 m using a force of 1000 N.

a Calculate the work done.

b Calculate the power of the man.

Extend

E1 A 7800 W lift took 10 seconds to move a 5000 N weight. How far did the lift move this weight?

SP9a Objects affecting each other

Specification reference: P9.1; P9.2

Progression questions

- What forces are there when two objects are touching?
- How can objects affect each other without touching?
- How are pairs of forces represented?

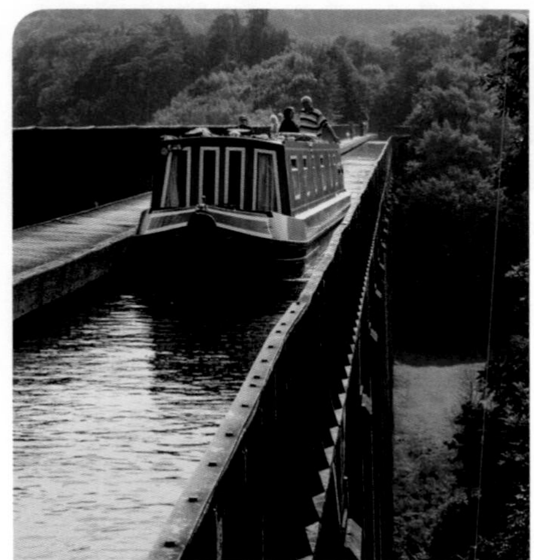

A What forces are acting on the narrowboat and on the aqueduct?

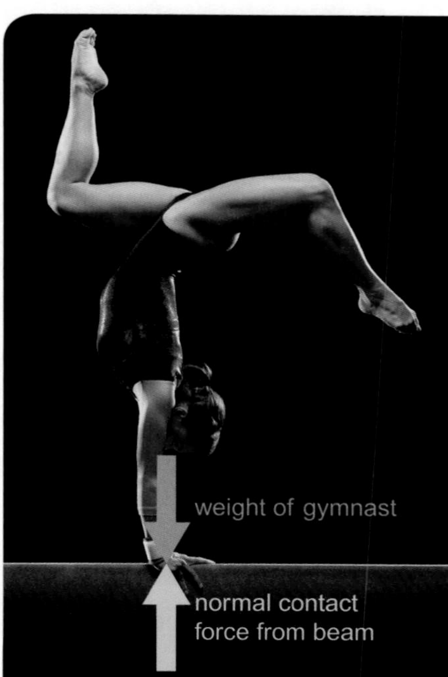

B The weight of the gymnast is balanced by the normal contact force from the beam acting upwards. The normal contact force acting on the gymnast is a reaction force due to the weight of the gymnast acting on the beam.

Objects can interact (affect each other) by exerting forces on each other. If the objects are touching, then the forces between them are **contact forces**. When you stand on the floor, there is an upwards force from the floor on you called the **normal contact force**. The narrowboat in photo A needs an engine to keep it moving because water resistance (a form of **friction**) slows it down. The force from the engine and water resistance are both contact forces.

The narrowboat is floating because of another contact force, called **upthrust**, from the water. The upthrust is balanced by a **non-contact force** called gravity. Gravity does not need to touch the boat to give it weight. The upthrust and the weight are both acting on the same object.

1 Look at photo A.

a How is the water affecting the narrowboat?

b How is the narrowboat affecting the water?

2 Describe the forces caused by the interaction of the aqueduct with the ground below it.

Gravity is a force that occurs between any two objects that have mass. The Moon stays in orbit around the Earth because the two bodies are attracting each other. The force from the Moon on the Earth is the same size as the force from the Earth on the Moon, but in the opposite direction. The gravitational forces between two objects with mass can be represented as **vectors** (arrows that show both direction and magnitude). These two forces are **action–reaction forces** (pairs of forces acting on *different* objects, in opposite directions).

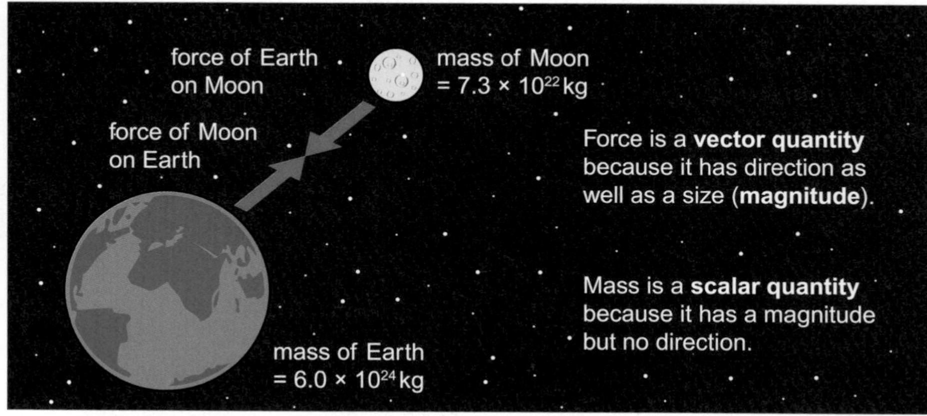

C The gravitational forces between the Earth and the Moon can be shown using arrows, where the length of the arrow represents the size of the force.

The space around an object where it can affect other objects is called a **force field**. The Moon and the Earth affect each other because the Moon is within the Earth's **gravitational field** and vice versa.

Other forces that affect objects they are not in direct contact with are **magnetism** and **static electricity**. A **magnet** can attract objects made from **magnetic materials** including iron, nickel and cobalt. A magnet can attract or repel another magnet. The space around a magnet where it can affect other materials is called the **magnetic field**.

An object charged with static electricity has an **electric field** (**electrostatic field**) around it. The electric field can affect objects within it. Two objects with the same charge that are close to each other produce a pair of forces that are equal in size and acting in opposite directions.

D Foam packing pellets can pick up a charge of static electricity and become attracted to things around them.

5 Draw a diagram with arrows to show the forces between two charged objects that are:

a attracting each other

b repelling each other.

Exam-style question

Two plastic rods with identical charges of static electricity are suspended near each other. Describe the forces between the two rods, and between the rods and the Earth. *(4 marks)*

3 Describe the effect of the gravitational force from the Sun on the Earth.

4 The force from the Sun on the Earth is 3.5×10^{22} N. What is the force of the Earth on the Sun?

Did you know?

The gravitational force between two people standing next to each other is about 0.000 003 N.

Checkpoint

How confidently can you answer the Progression questions?

Strengthen

S1 Name three contact forces and three non-contact forces.

Extend

E1 A heavy box is resting on a table. Describe how it interacts with the table and with the Earth.

E2 Describe one similarity and two differences between a gravitational field and a magnetic field.

SP9b Vector diagrams

Specification reference: H P9.3; H P9.4; H P9.5

Progression questions

- H What is a free body force diagram?
- H How and why do we resolve forces?
- H How do all the forces on a single body combine to affect it?

H

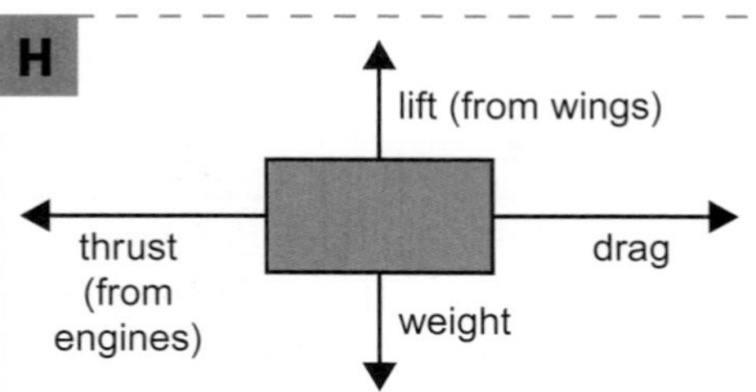

A A free body force diagram for an aeroplane. The arrows represent force vectors. The direction of the arrow shows the direction of the force and the length of the arrow represents its size. The simplest aeroplane to sketch is a box!

Every object usually has more than one force acting on it. If the forces are equal in size but in opposite directions, the **resultant force** (or **net force**) is zero. The forces on the object are said to be balanced.

When an aeroplane is flying at a constant velocity and height, the horizontal forces and vertical forces on it are balanced. The resultant (net) force on the aeroplane is zero. You can show the forces on the aeroplane using a **free body force diagram**, as shown in diagram A.

1 Describe the resultant force on a car that is slowing down.

2 Draw a free body force diagram to show:

a the vertical forces on a person sitting on a chair

b the forces on a car travelling at a constant speed.

The forces on an object are not always acting along the same line. When an aeroplane makes a cross-wind landing, the force from the aeroplane's engines is pushing it forwards, but there is also a force from the wind pushing it sideways. We can find the resultant force using a **scale diagram**.

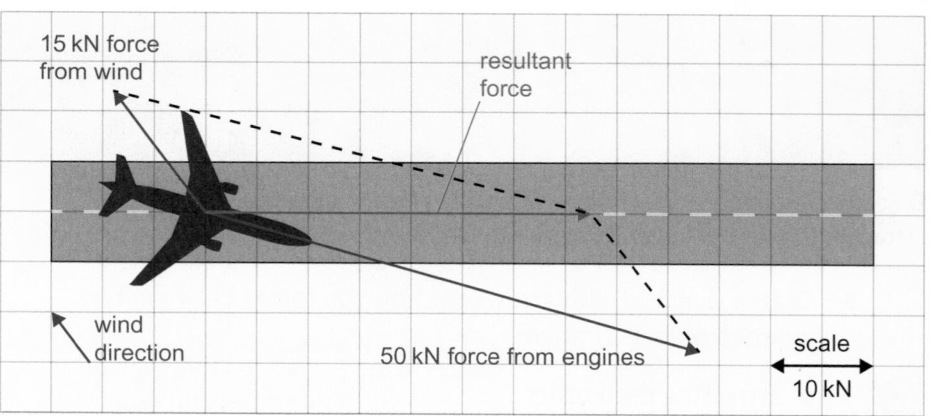

- Draw arrows at the correct angles to represent the forces. The length of each arrow should represent the size of the force.
- Draw lines to make a parallelogram.
- The resultant force is the diagonal of the parallelogram. Measure this arrow to work out the size of the resultant force.

B A scale diagram can be used to work out a resultant force.

Did you know?

Many airports have several runways pointing in different directions. This allows the pilot to use the one nearest to the wind direction to reduce the component of the wind acting across the runway.

3 In Figure B the force from the engines is 50 kN and the force from the wind is 15 kN at an angle of 150° from the direction the aeroplane is pointing. Draw a scale diagram to work out the resultant of these two forces.

H

The aeroplane in photo C is climbing at a steep angle. The thrust from its engines is helping it to climb. We can work out how much of the thrust is pushing the aeroplane upwards and how much is pushing it forwards by **resolving** the thrust force into two **component** forces at right angles to each other. This can be done using a scale diagram, as shown in diagram D.

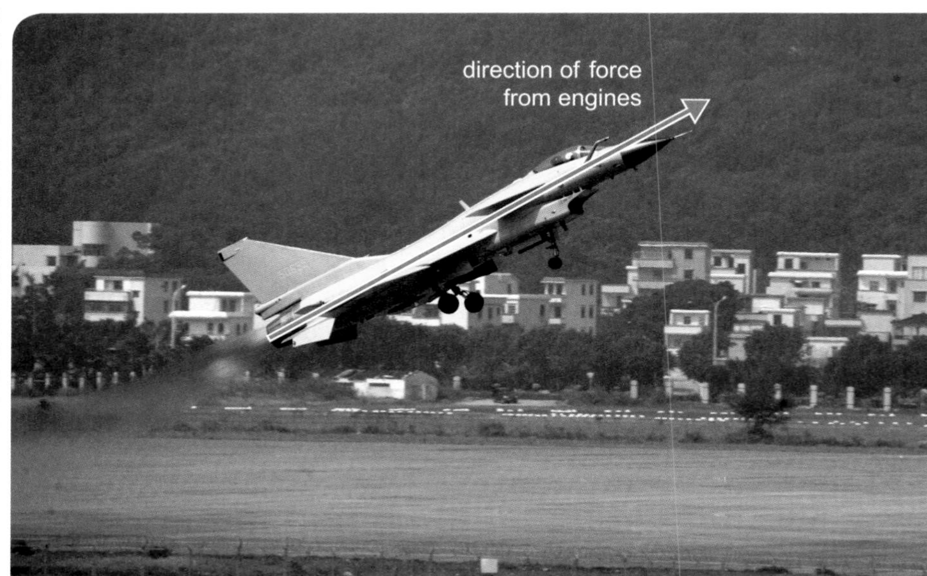

C The aeroplane is at an angle of 30° to the ground. The engines generate 200 kN of thrust.

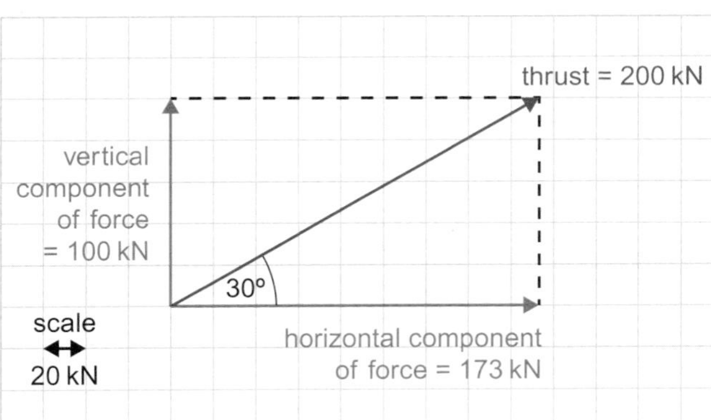

- Draw a force arrow to scale at the correct angle.
- Draw a rectangle with the sides in the directions you are interested in (e.g. horizontal and vertical).
- The resolved forces are the sides of the rectangle.

D A scale diagram can be used to resolve the forces on an aeroplane.

 **5** An aeroplane is climbing at an angle of 50° to the horizontal, with 200 kN of thrust. Draw a scale diagram to find the vertical component of this thrust.

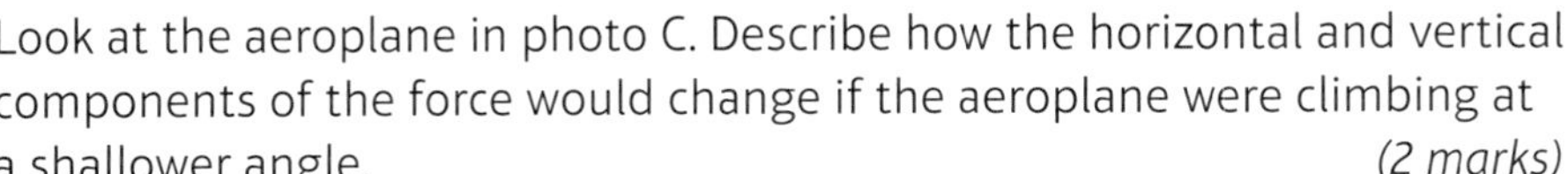

Exam-style question

Look at the aeroplane in photo C. Describe how the horizontal and vertical components of the force would change if the aeroplane were climbing at a shallower angle. *(2 marks)*

 **4 a** Which force, affecting the horizontal movement of the aeroplane, has not been included in diagram B?

 **b** The aeroplane is moving at a constant velocity. Explain how the size and direction of the force from part 4a compares to the resultant force you worked out in question 3.

Checkpoint

How confidently can you answer the Progression questions?

Strengthen

S1 There is a forwards force of 20 N on a toy car and a backwards force of 5 N.

a Calculate the resultant force.

b Explain the effect of the resultant force.

c Draw a free body force diagram to represent all of the forces on the toy car.

Extend

E1 A model rocket takes off at an angle of 20° from the vertical. Its thrust is 50 N. Draw a scale diagram to work out the horizontal and vertical components of the thrust.

SP9c Rotational forces

Specification reference: P9.6P; P9.7P; P9.8P; P9.9P

Progression questions

- How do you calculate the turning effect of a force?
- How can you use moment calculations to work out if two rotational forces will balance?
- How do levers and gears transmit the rotational effects of forces?

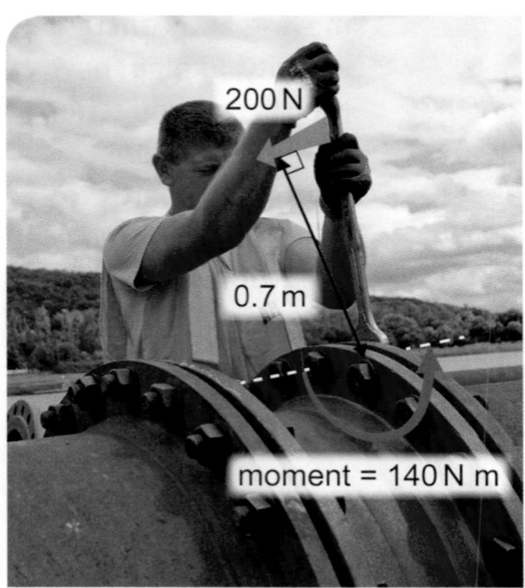

A a large spanner being used to tighten a bolt on a pipeline

 **2** Look at photo A. A force of 100 N is being applied at a distance of 0.5 m normal to the force. Calculate the moment.

The spanner in photo A is producing a turning force on the nut attached to the bolt on the pipeline. A turning force is called a **moment**.

 **1 Describe two other situations where a force produces a moment.**

The moment of a force depends on the size of the force and where the force is applied. The greater the force, and the further it is applied from the **pivot**, the greater the moment. The distance between the force and the pivot is measured **normal** (at right angles) to the direction of the force.

Moment, force and distance are linked by the equation below. The unit for moment is **newton metres** (**N m**).

moment of a force (N m) = force (N) × distance normal (perpendicular) to the direction of the force (m)

Diagram B shows a steelyard used for weighing goods. When the anti-clockwise moment from the goods is equal to the clockwise moment from the weights, the steelyard will balance and the weight of the goods can be worked out. When it is balanced, the steelyard is said to be **in equilibrium**.

When a system involving rotational forces is in equilibrium:

the sum of clockwise moments = the sum of anti-clockwise moments

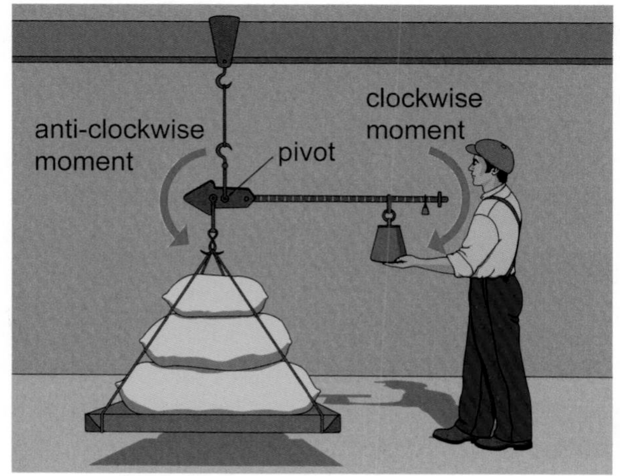

B The steelyard allows things to be weighed without the user having to lift weights as heavy as the goods.

Worked example

In diagram B the sacks are hanging from a point 0.1 m from the pivot. They are balanced by a weight of 300 N hanging 1 metre from the pivot and a weight of 20 N hanging 1.2 m from the pivot. Calculate the weight of the sacks.

sum of clockwise moments = 300 N × 1 m + 20 N × 1.2 m

= 300 N m + 24 N m = 324 N m

sum of clockwise moments = sum of anti-clockwise moments

324 N m = weight × 0.1 m

$$\text{weight} = \frac{324\text{ N m}}{0.1\text{ m}} = 3240\text{ N}$$

The horizontal beam in the steelyard is a **lever** – a bar that pivots about a point and is used to transfer a force. Diagram C shows another example of a lever.

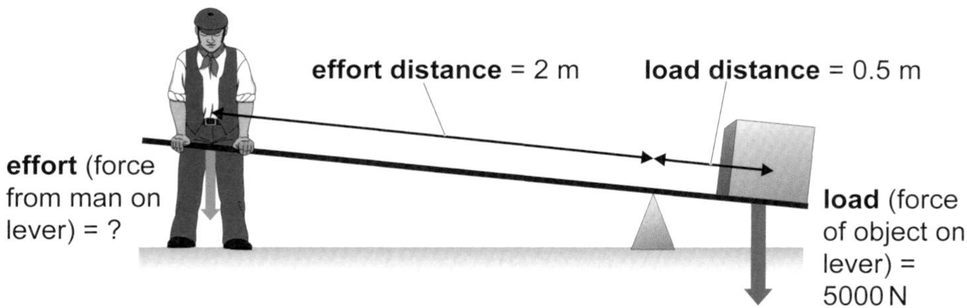

C using a lever to lift a load

3 Look at diagram C. The man is pushing down just hard enough to balance the load.

9th **a** Calculate the effort force from the man.

8th **b** What will happen if the man pushed down with a greater force than you calculated in your answer to part a?

9th **4** Look at diagram C again. The man uses a longer lever, so that the effort distance is 3 m. The load distance stays the same. Calculate the force needed to just balance the load.

The rotational effect of a force can also be transmitted by **gears**. Photo D shows the gears inside a water mill. Gear A is driven by the water wheel. The rotation is passed on to gear B by the interlocking teeth, and this drives the milling stones. Gear A has around eight times as many teeth as B, so each complete turn of A turns B by eight complete turns.

8th **5** Gear X has 10 teeth and is connected to gear Y which has 40. How many times will:

a gear Y turn for each turn of gear X

b gear X turn for each turn of gear Y?

D These gear wheels are part of a water mill.

Exam-style question

A screwdriver is being used as a lever to open a tin of paint. It pivots about the edge of the paint tin. A force of 10 N is applied at 15 cm from the pivot. The force on the lid is 0.5 cm from the pivot. Calculate the force on the lid. *(4 marks)*

Did you know?

Until the beginning of the 20th century, steelyards three stories high were used to weigh whole wagon-loads of goods.

Checkpoint

How confidently can you answer the Progression questions?

Strengthen

S1 Explain why it is easier to undo a stiff nut if you use a longer spanner.

S2 Joe has a weight of 600 N and sits 1.5 m from the centre of a see-saw.

a Calculate Joe's moment.

b Sally has a weight of 400 N. Calculate how far she must sit from the centre of the see-saw to balance Joe.

Extend

E1 Explain why a lever is useful to help you to move a heavy object. Include a calculation to illustrate your answer.

Walking uphill

Al walks directly up a hill and he takes 12 minutes to get to the top. Bev walks up the same hill on a shallower path that zig zags as it goes up. She takes 15 minutes to get to the top.

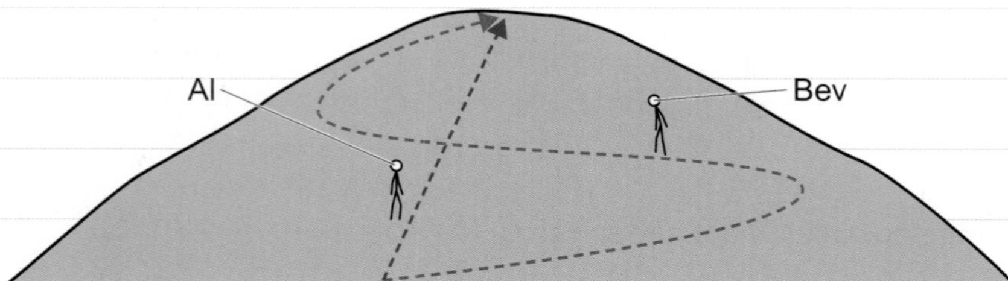

Explain who has exerted the greater power, and who has transferred more energy while getting to the top of the hill. Include any assumptions you make in your answer. **(6 marks)**

Student answer

Power is the rate of doing work, and is measured in watts. In climbing the hill, they are doing work against gravity, and the energy transferred is the force (their weight) multiplied by the distance moved in the direction of the force (up the hill) [1]. *If they both have the same weight, as they both have climbed the same distance, in theory they have both transferred the same amount of energy* [2]. *Al exerted the greater power because he gained the height in a shorter time* [3].

However, this answer may not be correct if their weights are not the same. It also assumes that the human body is totally efficient, which it is not. As Bev was walking for longer, she will waste more energy and so will have transferred more energy altogether [4].

[1] The answer correctly describes what power and work mean.

[2] This states that both have transferred the same amount of energy if we make an assumption about their weights.

[3] This part of the answer explains who has transferred more power.

[4] The final section points out that the answer relies on certain assumptions that may not be correct.

Verdict

This is a strong answer. It answers all the points in the question and also explains the assumptions that had to be made in reaching the conclusion. The points in the answer are given in a logical order and linked together with scientific ideas.

Exam tip

Make sure you understand key terms and can use them correctly. In this example, the key terms are power and energy. It can be useful to circle the key terms in the question to help you think about what they mean as you prepare your answer.

Paper 2

SP10 Electricity and Circuits / SP11 Static Electricity

An incubator has many circuits to keep a premature baby alive. It monitors temperature, blood oxygen concentration, heart rate and breathing rate.
It can make automatic adjustments and alert staff to problems. The oxygen concentration inside the incubator may be higher than normal, which could allow a fire to break out more easily. So the incubator is earthed to prevent a build up of static electricity.

In this unit you will learn about how electricity is supplied and used in different circuits, and about static electricity.

The learning journey

Previously you will have learnt at KS3:

- about electric current and voltage
- about series and parallel circuits
- that conductors have low resistance and insulators have high resistance
- about electric fields, how objects become charged and how charged objects behave.

In this unit you will learn:

- about current, charge and potential difference
- how to calculate resistance, power and energy transferred
- about components with changing resistance
- about the UK domestic electricity supply and electrical safety features in homes
- how earthing works and why it is important.
- about the shape and size of electric fields and how they explain some phenomena caused by static electricity.

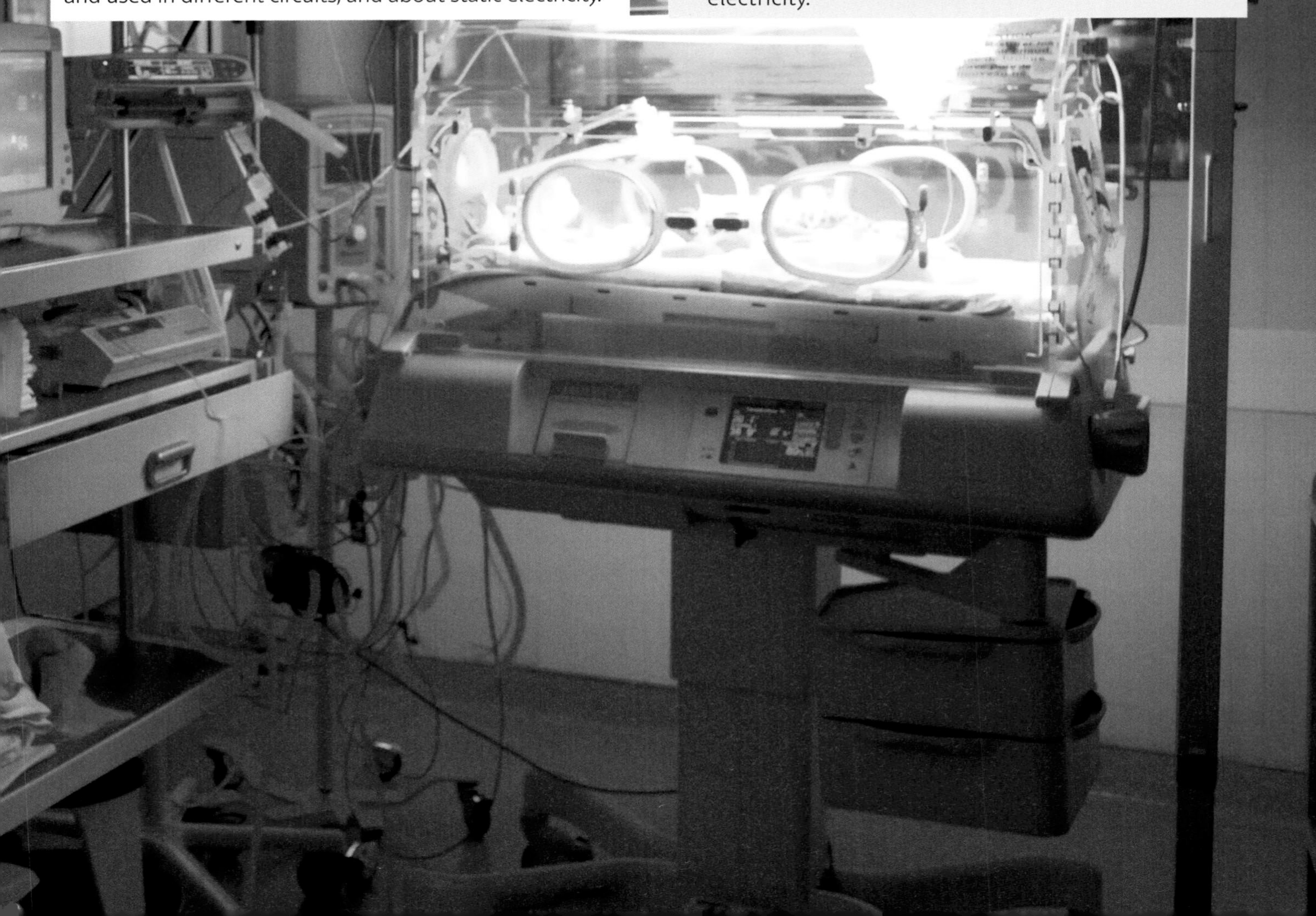

SP10a Electric circuits

Specification reference: P10.1; P10.2; P10.3

Progression questions

- How does the structure of atoms affect the flow of electric current?
- What are the names and symbols of components used in electric circuits?
- What are the differences between series and parallel circuits?

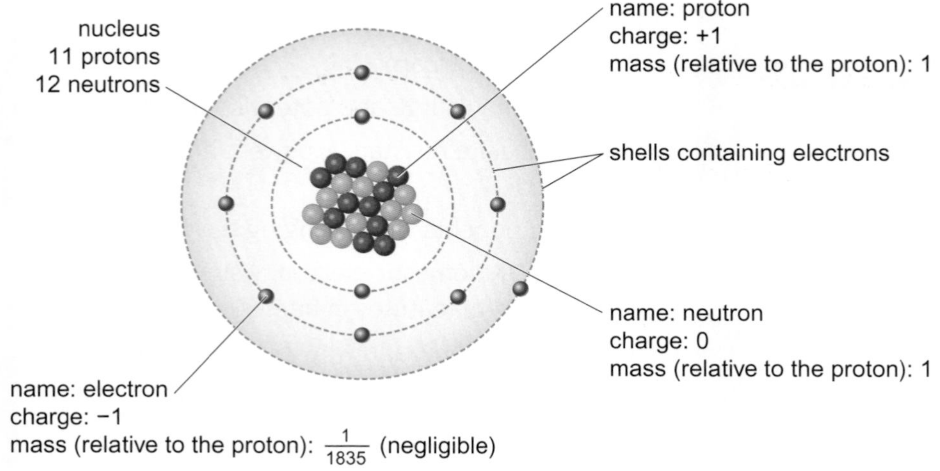

A the structure of a sodium atom (not to scale)

Diagram A shows a sodium **atom**. The atom has a central **nucleus** of positively charged **protons** and uncharged **neutrons**. Both these particles have a similar mass.

The **electrons** are found at different distances from the nucleus, in **shells**. Electrons are much smaller than protons or neutrons. Each electron has a negative charge, equal but opposite to the charge on a proton. The number of electrons is equal to the number of protons and so overall an atom is uncharged.

 **1** Draw a table of the particles in the atom, giving their charge, mass (relative to the proton) and their location in the atom.

 **2** Draw a diagram of a carbon-14 atom with 6 protons and 8 neutrons.

Current in metals

Sodium is a metal. Look at diagram A. There is one electron in the outer shell that is only weakly attracted to the nucleus. All metals have electrons like this, including copper which is used for electrical wiring. These electrons can easily be removed, so a metal wire has many 'free' electrons. When a **battery** is attached to the wire the **voltage** 'pushes' the free electrons around the circuit. The electrons are negatively charged so they move towards the positive terminal of the battery.

When working with circuits, however, the conventional direction of current is used. Conventional current direction goes from the positive terminal to the negative terminal of the battery.

Did you know?

The conventional direction of current is thanks to Benjamin Franklin, an American scientist. He chose the direction long before the electron was discovered.

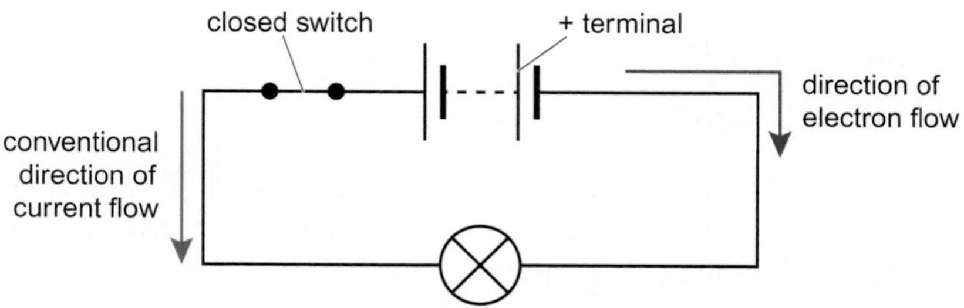

B electron flow and conventional current

Series and parallel circuits

Circuit diagrams are used to show the **components** and the junctions in a circuit. Table C shows some common circuit symbols.

Components in circuits can be connected in **series** or **parallel**. In series circuits there is just one route the current can take around the circuit. In parallel circuits there are junctions that allow the current to take different routes. Diagram D shows circuit diagrams for both series and parallel circuits.

Circuit symbol	Component	Circuit symbol	Component
	switch (open)		lamp
+ −	cell	A	ammeter
	battery	V	voltmeter

C table of circuit symbols

In the series circuit, lamps cannot be switched on and off individually, and if one lamp fails they will all switch off. In the parallel circuit each lamp can be switched separately.

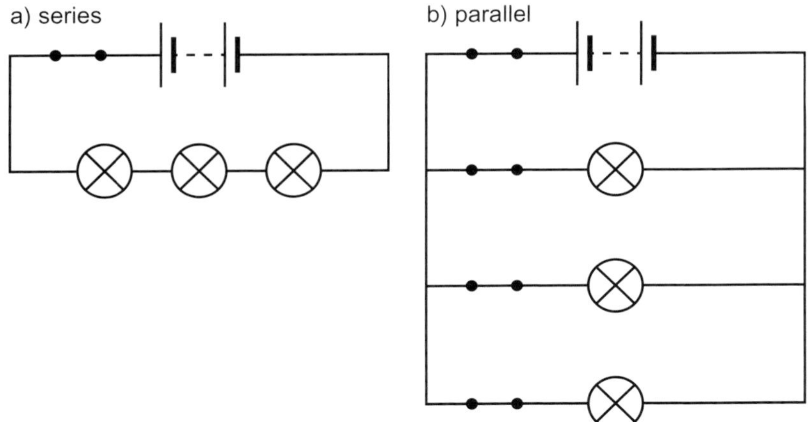

D three lamps connected in a) series and b) parallel

3 Draw a series circuit with:

 **a** 2 lamps

 **b** a switch and 4 lamps.

 **4** Draw a circuit with 2 lamps in parallel.

 **5** Draw a circuit containing 3 lamps, one in series and two in parallel. Include 3 switches (both open and closed) to show how two of the lamps could be lit and one of them unlit.

Exam-style question

Look at diagram D. Explain what happens in each circuit if all the lamps are on and then one lamp breaks. *(4 marks)*

Checkpoint

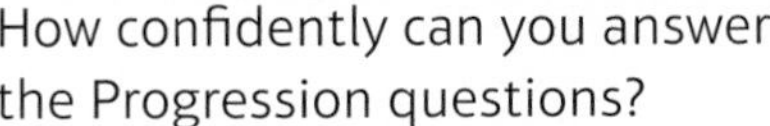

How confidently can you answer the Progression questions?

Strengthen

S1 Draw a circuit diagram with a cell, two lamps in parallel and two switches so that each lamp can be turned off separately.

S2 Add labelled arrows to your diagram from **S1** showing the directions of conventional current and electron flow.

Extend

E1 Describe the way that the protons, neutrons and electrons are arranged in an atom and explain why having free electrons means that metals can conduct electricity.

E2 Compare and contrast series and parallel circuits.

SP10b Current and potential difference

Specification reference: P10.4; P10.7; P10.10; P10.11

Progression questions

- How is electric current measured?
- What happens to the electric current at a junction in the circuit?
- What is potential difference and how do you measure it?

Current

Electric current is measured in units called **amperes** (often shortened to **amps**, A), using an **ammeter**. An ammeter is connected in series to measure the current passing through a component or circuit.

The total amount of current stays the same on its journey around the circuit. The current leaving the positive terminal of the battery is the same as the current arriving at the negative terminal. This is because current is **conserved**. In a parallel circuit, current splits at a junction to travel along different branches, but the total amount entering the junction is the same as the total amount leaving.

Did you know?

There are several different species of fish called electric rays. They can use electricity to stun or kill their prey. The voltage used can be as high as 220 V.

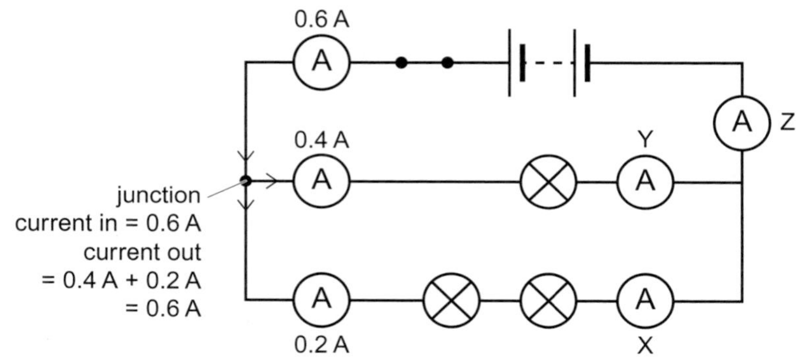

A Currents are conserved at the junctions.

1 In the circuit in diagram A, what is the reading on ammeter:

 **a** X

 **b** Y

 **c** Z?

 **2** A current of 2 A passes through a circuit with 5 identical lamps in parallel. What is the current through each lamp?

Potential difference

You need a **potential difference (p.d.)** to 'push' current around an electric circuit. Potential difference is also called voltage. Diagram B shows marbles on a ramp. This can be used as a model for potential difference. On a flat surface the marbles won't move but when there is a height difference, they can roll down the ramp. Similarly, electrons will flow when a potential difference is applied across a component.

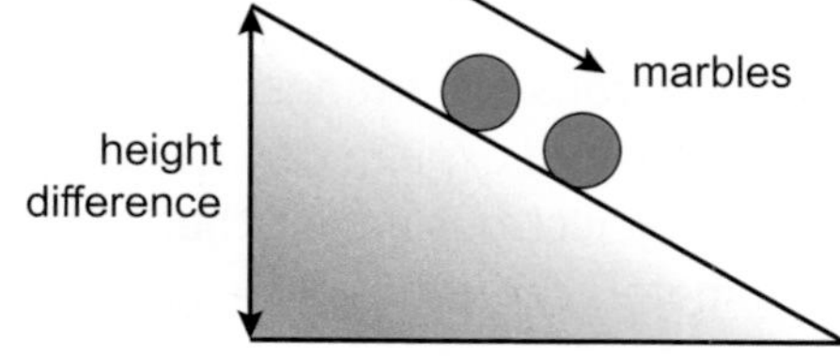

B A height difference applied to the ramp can make marbles roll.

A circuit contains electrons all the way round. For a current to flow, the circuit must be closed and contain a source of potential difference (such as a **cell** or battery). The electrons all move together when a current flows.

3 Explain what happens to the marbles when the height difference in diagram B is increased.

4 For a current to flow, why must there be:

a a potential difference

b a closed circuit?

The bigger the potential difference across a component the bigger the current.

In a parallel circuit, the potential difference across each branch of the circuit is the same. When there is more than one component in a branch of a circuit, the potential differences across all the components add up to give the total potential difference supplied by the cell or battery.

Potential difference is measured in **volts,** V, using a **voltmeter.** A voltmeter is always connected in parallel to measure the potential difference across a component or circuit.

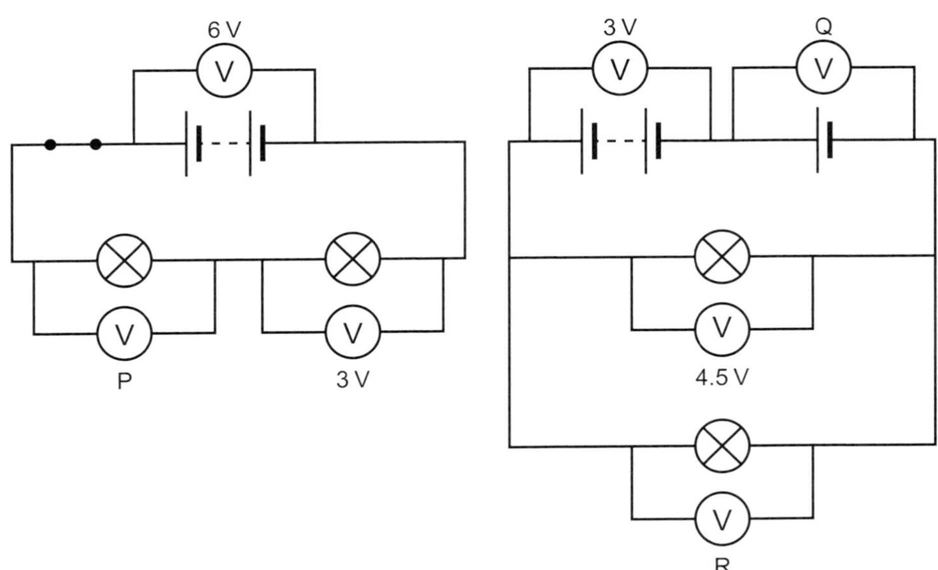

C A voltmeter is always connected in parallel with components.

5 In the circuits in diagram C, calculate the potential difference measurements on voltmeters P, Q and R.

Exam-style question

Describe how you would measure the current and potential difference in a circuit with one lamp. You may draw a circuit diagram to help you. *(2 marks)*

Checkpoint

How confidently can you answer the Progression questions?

Strengthen

S1 List the symbols used in the circuits in diagram D.

S2 What are the readings on meters A and B?

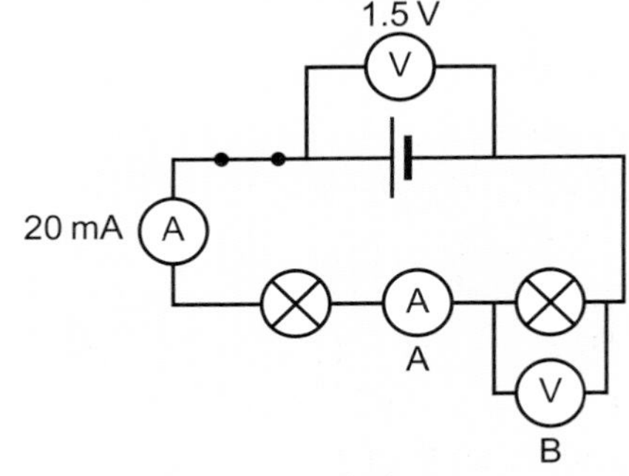

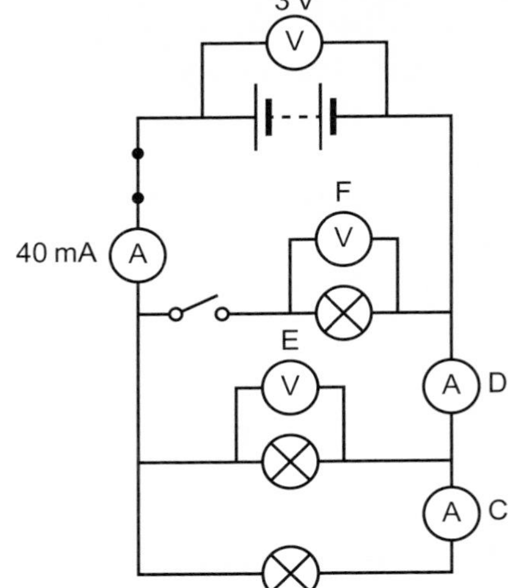

D Two circuits, in which the lamps are all identical.

Extend

E1 State and explain each of the meter readings A to F in diagram D.

SP10c Current, charge and energy

Specification reference: P10.5; P10.6; P10.8; P10.9

Progression questions

- What is a coulomb?
- What is the connection between the electric current and the amount of charge that flows in a circuit?
- What is the equation that relates electric charge, potential difference and the energy transferred in a circuit?

A A defibrillator uses an electric charge to restart a heart after a heart attack. Defibrillators need to build up a charge of about 160 mC before being used to 'shock' a heart back into action.

Moving charged particles form an electric current. Electric **charge** is measured in **coulombs** (C). One coulomb is the charge that passes a point in a circuit when there is a current of 1 amp for 1 second.

In metals the current is a flow of electrons. Each electron has a very tiny negative charge, just -1.6×10^{-19} C (less than one millionth of a millionth of a millionth of a coulomb).

The size of the current at any point in a circuit tells you how much charge flows past that point each second. Electric current is the **rate** of flow of charge.

1 Explain how a current flows in a metal (use the words electron and charge).

The charge that flows in a set time can be calculated using the equation:

$$\underset{(C)}{\text{charge}} = \underset{(A)}{\text{current}} \times \underset{(s)}{\text{time}}$$

This can also be written as:

$$Q = I \times t$$

where Q represents charge

I represents current

t represents time.

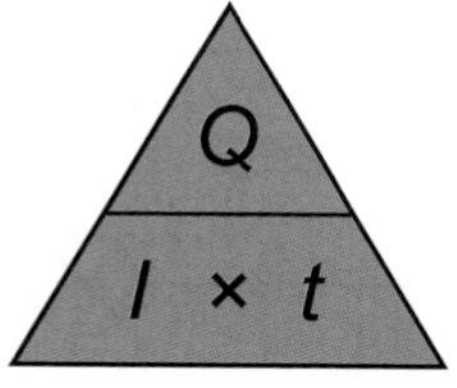

B This triangle can help you to rearrange the equation.

Worked example W1

The current in a lamp is 0.6 A. How much charge flows through it in 1 minute?

$$Q = I \times t$$
$$= 0.6\,\text{A} \times 60\,\text{s}$$
$$= 36\,\text{C}$$

Worked example W2

There is a current of 0.3 A in a circuit. How long will it take for 36 C to flow past a point in the circuit?

$$t = \frac{Q}{I}$$
$$= \frac{36\,\text{C}}{0.3\,\text{A}}$$
$$= 120 \text{ seconds}$$

2 A current of 2 A is switched on for 8 s. Calculate how much charge flows.

3 The current in a lamp is 0.5 A. Calculate how long it will take for 10 C of charge to flow through the lamp.

Energy and charge

Diagram C shows how energy is transferred in a circuit. The cell transfers energy to the charge, and so the charge then has the potential to transfer energy to other components in the circuit. The charge has 'potential energy'.

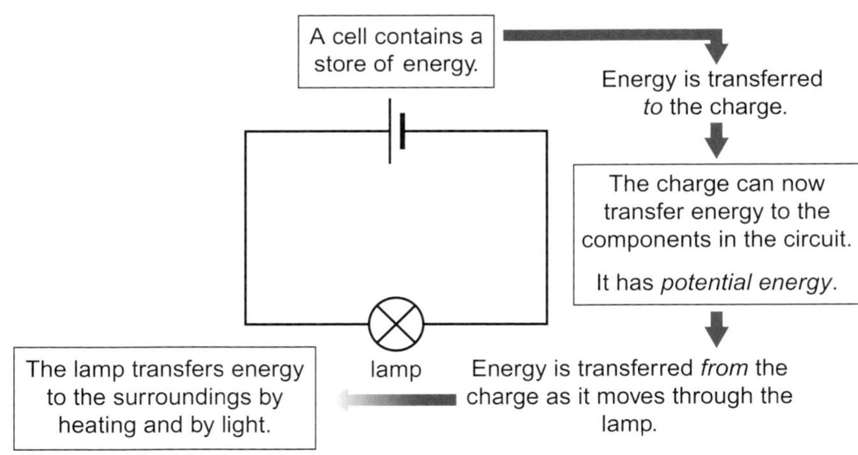

A cell contains a store of energy.
Energy is transferred *to* the charge.
The charge can now transfer energy to the components in the circuit.
It has *potential energy*.
lamp
Energy is transferred *from* the charge as it moves through the lamp.
The lamp transfers energy to the surroundings by heating and by light.

C how energy is transferred in a circuit

The potential *difference* of a cell is the amount of potential energy the cell transfers to each coulomb of charge flowing through it. There is a potential difference of 1 volt when there is a transfer of 1 joule of energy to each coulomb of charge (1 volt = 1 joule per coulomb). These quantities are related by the following equation:

energy transferred (J) = charge moved (C) × potential difference (V)

This can also be written as:

$E = Q \times V$

where E represents energy transferred

Q represents charge

V represents potential difference.

E

$Q \times V$

D

Worked example W3

The potential difference across a lamp is 1.5 V. When the circuit is switched on, 600 J of energy is transferred in the lamp. How much charge flowed through the lamp?

$$Q = \frac{E}{V}$$

$$= \frac{600\text{ J}}{1.5\text{ V}}$$

$$= 400\text{ C}$$

4 Calculate how much energy is transferred when 8 C of charge flows through a potential difference of 3 V.

5 150 J of energy is transferred when 50 C of charge flows through a wire. Calculate the potential difference across the wire.

Exam-style question

A mobile phone battery provides a potential difference of 4 V. Explain what is meant by 'a potential difference of 4 V'. *(2 marks)*

Checkpoint

How confidently can you answer the Progression questions?

Strengthen

S1 The current in a circuit is doubled and it is switched on for three times as long. Explain the change in the amount of charge that flows in the circuit.

S2 Explain what happens to the charge flowing in a circuit when the cell is replaced by a cell with double the potential difference.

Extend

E1 The label on a 12 V car battery says that it will give a current of 44 A for 1 hour. Calculate the charge and the potential energy stored in the battery.

SP10d Resistance

Specification reference: P10.12; P10.13; P10.14; P10.15; P10.16

Progression questions

- What is electrical resistance?
- What is the connection between voltage, current and resistance?
- What are the different effects of adding resistors in series and parallel?

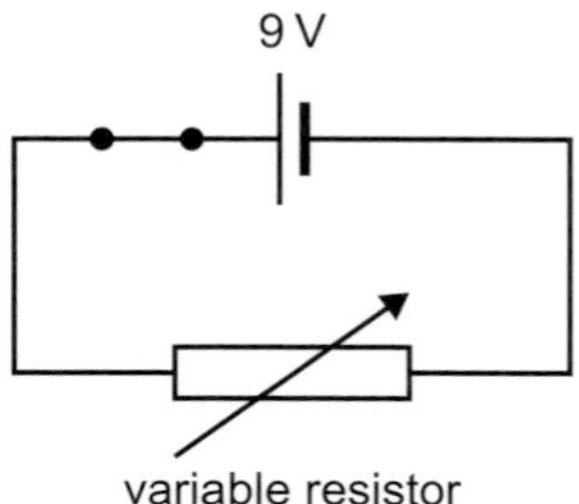

A When the variable resistor in this circuit is used to increase the resistance, the current decreases.

 **1** In diagram A the resistance is 600 Ω. Calculate the current.

 **2** In diagram A, explain what happens to the current when the resistance is decreased.

Some wires and components need a larger potential difference to produce a current through them than others. This is because they have a large electrical **resistance**. Resistance is measured in units called **ohms** (Ω). The resistance of a wire, a component or a circuit is calculated using the equation:

potential difference (V) = current (A) × resistance (Ω)

This can also be written as:

$V = I \times R$

where V represents potential difference

I represents current

R represents resistance.

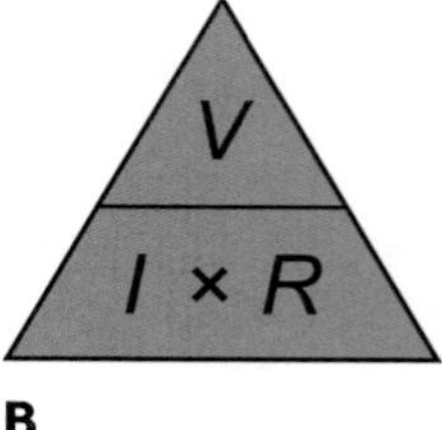

B

Worked example W1

Calculate the resistance in circuit A when the current is 0.3 A.

$$R = \frac{V}{I} = \frac{9\,\text{V}}{0.3\,\text{A}} = 30\,\Omega$$

Resistors in series

When resistors are connected in series the total resistance of the circuit is increased because the pathway becomes harder for current to flow through. The potential difference from a cell is shared between the resistors, but it may not be shared equally. There will be greater potential difference across resistors with higher resistances.

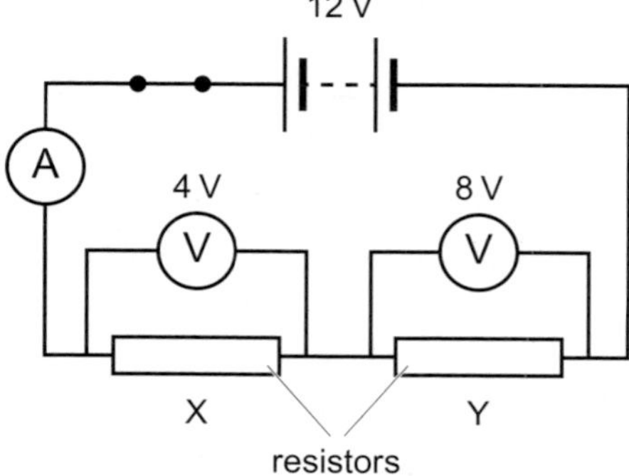

C Resistors X and Y are connected in series. Adding resistors in series increases the total resistance of the circuit.

Worked example W2

In circuit C, X has a resistance of 20 Ω. Calculate: **a)** the current in the resistors and **b)** the resistance of Y.

a)

$$I = \frac{V}{R}$$

$$= \frac{4\,\text{V}}{20\,\Omega}$$

$$= 0.2\,\text{A}$$

b)

$$R = \frac{V}{I}$$

$$= \frac{8\,\text{V}}{0.2\,\text{A}}$$

$$= 40\,\Omega$$

For Y the current is the same.

3 In diagram C, X and Y are changed for two 30 Ω resistors. The potential difference across each is 6 V. Calculate:

a the current

b the total resistance.

4 In diagram C, X is 20 Ω and Y is changed to 100 Ω. Compare and contrast the current through, and the potential difference across, X and Y.

Parallel circuits

When resistors are connected in parallel the total resistance of the circuit is less than the resistance of the individual resistors. This is because there are now more paths for the current.

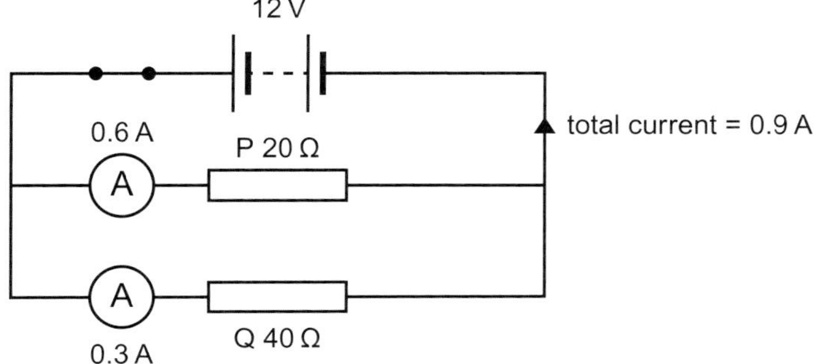

D Resistors P and Q have the same potential difference across them, but the current in resistor P is larger. This is because P has less resistance.

5 Two 30 Ω resistors are connected in place of P and Q in diagram D. Explain whether the total resistance of this circuit is larger or smaller than 30 Ω.

Testing and measuring

The variable resistor in diagram E is used to change the current in the circuit. Measurements of the current and potential difference are recorded and the resistance of Z is calculated.

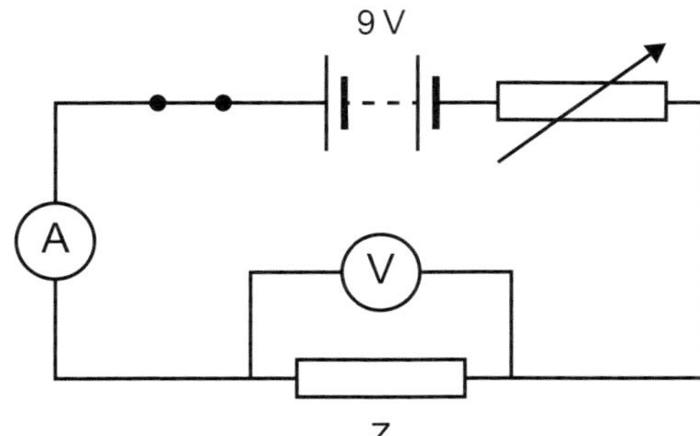

E A circuit like this one can be used to check whether a resistor has the correct value or to measure an unknown resistance.

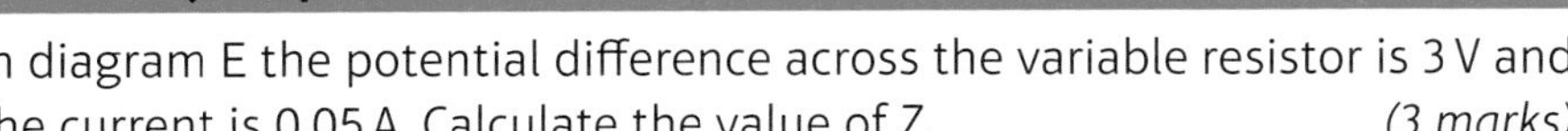

Exam-style question

In diagram E the potential difference across the variable resistor is 3 V and the current is 0.05 A. Calculate the value of Z. *(3 marks)*

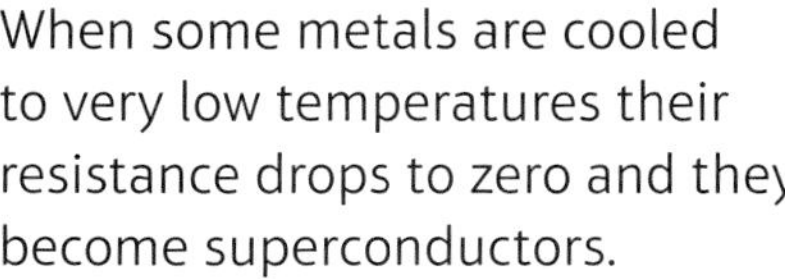

Did you know?

When some metals are cooled to very low temperatures their resistance drops to zero and they become superconductors.

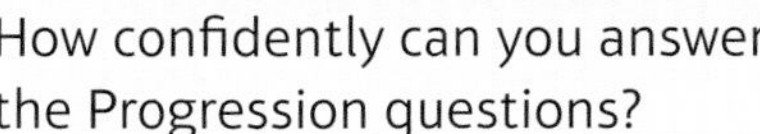

Checkpoint

How confidently can you answer the Progression questions?

Strengthen

S1 The resistance of the circuit in diagram A is increased. Use the equation for voltage, current and resistance to help you describe what will happen to the current in the circuit.

S2 You have three 100 Ω resistors. Draw diagrams to explain how they should be connected to give the maximum and the minimum total resistance.

Extend

E1 In diagram C, X and Y are changed for two 200 Ω resistors. Using $V = I \times R$, show that the total resistance of the circuit is 400 Ω.

E2 Two 200 Ω resistors are connected in place of P and Q in diagram D. Calculate the total resistance of the circuit. (*Hint:* work out the total current.)

SP10e More about resistance

Specification reference: P10.18; P10.19; P10.20; P10.21

Progression questions

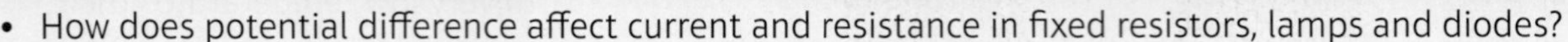

- How does potential difference affect current and resistance in fixed resistors, lamps and diodes?
- How do light intensity and temperature affect resistance in LDRs and thermistors?
- How are circuits used to explore resistance in lamps, diodes, thermistors and LDRs?

1 Look at graph A.

a State what a fixed resistor is.

b For a fixed resistor, if the potential difference increases by 20%, by what percentage will the current increase?

2 How can you tell from the graph that, for a filament lamp current is not directly proportional to potential difference?

Did you know?

The first diodes were made over 100 years ago. They were called cat's-whisker diodes because they used a thin wire touching a crystal. In 2015 scientists reported making a diode from a single molecule.

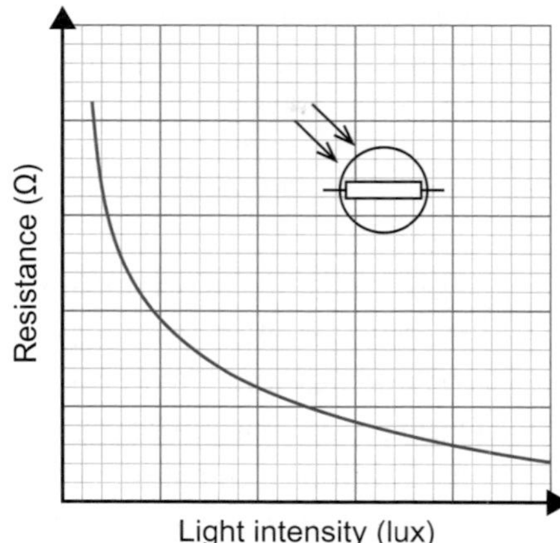

B The resistance of an LDR changes with light intensity.

Graph A shows that when potential difference changes across a fixed resistor, the current changes by the same percentage. The two variables are in **direct proportion**, and the graph forms a straight line going through the origin. This happens because the resistance stays the same.

Other components, such as filament lamps and **diodes** (also shown in graph A), have resistances that change when potential difference changes.

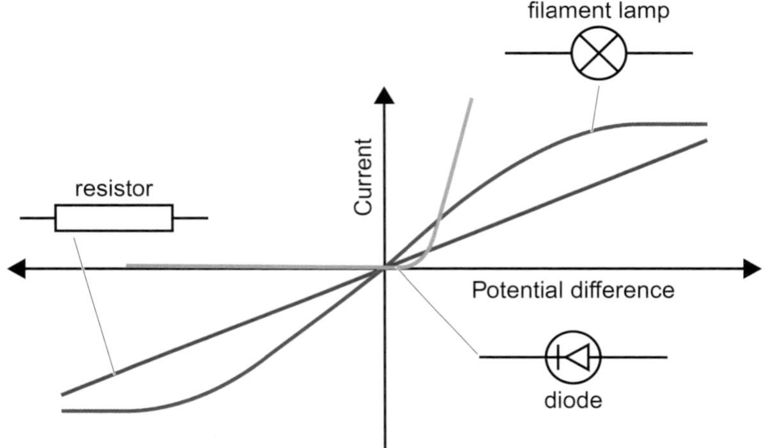

A graph of current against potential difference for a fixed resistor, filament lamp and diode

A potential difference across a filament lamp causes a current to flow through it. The current causes the filament to heat up and glow. The greater the potential difference, the more current flows and the hotter and whiter the filament gets. However, as it heats up, the filament's resistance increases. This means that when the potential difference changes, the current does not change by the same percentage (the two variables are not in direct proportion).

A diode has a low resistance if the potential difference is in one direction but a very high resistance if the potential difference is in the opposite direction. This means that current can only flow in one direction.

A **light-dependent resistor** (**LDR**) has a high resistance in the dark but the resistance gets smaller when the light intensity increases.

3 Describe how the resistance of an LDR changes with increasing light intensity.

Thermistors have high resistances at low temperatures but as the temperature increases the resistance decreases.

4 Write down the name and symbol of the five components mentioned so far on these two pages.

5 Why does the current flowing through a thermistor increase with increasing temperature?

The circuit in diagram D can be used to explore how the resistance of the lamp changes as the potential difference across the lamp is changed. The current through the lamp, measured by the ammeter, is recorded for different values of the potential difference measured on the voltmeter.

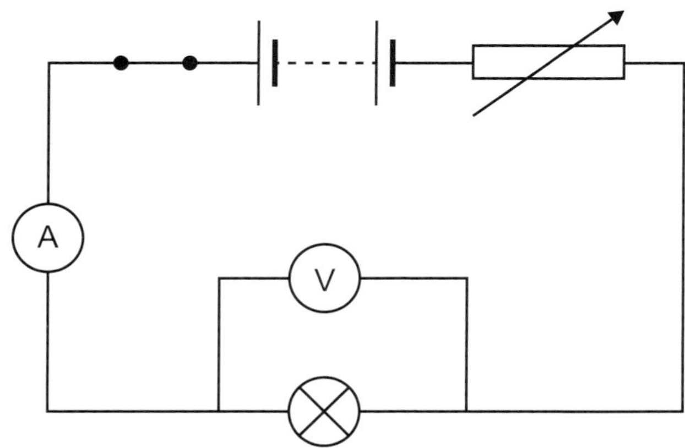

D A circuit used to explore variation in resistance of a lamp. It can easily be adapted to explore resistance in a diode, thermistor or LDR.

6 **a** Explain how you would use the circuit in diagram D to find out how the resistance of a lamp changes with potential difference.

b Predict what will happen.

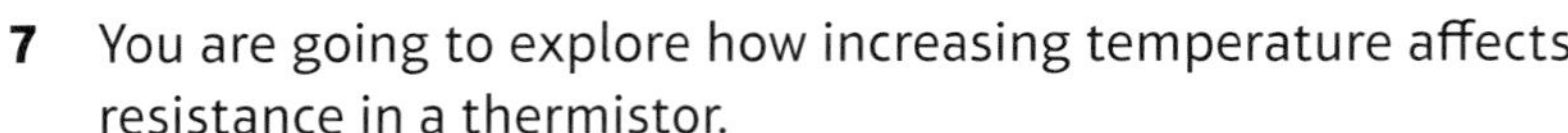
7 You are going to explore how increasing temperature affects resistance in a thermistor.

a List the apparatus you need.

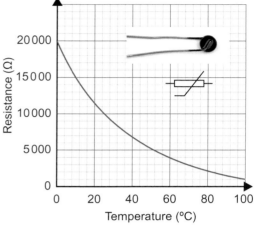
b Draw a circuit diagram of your set-up.

c Suggest the measurements you would make.

Exam-style question

A circuit similar to the one in diagram D is set up. This time the lamp in the circuit is replaced by an LDR. A light source which can be brightened and dimmed is shone onto the LDR. Explain what will happen when the light intensity of the light source is changed. *(2 marks)*

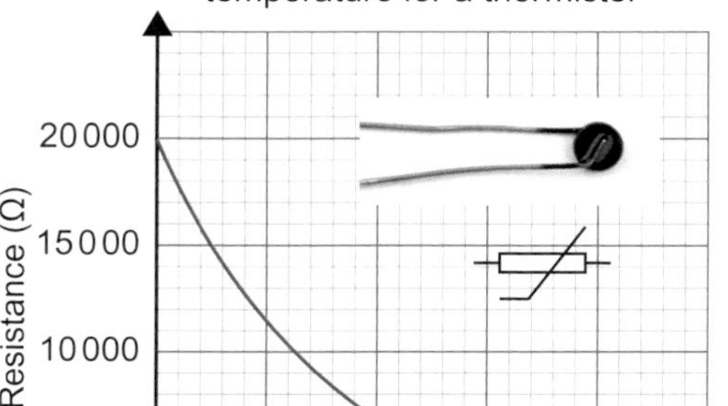

C The resistance of a thermistor changes with temperature.

Checkpoint

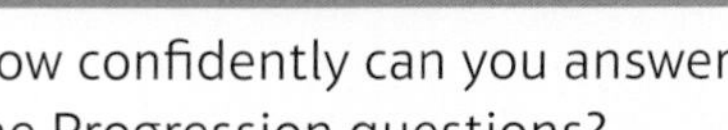

How confidently can you answer the Progression questions?

Strengthen

S1 Look at graph A. Compare and contrast the graph for the fixed resistor with that of:

a the filament lamp

b the diode.

S2 Explain which components could be used to switch a heater and a light on in an hospital incubator.

Extend

E1 For the thermistor shown in graph C, at what temperature is it most sensitive to changes in temperature? (*Hint*: find the biggest change in resistance for a 1 degree change in temperature - one quarter of a square on the temperature axis.)

SP10e Core practical – Investigating resistance

Specification reference: P10.17

Aim

Construct electrical circuits to:

a investigate the relationship between potential difference, current and resistance for a resistor and a filament lamp

b test series and parallel circuits using resistors and filament lamps.

A Circuits are built up from different components.

B Older style light bulbs are filament lamps. They give out light when the electricity flowing through them makes the filament so hot that it glows white. The lamp shown in the photo is not quite so hot, but you can clearly see the filament.

Most machines around us rely on electricity in some way. The circuits inside computers, cars and phones can be quite complex, but they are all built up from simpler components. Engineers who design the circuits need to know the characteristics of different components. For example, resistors are used to control the amount of current flowing in part of a circuit, but not all components keep the same resistance if the potential difference across them changes.

Your task

You will construct a circuit to investigate the link between potential difference, current and resistance for a resistor and a filament lamp.

You will then find out what happens to the current through filament lamps when they are used in series and parallel circuits.

Method

Investigating resistance

A Set up circuit C. Use a power pack that can provide different potential differences. Ask your teacher to check your circuit before you switch it on.

B Set the power pack to its lowest voltage (potential difference) and switch on. Write down the readings on the ammeter and voltmeter and then switch the power pack off.

C Repeat step B for five different voltage settings, up to a maximum of 6 V.

D Replace the resistor in the circuit with two filament lamps, and repeat steps B and C.

Filament lamps in series and parallel circuits

E Set up circuit D. Ask your teacher to check your circuit before you switch it on.

F Set the power pack to its lowest voltage. Write down the readings on the ammeter and the voltmeters. Repeat with the power pack set to provide different voltages, up to a maximum of 6 V.

G Now set up circuit E and ask your teacher to check it. Repeat step F for several different voltage settings.

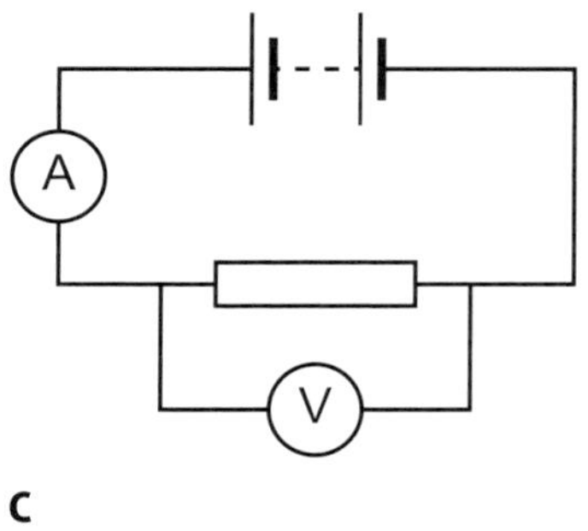

C

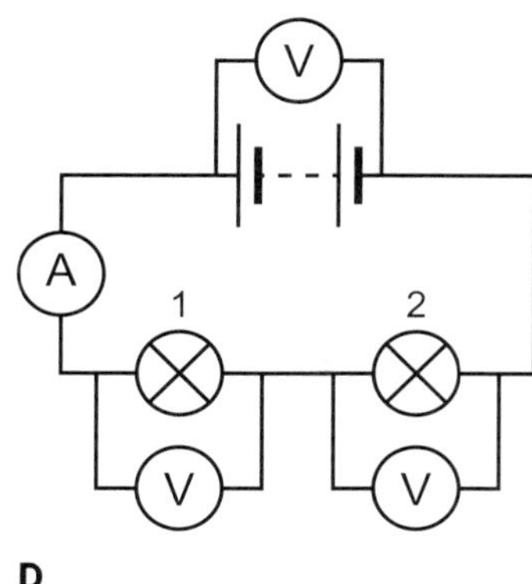

D

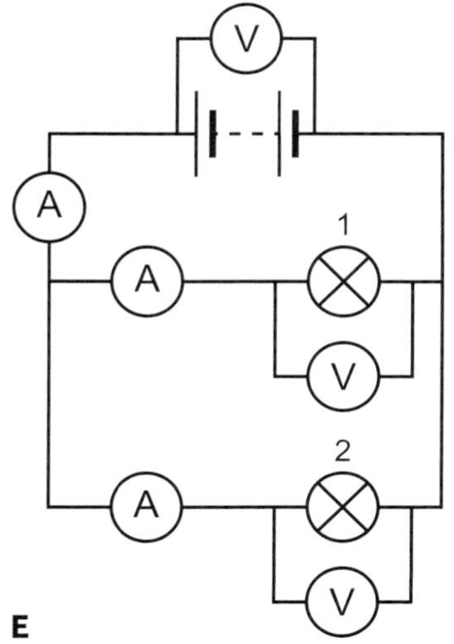

E

Exam-style questions

1 State the units for measuring resistance. *(1 mark)*

2 a Explain why the ammeter in diagram C is in series with the resistor (and not in parallel). *(2 marks)*

b Explain why the voltmeter is in parallel with the resistor (and not in series). *(2 marks)*

3 Table F shows a set of results from the 'Investigating resistance' investigation. Plot a graph to present these results, showing both sets of results on the same axes. Draw a curve of best fit through the points for each component. *(6 marks)*

4 Use the readings at 1 V and at 6 V given in table F to calculate the resistance of:

a the resistor *(3 marks)*

b the filament lamp. *(2 marks)*

5 Use your graph from question 3 and your answers to question 4 to write conclusions for the investigation with:

a the resistor *(3 marks)*

b the filament lamp. *(3 marks)*

c Describe how you could find out if your conclusions are applicable to all resistors or to all filament lamps. *(3 marks)*

6 Tables G and H show some results from the investigation on filament lamps in series and parallel circuits.

a Explain what the current readings (x) and (y) would be. *(2 marks)*

b Explain what the current reading (z) would be. *(2 marks)*

7 a Use the information in tables G and H to calculate the overall resistance of:

i circuit D *(2 marks)*

ii circuit E. *(2 marks)*

b Describe how two bulbs can be put in a circuit to give the lowest possible overall resistance. *(1 mark)*

Potential difference (V)	Current (A)	
	resistor	filament lamp
0	0	0
1	0.2	0.12
2	0.4	0.23
3	0.6	0.33
4	0.9	0.41
5	1.0	0.47
6	1.2	0.53

F

	Potential difference (V)		
	power pack	lamp 1	lamp 2
Series (D)	4	2	2
Parallel (E)	4	4	4

G

	Current (A)		
	power pack	lamp 1	lamp 2
Series (D)	0.23	(x)	(y)
Parallel (E)	(z)	0.41	0.41

H

SP10f Transferring energy

Specification reference: P10.22; P10.23; P10.24; H P10.25; P10.26; P10.27

Progression questions

- What are the advantages and disadvantages of the heating effect of a current?
- How can the energy transfer that causes the heating effect be explained?
- H How can unwanted energy transfer be reduced in wires?

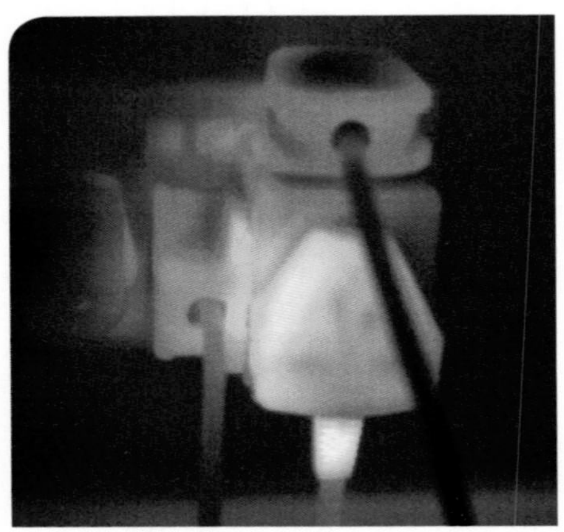

A This thermal image shows that plugs and wires are heated by the currents passing through them.

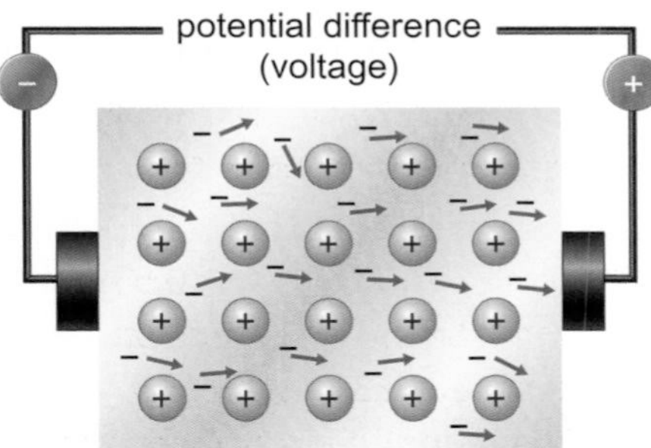

B Inside a resistor, free electrons (shown by –) move through a lattice of positive ions.

 **3** Using the model of a lattice of ions, suggest why a larger current makes a resistor hotter.

 **4** Explain why thick copper cables are used to carry electricity from electrical substations to homes.

All circuits have some resistance, so they warm up when there is a current. When a current passes through a resistor, energy is transferred because electrical **work** is done against the resistance. The energy is transferred by heating and the resistor becomes warm.

The heating effect is useful in an electric heater or a kettle. It is not useful in a computer or in plugs and wires because it means that useful energy is being transferred from the circuit by heating, and spread out or **dissipated**. The surroundings gain thermal energy.

 **1** Name four appliances where the heating effect of a current is useful.

 **2** Describe an example of when the heating effect of a current is not useful.

A model of resistance

Diagram B shows the structure inside a resistor. As the electrons flow through the lattice of vibrating ions, they collide with the ions. The more collisions they make with the ions, the harder it is for them to pass through, so the higher the electrical resistance. When the electrons collide with the ions, they transfer energy to them.

H Reducing resistance

Resistance in circuits can be reduced by using wires made from metals with low resistance, such as copper. Thicker wires also have lower resistance. Resistance can also be decreased by cooling metals so that the lattice ions are not vibrating as much.

C The wires carrying electricity in this photo are thick wires made of aluminium which has a low resistance. When the resistance is lower, less energy is transferred by heating and less energy is dissipated.

Calculating the energy transferred

energy transferred (J) = current (A) × potential difference (V) × time (s)

This can also be written as:

$E = I \times V \times t$

Worked example W1

A 12 V battery supplies a current of 0.3 A to a heater for 8 minutes. Calculate the energy that is transferred in heating up the heater and the surroundings.

$E = I \times V \times t$

$= 0.3\,\text{A} \times 12\,\text{V} \times 8\text{ minutes}$

$= 0.3\,\text{A} \times 12\,\text{V} \times (8 \times 60\,\text{s})$

$= 1728\,\text{J}$

Note: for the equation to work, time has to be in seconds and so minutes need to be converted to seconds here.

Worked example W2

The current in a lamp is 0.4 A when it is connected to a 230 V supply. Calculate how long it takes to transfer 2000 J of energy.

$t = \frac{E}{I \times V}$

$= \frac{2000\,\text{J}}{0.4\,\text{A} \times 230\,\text{V}}$

$= 21.7\,\text{s}$

5 Calculate the energy transferred when a TV using a 230 V supply and current of 0.9 A is switched on for one minute.

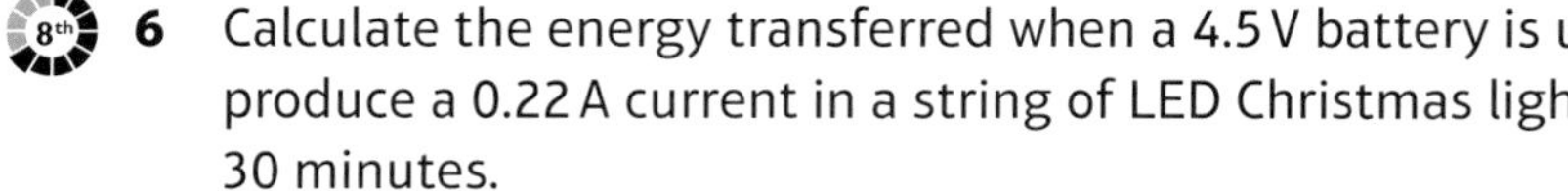

6 Calculate the energy transferred when a 4.5 V battery is used to produce a 0.22 A current in a string of LED Christmas lights for 30 minutes.

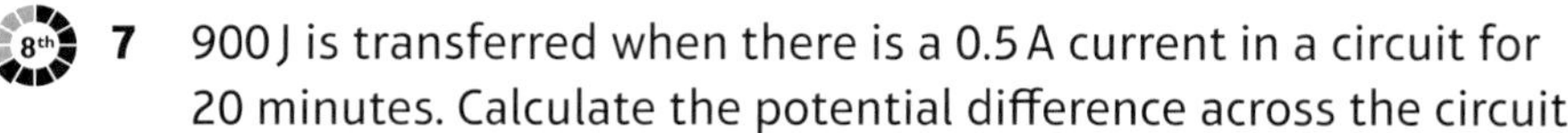

7 900 J is transferred when there is a 0.5 A current in a circuit for 20 minutes. Calculate the potential difference across the circuit.

Exam-style question

To boil the water in a kettle requires 325 kJ of energy. The electric kettle transfers this energy in 1 minute and 49 s. Explain why it will actually take longer than this to boil the water. *(2 marks)*

Did you know?

Computer core chips are cooled by a fan. If the fan stops, the electric currents inside the chips overheat them. They can even catch fire.

Checkpoint

How confidently can you answer the Progession questions?

Strengthen

S1 The label on an extension lead says '230 V max 13 A (unwound)'. Suggest why the lead must be unwound if it is carrying the maximum current.

Extend

E1 When the temperature of a material increases, the ions in the lattice vibrate more. Suggest what effect this will have on the electrical resistance of the material.

E2 Compare the energy transferred by a 3 V LED with a current of 20 mA and a 2.5 V filament lamp with current of 320 mA.

SP10g Power

Specification reference: P10.28; P10.29; P10.30; P10.31

Progression questions

- What is power and what units are used to measure it?
- How is power related to the energy used in joules?
- How can you calculate power when you know current, potential difference and/or resistance?

A This oven takes an hour to roast a chicken. There are 28 halogen lights in the downstairs rooms of this house.

In photo A, it takes about an hour to roast the chicken. This transfers about the same amount of energy as using the 28 halogen ceiling lights for 2 hours.

The energy transferred by an electric current depends on the time taken, so it is often more useful to compare the **power** of the appliances. Power is the energy transferred per second. This is often shown on appliances as the **power rating**. Power is measured in **watts** (W). 1 W is a transfer of 1 joule per second.

To calculate the power, use the equation:

$$\underset{(\text{W})}{\text{power}} = \frac{\underset{(\text{J})}{\text{energy transferred}}}{\underset{(\text{s})}{\text{time taken}}}$$

This can also be written as:

$$P = \frac{E}{t}$$

where P represents power

E represents energy transferred

t represents time taken.

Worked example W1

Calculate the energy transferred by a 800 W microwave in 1 minute.

$E = P \times t$

$= 800\,\text{W} \times 60\,\text{s}$

$= 48\,000\,\text{J}$

B The Blackpool Tower originally had 10 000 filament lights, but now has 25 000 LED lights because they use less power and decrease the electricity bill.

1 The power rating of the oven in photo A is approximately 3kW.

7th **a** Calculate the energy used by the oven in one hour.

8th **b** Calculate the energy used by one halogen light in 2 hours.

8th **c** Calculate the power of one halogen light.

8th **2** A kettle transfers 540 000 J of energy in 3 minutes. Calculate the power of the kettle.

Calculating electrical power

The power transfer in a component or appliance is proportional to the potential difference across it and the current through it. This means that:

$$\underset{(W)}{\text{electrical power}} = \underset{(A)}{\text{current}} \times \underset{(V)}{\text{potential difference}}$$

This can also be written as:

$$P = I \times V$$

where P represents power

I represents current

t represents voltage.

3 Calculate the power of a wheelchair motor that has a current of 20 A and a potential difference of 12 V.

4 Calculate the current in an electric kettle with a power rating of 3 kW and a potential difference of 230 V.

In *SP10d Resistance* you learned the equation $V = I \times R$. Using this to substitute for V in the equation $P = I \times V$ gives a new equation for power:

$$P = I \times I \times R \text{ or } P = I^2 \times R$$

$$\underset{(W)}{\text{electrical power}} = \underset{(A^2)}{\text{current}^2} \times \underset{(\Omega)}{\text{resistance}}$$

Worked example W2

An electric cable has a resistance of 900 Ω and a current of 3 A through it. Calculate the power transferred in kilowatts.

$$P = I^2 \times R$$

$$= (3\text{ A})^2 \times 900\ \Omega$$

$$= 9\text{ A}^2 \times 900\ \Omega$$

$$= 8100\text{ W}$$

$$= 8.1\text{ kW}$$

5 A 46 W electric blanket has a resistance of 1150 Ω. Calculate the current in the blanket.

Exam-style question

Calculate the energy transferred by a 5 W spotlight in 2 hours. *(2 marks)*

Did you know?

The Three Gorges Dam in China is the largest power station in the world. It produces 22 500 MW of hydroelectric power from the Yangtze River.

Checkpoint

How confidently can you answer the Progression questions?

Strengthen

S1 Describe what is meant by the power of a 650 W electric toaster.

S2 State three ways the power of an appliance can be worked out.

Extend

E1 500 W of power can be supplied to a building at 250 V or at 1000 V. Calculate the current in the cable in each case.

E2 If the cable has a resistance of 100 Ω, calculate the energy transferred by heating the cable and the surroundings in each case. Compare your answers and suggest whether it is more efficient to use 250 V or 1000 V.

SP10h Transferring energy by electricity

Specification reference: P10.32; P10.33; P10.34; P10.35; P10.36; P10.42

Progression questions

- How is energy transferred from electrical cells or batteries to motors and heating devices?
- What is the difference between direct and alternating, for both current and voltage?
- What is the voltage and frequency of the UK domestic electricity supply?

In photo A some of the energy stored in the battery is transferred by electricity to the motor, where it is transferred to a store of kinetic energy in the fan. Some energy will also be transferred by heating the wires, the motor and the surroundings. In the end, all the energy will be dissipated by heating, making the surroundings a little warmer (increasing their store of **thermal energy**).

The gloves in photo B contain wire that has a high resistance. Energy stored in the battery is transferred by electricity to the high resistance wire where it is transferred by heating to a store of thermal energy in the wire. The energy is then transferred by heating to the gloves and hands of the wearer, and eventually dissipates to the surroundings.

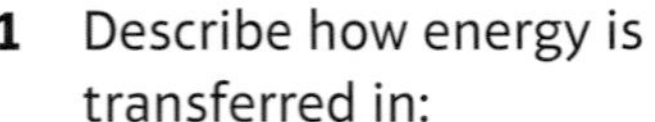

1 Describe how energy is transferred in:

a a battery operated toothbrush

b a battery operated cup that warms drinks.

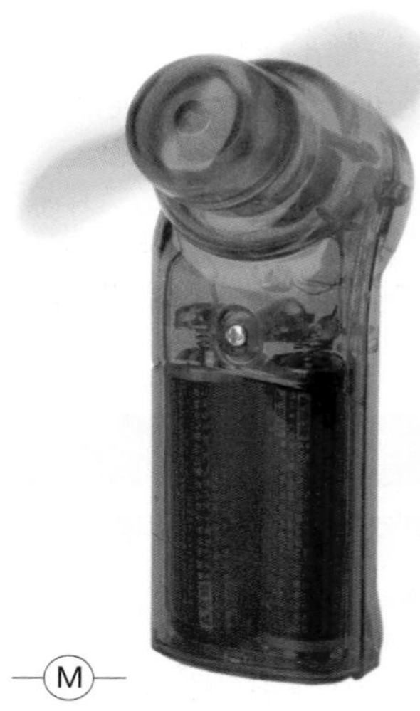

A This battery-operated fan contains an electric motor.

B These gloves warm your hands – they use a battery-operated heating circuit.

Did you know?

In the 1880s and 90s there was a 'War of the currents' in the USA between the Edison Electric Light Company and the Westinghouse Electric Company over whether mains electricity should be d.c. (direct current) or a.c. (alternating current). The 'war' was eventually won by Westinghouse and alternating current. The difference is explained on the next page.

Appliances that need a large amount of power use **mains electricity**. In a power station energy is transferred from a store of kinetic energy (such as a turbine) by electricity. The electricity is carried to our homes through a network of wires and cables known as the **national grid**.

In our homes, appliances use the energy transferred by electricity in various ways. For example, the motor in a washing machine transfers energy to kinetic energy in the washing machine drum.

2 Describe the energy transfers in:

a mains operated hair straighteners

b a mains operated electric drill.

Direct voltage and alternating voltage

Cells and batteries have a positive and a negative terminal and the direction of the movement of charge stays the same. This is called **direct current** (**d.c.**).

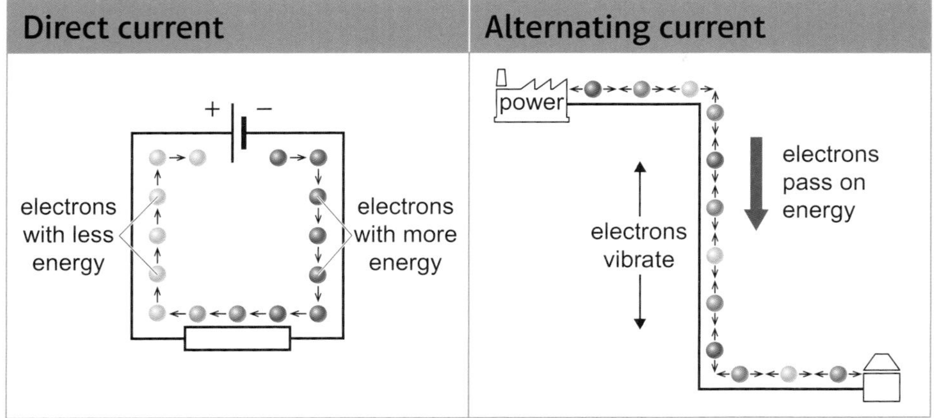

C The direction of the current and the movement of charge stays the same in direct current but changes in alternating current.

Mains electricity is produced using generators that rotate, causing the direction of the current to keep changing. This is called **alternating current** (**a.c.**). The voltage also changes, increasing to a peak voltage then decreasing to zero. It then increases to a peak in the opposite direction before decreasing back to zero. This cycle then repeats. In the UK there are 50 of these cycles per second or, in other words, the frequency of the mains supply is 50 **hertz** (Hz). The voltage is constantly changing but the average effect is the same as a d.c. voltage of 230 V.

3 A mobile phone is provided with 5 V d.c. using a charger plugged into the mains supply. State two changes the charger makes to the mains electricity.

4 Compare a.c. and d.c. in terms of:

a the movement of charges **b** the voltage.

Power rating of domestic appliances

The power rating of an appliance is measured in watts (W). A kettle with a power rating of 3 kW transfers 3000 joules of energy each second (from the mains electricity supply to a store of thermal energy in the water).

5 What is the relationship between energy transferred and the power rating?

6 A hairdryer has a power rating of 1200 W. Calculate the amount of energy it transfers every hour. Show your working.

Exam-style question

An electric deep fat fryer uses the UK mains electricity supply. Describe all the energy transfers that take place when the fryer is used. *(3 marks)*

Checkpoint

How confidently can you answer the Progression questions

Strengthen

S1 Describe how energy is transferred in a battery operated hairdryer.

S2 Draw a table to compare the voltage and the movement of charge in UK mains electricity and a 12 V car battery.

Extend

E1 Explain an advantage of using the UK mains supply instead of a 12 V battery for an electric kettle.

SP10i Electrical safety

Specification reference: P10.37; P10.38; P10.39; P10.40; P10.41

Progression questions

- What is the difference between the live and the neutral wires?
- How do earth wires and fuses make circuits safer?
- What are the potential differences between the live, neutral and earth wires?

In the UK, appliances are connected to the mains electricity with a 3-pin plug as shown in photo A.

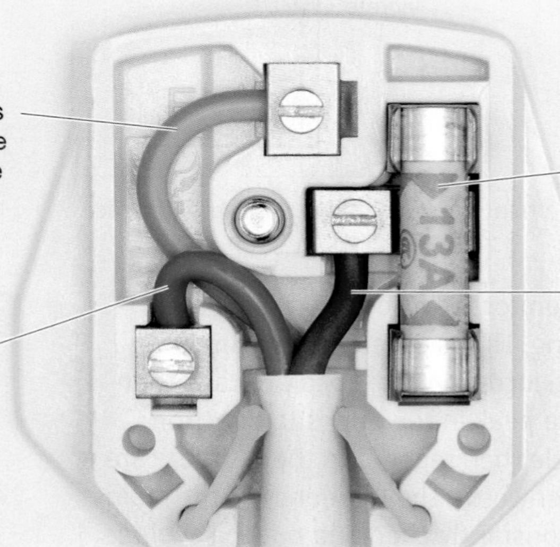

earth wire – connects the metal parts of the appliance to a large metal spike or metal tubing that is pushed into the ground. It is for safety and is at 0V if the circuit is correctly connected.
neutral wire – the return path to the power station. If the circuit is correctly connected it is at a voltage of 0V.
fuse – a safety device marked with the current it can carry. Usually 3A, 5A or 13A.
live wire – connects the appliance to the generators at the power station. The voltage on this wire is 230V.
13A

A A 3-pin plug is designed to safely connect appliances to mains electricity.

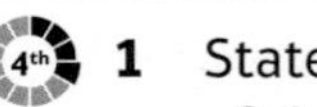

1 State the colours of each of the wires in a plug.

2 Calculate the potential difference between the following pairs of wires.

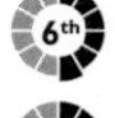

a live and neutral

b live and earth

c neutral and earth

3 An iron uses a current of 4 A. Fuse ratings are usually 3 A, 5 A or 13 A. Explain which fuse is best to use in the plug for the iron.

Did you know?

An a.c. electric current of just 0.1 A through the heart is enough to stop it and kill you. The voltage needed to produce this current depends on the resistance of the path through your body. This resistance could be as low as 1 kΩ for wet skin but 500 kΩ for dry skin.

Safety features

Switches are connected in the live wire of a circuit. When they are off, no current goes through the appliance.

A fuse is a tube with a thin wire inside. The current passes through the wire and the wire gets hotter. If the current exceeds a certain value the wire melts. This breaks the circuit and stops the current.

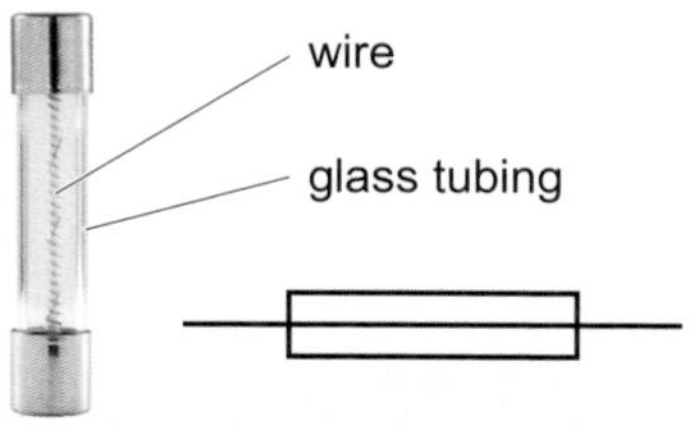

wire
glass tubing

B The fuse melts before wiring or parts of an appliance can overheat. Once the fault is fixed, a new fuse can be fitted.

If a faulty appliance draws too much current, it can caused overheating of the wiring in either the walls or in the appliance. This can cause fires. A fuse stops this from happening. If an appliance develops a fault, its metal parts can be at a high voltage. If you touched the metal you might get dangerous electric shock – a current would flow through you into the ground (which is at 0 V). For this reason, the metal parts of appliances are connected to the earth wire so that this current goes into the ground instead of through you.

If a fault causes the live wire to touch a metal part, it makes a very low resistance circuit between 230 V and 0 V (the earth). This causes a very large current to flow to the earth, which heats up the wire and could cause a fire. If this happens, the current blows the fuse and cuts off the mains electricity supply.

Circuit breakers can be an alternative to fuses. They detect a change in the current and safely switch off the supply.

C Circuit breakers are a type of automatic switch that stop current flowing if there is a problem in the circuit, such as too much current or current flowing in the wrong wires.

One advantage of circuit breakers is that once a fault is fixed they can be switched back on again, whereas a fuse has to be replaced. Another advantage of some types is that they work very quickly, so can save lives. A fuse takes some time to melt and will not prevent you getting a shock if, for example, you touch a live wire.

 **7** Compare and contrast circuit breakers and fuses as safety devices.

Exam-style question

Why is the earth wire needed in a plug? *(2 marks)*

 **4** Explain why a current would flow though you if you touched a metal part of a faulty appliance that did not have an earth wire.

5 Explain why the fuse must be in the live wire and not the neutral wire of a circuit.

 **6** Suggest what could happen if a fuse were replaced by a piece of ordinary wire and a fault caused the live wire to touch the metal case of an oven.

Checkpoint

How confidently can you answer the Progression questions?

Strengthen

S1 Describe the functions and voltage of the three wires found in a plug.

S2 Write a sentence or two for each of the following safely devices to explain how they work: fuses, the earth wire, circuit breakers.

Extend

E1 All appliances could be fitted with 13 A fuses. Explain the advantages of replacing them with lower value fuses and how you would work out which fuse to use.

SP11a Charges and static electricity

Specification reference: P11.1P; P11.2P; P11.3P; P11.4P

Progression questions

- What sort of charges are there on the particles in an atom?
- How can an insulator become charged using friction?
- How can an insulator gain an induced charge?

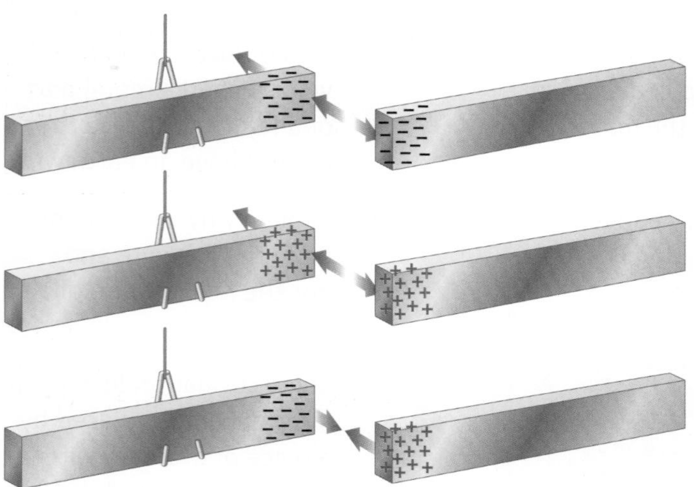

A Hanging rods with opposite charges attract each other and will move towards one another due to the force between them. Rods with the same charges repel.

Plastics, such as acetate and polythene, can collect a charge because they are **insulators.** The charges cannot flow through the plastic into other materials. When you rub an **acetate rod** with a dry duster some of the electrons move from the acetate onto the duster. Since electrons are negatively charged, the duster gains a negative **charge** and this leaves the acetate rod with a positive charge**.** When you rub a polythene rod with a duster, electrons move from the duster onto the polythene and the opposite situation occurs.

1 A polythene rod is rubbed with a dry duster. State what charge will be left on:

 **a** the polythene

 **b** the duster.

There is a force of attraction between rods with opposite charges. If they can move, they will move towards each other. Rods with the same charge repel each other.

2 Explain what happens when the experiment in diagram A is done with the following pairs of rods, each of which has been charged by rubbing with a duster:

 **a** two acetate rods

 **b** two polythene rods

 **c** an acetate rod and a polythene rod.

 **3** Explain why two plastic rulers rubbed with a cloth will repel each other.

B These hair strands all have the same charge.

In photo B the child's hair has become charged. This is an example of **static electricity**. The charge is not able to flow away to the surroundings. Since all the hairs have the same charge, they repel each other.

 **4** Explain how the electric charge on a person causes their hair to behave as shown in B.

Charging by induction

A charged object can affect the distribution of charges on an uncharged object. For example, if a negatively charged balloon is brought near a wall, the negatively charged electrons in the wall are repelled. This causes the surface of the wall to become positively charged. No charges are transferred from the balloon and so we say that the positive charge on the wall has been induced. This is called charging by **induction**.

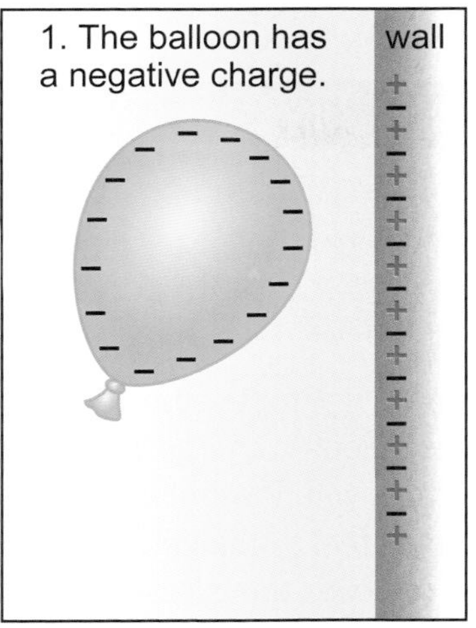

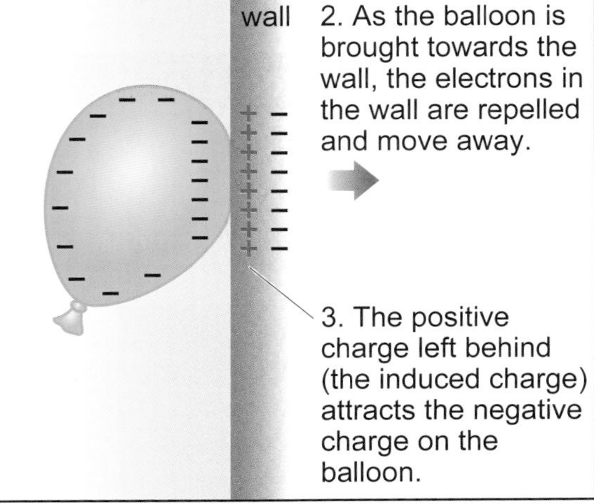

wall
2. As the balloon is brought towards the wall, the electrons in the wall are repelled and move away.
3. The positive charge left behind (the induced charge) attracts the negative charge on the balloon.

C A charged balloon can induce a charge in a wall.

5 **a** Look at diagram C. Explain why the surface of the wall becomes positively charged.

b How is this different to how the balloon became charged by rubbing it with a duster?

Did you know?

This comb has become charged by being brushed through dry hair and can now induce a charge in a thin stream of water.

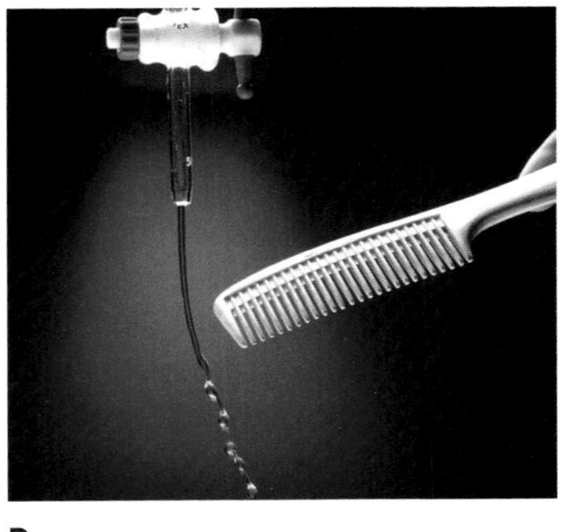

D

6 Explain how rubbing a plastic ruler on a sleeve and holding it over small pieces of paper makes the paper jump up and stick to the ruler.

7 Look at photo D in *SP9a Objects affecting each other*. Explain why the polystyrene packing pellets are stuck to the person's arms.

Exam-style question

Two uncharged balloons are each hanging vertically from a thread, so that they are next to each other but not touching. Explain what will happen to the balloons:

a when both of them are given a negative charge *(2 marks)*

b when only one of them is given a negative charge. *(3 marks)*

Checkpoint

How confidently can you answer the Progression questions?

Strengthen

S1 When you take off a jumper, it can become charged. Explain why taking off a jumper sometimes makes your hair stick up.

S2 A balloon is charged by rubbing it with a duster. The balloon is then attracted to a positively charged rod. Explain what charge is on the balloon and what will happen if it is brought close to a negatively charged rod.

Extend

E1 Describe what happens if an uncharged rod is hung up as in diagram A and a charged rod brought close to it.

E2 A mirror is rubbed with a dry cloth. It quickly becomes covered in a layer of dust. Explain how the dust is attracted and sticks to the mirror.

SP11b Dangers and uses of static electricity

Specification reference: P11.4P; P11.5P; P11.6P; P11.7P

Progression questions

- Why is earthing needed?
- How does earthing work?
- How do insecticide sprayers work?

A Lightning conductors are an important way of earthing buildings and protecting them from lightning.

When you walk across some types of carpet, you may end up with an unbalanced electric charge. If you then touch a conductor, such as a metal tap, electrons will flow between the tap and you. Sometimes there is a spark and you may feel a small electric shock. The electrons flow in whichever direction removes the excess charge and you become **discharged** or **earthed**.

1 Explain why after sliding out of a car seat, touching the metal door may give you an electric shock.

2 A boy has become charged and touches a metal door handle.

a Explain which way the electrons flow if the charge on him is positive.

b Explain why he will not be earthed if the door handle is plastic.

Static electricity builds up in clouds due to friction between particles of ice or water moved by air currents. When the charge is large enough, charged particles travel through the air between the cloud and the earth. This causes both lightning and thunder. Lightning can be dangerous and so to discharge clouds safely, buildings are earthed by having lightning conductors made of thick metal running down from their tops into the earth.

3 A storm cloud has a positive charge. Describe what happens to the charges when lightning from the cloud strikes a lightning conductor.

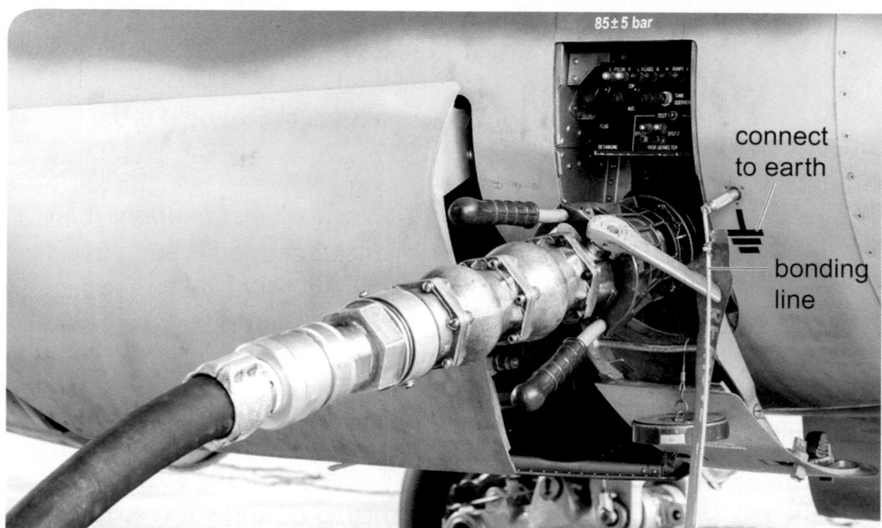

B The bonding line is attached while this aircraft is refuelled.

Sparks can be dangerous, for example when there is fuel vapour that could ignite. This can be a problem for refuelling aircraft, which often become charged when flying through the air. A charge can also build up when fuel flows through a pipe. To prevent a spark between a fuel pipe and an aircraft, a 'bonding line' is connected to earth the aircraft before refuelling begins.

4 What might happen if the bonding line in photo B were not attached?

At petrol filling stations the storage tanks, pipes and hoses are earthed. Cars are earthed through their tyres, which contain a form of carbon that makes them conducting, and you are earthed when you touch the metal of the car or the pump.

 **5** Why is it important to make sure that charge does not build up on vehicles at petrol stations?

Did you know?

Engineers who work with computers often wear a wire connecting them to earth so that static electric charge cannot build up and damage the computer chips.

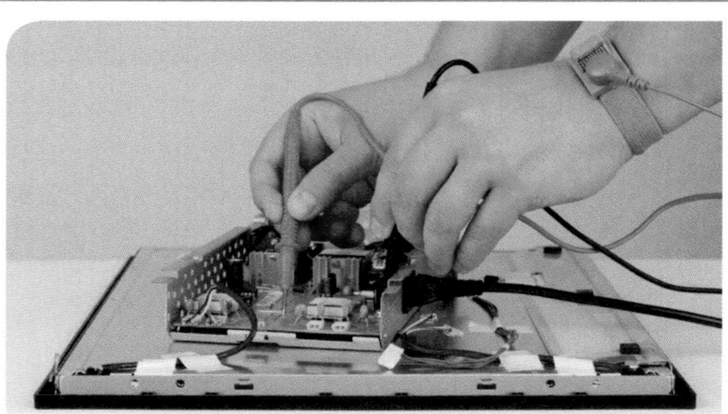

C

Uses of static electricity

Electrostatic spraying makes use of static electricity. For example, when spraying crops with insecticide, electrodes on the spray nozzle charge the spray droplets as they pass. The charged droplets spread out because they repel each other and then they are attracted to the plants by induction. This means that the spray spreads around the plant, even underneath it. Less spray falls on the ground and farmers don't need to use as much.

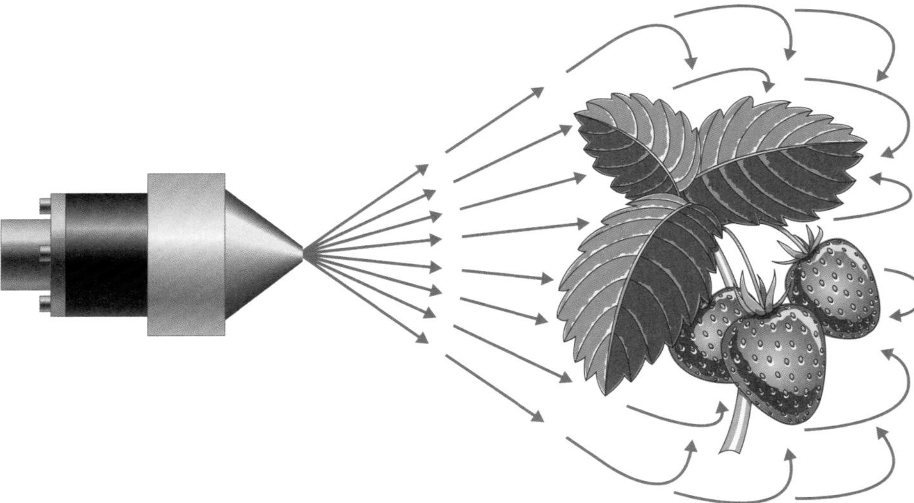

D Static electricity makes the droplets of insecticide spread out.

 **6** **a** Why does electrostatic spraying reduce the amount of spray needed?

 **b** Suggest two advantages of electrostatic spraying.

Exam-style question

When spray-painting a metal car door, the spray droplets are given a positive charge and the door is earthed. Explain why the door is earthed.

(2 marks)

Checkpoint

How confidently can you answer the Progression questions?

Strengthen

S1 What does earthing mean?

S2 State one disadvantage and one use of electrostatic charges.

Extend

E1 Explain how the risk of a fire is reduced when refuelling aircraft and cars.

E2 An electrostatic paint sprayer gives a more even coating of paint than using a brush. Use your knowledge of charges to explain why.

SP11c Electric fields

Specification reference: P11.8P; P11.9P; P11.10P

Progression questions

- What is an electric field?
- What do the field lines tell you about an electric field?
- How do electric fields help to explain static electrical effects?

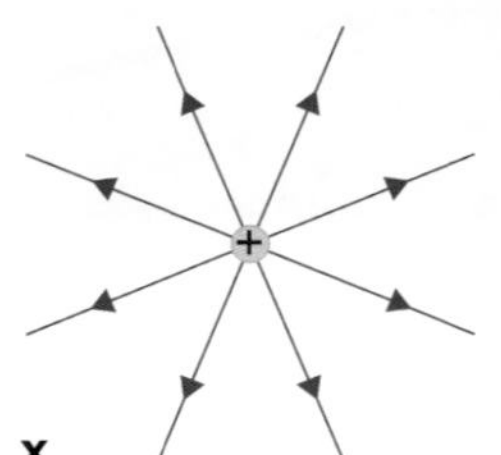

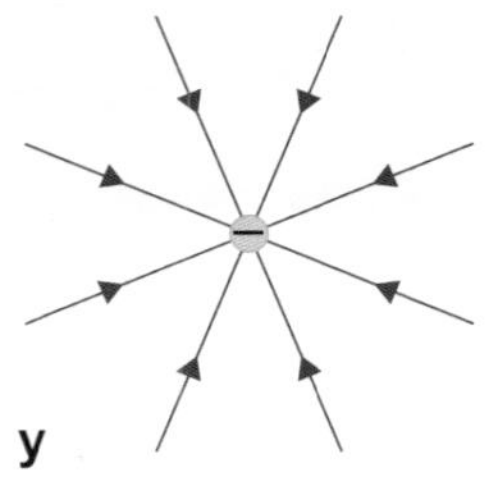

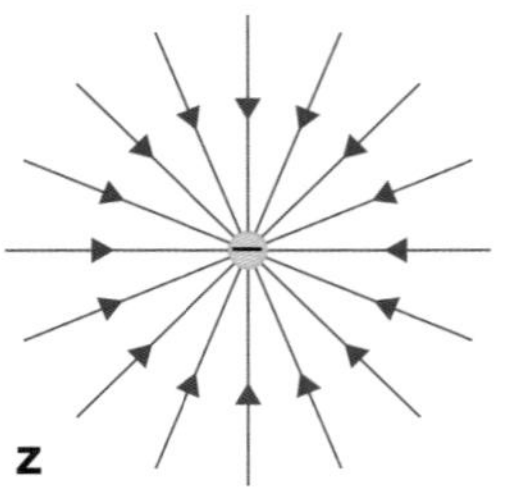

A an electric field around positive and negative point charges

A **force field** is the volume of space around an object in which another object can experience a force. A magnet has a force field around it called a magnetic field. A magnetic material will feel a force from the magnet if it is inside this magnetic field.

1 Which point charge in diagram A produces the strongest field?

2 Draw diagrams to show the difference between a weaker and a stronger positive point charge.

3 Using diagram A, describe the direction a positive point charge would move if it were placed:

a to the left of the positive point charge in x

b to the left of the negative point charge in z.

A charged object has a force field around it called an **electric field** (or **electrostatic field**). Another charged object placed in the field will experience a force. The drawings in diagram A show how we represent electric fields around a charge that is at a single point (called a **point charge**). The lines are called **field lines** and they:

- never cross
- show where the field is strongest (which is where the field lines are closest together).
- show the direction of the force on a charge in the field
- start on a positively charged object, for example the point charge, and end on a negative charged object. If there is only one object they keep going, becoming more widely spaced

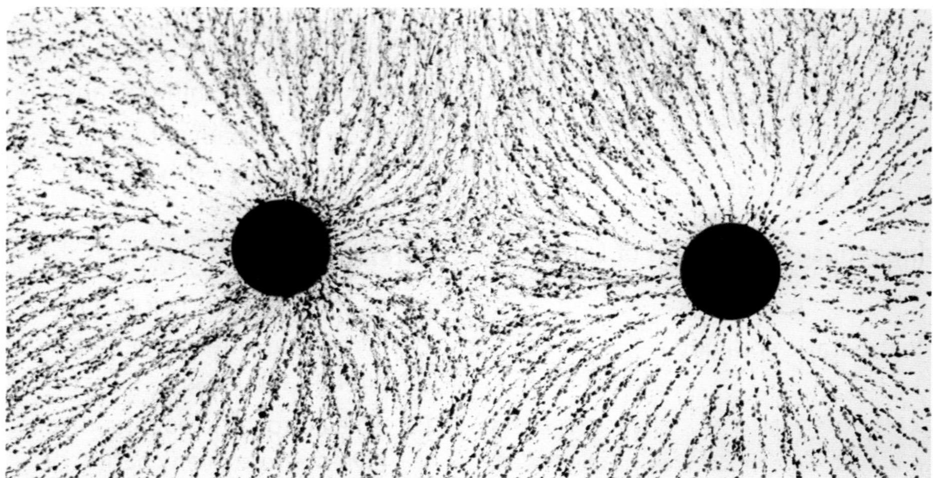

B The black circles are electrodes, each charged with the same high, positive electric charge. Black specks of pepper have lined up with the electric field so that they show the pattern of the electric field lines.

Photo B shows specks of pepper floating in cooking oil. The electric charge on an electrode causes charge to be redistributed in each pepper speck. A positive charge is induced on one end and a negative charge on the opposite end of each speck. These induced charges on each speck are in the electric field around each electrode, and so they feel a force. The forces are strong enough to line the speck up with the electric field, with the negative end of the speck towards the electrode.

In photo B, the pepper specks form field lines. Where they are more concentrated shows where the electric field is stronger. The electric field between two parallel plates, however, is **uniform** – it is the same in all places between the two points, as shown in diagram C.

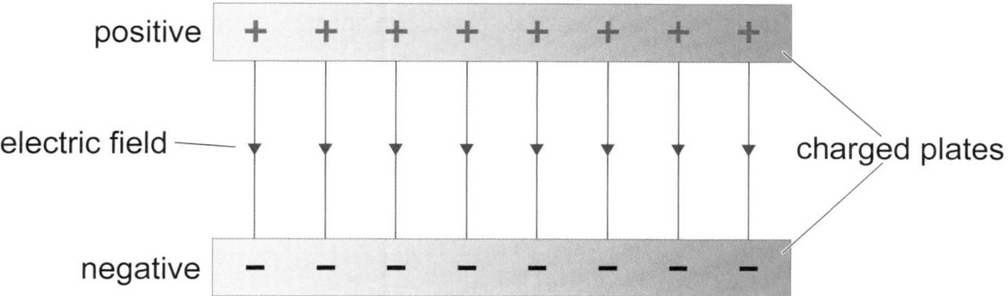

C The electric field between two parallel plates is uniform.

Did you know?

Power lines carry a voltage of 400 000 V and the ground is at 0 V so an electric field is created between the power lines and the ground. This field causes mercury atoms in fluorescent tubes to give out UV light. The UV light strikes the coating of the tube and makes it glow. This effect was used by Richard Box to create this piece of art near Bristol, called *Field*.

D

5 Describe how a positive charge would move if it were placed between the charged parallel plates.

6 What can you say about the strength of the electric field at different points between the parallel plates in diagram C?

Exam-style question

A positively charged balloon is held above a small piece of paper. Explain the effect of the balloon's electric field on the piece of paper. *(3 marks)*

4 Describe what would happen if a grain of pepper with an unbalanced positive charge were placed close to the negative electrode shown in diagram C.

Checkpoint

How confidently can you answer the Progression questions?

Strengthen

S1 Copy the diagram of the parallel plates from diagram C and then draw a second diagram to show the electric field when the charge on each plate is doubled.

S2 Add a rectangular speck of dust to the electric field in the first drawing and use labels to explain why it lines up in the electric field.

Extend

E1 Explain why electric field lines cannot cross.

E2 Photo B shows the field between two identical electric charges. Compare this with the magnetic field pattern for two bar magnets aligned with identical poles, and then with opposite poles. Use this information to help you draw the expected electric field pattern between two opposite electric charges attracting.

Electrical safety

A 3-pin plug is used to connect a television to the mains electrical supply. In the plug there is a 5 amp fuse and an earth connection. Explain how these safety features work to make using the television safer. **(6 marks)**

Student answer

The 5 amp fuse is made of thin wire which melts if a current greater than 5 amps passes through it [1]. *The earth connection joins any metal parts of the television to a large metal spike in the ground, through the plug. If there is a fault which connects the metal parts of the television to a high voltage, then a high current passes through the earth wire to the ground* [2]. *The current also flows through the fuse and the wire melts* [3]. *This disconnects the television so that nobody can get an electric shock by touching it* [4].

[1] This part of the answer describes how the fuse works, but the information about how this makes using the television safer is at the end of the answer. It would be better to have all the information about the fuse in one paragraph, and all the information about the earth wire in another.

[2] This explanation should have included the fact that the earth wire makes using the television safer by taking charge away from the metal parts of the television and so preventing someone getting an electric shock by touching it.

[3] This sentence would be clearer if the student had stated that the wire that melts is the one inside the fuse.

[4] The main purpose of a fuse is to protect against the wiring overheating and causing a fire. The fuse will protect people from electric shocks only if the live wire touches the outer casing of the television, in which case a large additional current flows and the fuse wire melts.

Verdict

This is an acceptable answer. It shows an understanding of how a fuse works and how the earth connection works. The use of scientific language is good with the terms current and voltage used correctly.

This answer could be improved by explaining that the earth wire is the primary safety feature for preventing electric shocks and does this by removing any build-up of charge (or current flowing through) the parts of the television that someone could touch. This answer could also be improved by including an explanation of how a fuse improves safety – if a fault causes a high current, it breaks the circuit so that overheating does not cause fire or damage to the television.

Exam tip

When a question asks for a series of points it is often a good idea to pull the question apart and make a list. In this question you are being asked about two safety features and how each one works. A planning list might look like this:

earth { *how it works* / *how it makes the television safer* } *fuse*

Paper 2

SP12 Magnetism and the Motor Effect / SP13 Electromagnetic Induction

The aurora borealis occurs when charged particles flowing out from the Sun become trapped by the Earth's magnetic field and enter the Earth's atmosphere near the North Pole. These particles collide with atoms in the atmosphere, and the atoms gain energy. This energy is then emitted by the atoms as light.

In this unit you will learn about magnetic fields and how they are used to produce forces and to change the voltage of electricity supplies.

The learning journey

Previously you will have learnt at KS3:

- how to plot the shape of a magnetic field and that the Earth has a magnetic field
- that electric currents cause magnetic fields, including in electromagnets and motors.

In this unit you will learn:

- about permanent and induced magnets, and how to represent a magnetic field
- about the magnetic field around a current in a wire and the factors that affect it
- how the fields from the individual coils in a solenoid interact
- how to use the power equation for transformers
- how transformers are used in the national grid
- **H** how to use the turns ratio equation for transformers
- **H** how a current can be induced in a wire and the factors that affect it
- **H** how to work out the size and direction of the force on a wire carrying a current in a magnetic field.

SP12a Magnets and magnetic fields

Specification reference: P12.1; P12.2; P12.3; P12.4; P12.5; P12.6

Progression questions

- How are magnets used?
- What shape are magnetic fields and how can they be plotted?
- What is the evidence that the Earth has a magnetic field?

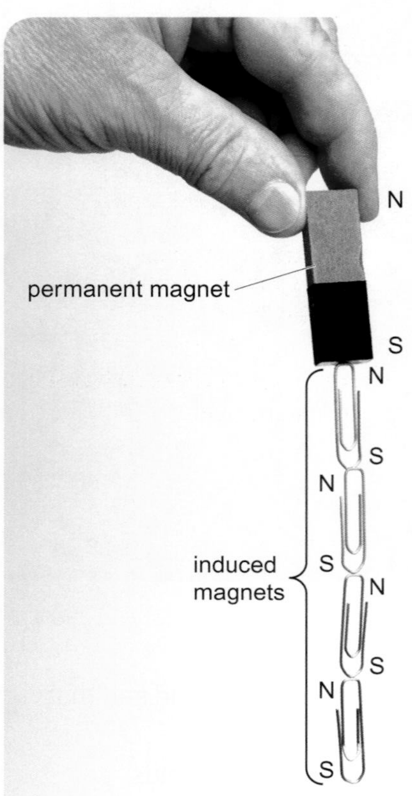

A A permanent magnet can turn objects made from magnetic materials into induced magnets.

A bar magnet is a **permanent magnet** because it is always magnetic. A magnet can attract **magnetic materials**. These include the metals iron, steel, nickel and cobalt. The space around a magnet where it can attract these materials is called the **magnetic field**.

A bar magnet has two ends, one called a north-seeking pole and one called a south-seeking pole (usually called the north pole and south pole for short). If two magnets are placed close to each other, the north pole on one magnet attracts the south pole on the other. If two north poles or two south poles are put close together, the magnets repel each other.

4th **1** Suggest what material the paper clips in photo A are made from. Explain your answer.

When a piece of magnetic material is in a magnetic field it becomes a magnet itself. This is called an **induced magnet**. It stops being magnetic when it is taken out of the field again.

5th **2** In photo A, how would the induced magnetism in the paper clips be different if the magnet were held the other way up?

Magnets are used in electric motors, generators, loudspeakers and other electrical devices. They are also used for simpler things such as door latches and knife holders.

The shape of a magnetic field can be found using **plotting compasses**. We represent magnetic fields using lines that show how a single north pole would move (from north to south). The field is strongest where the lines are closest together.

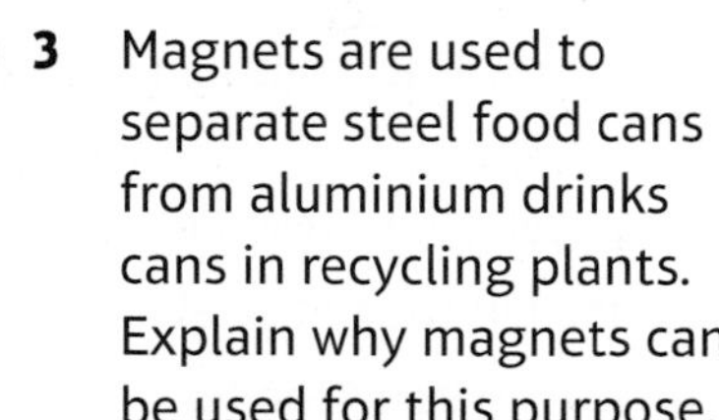

5th **3** Magnets are used to separate steel food cans from aluminium drinks cans in recycling plants. Explain why magnets can be used for this purpose.

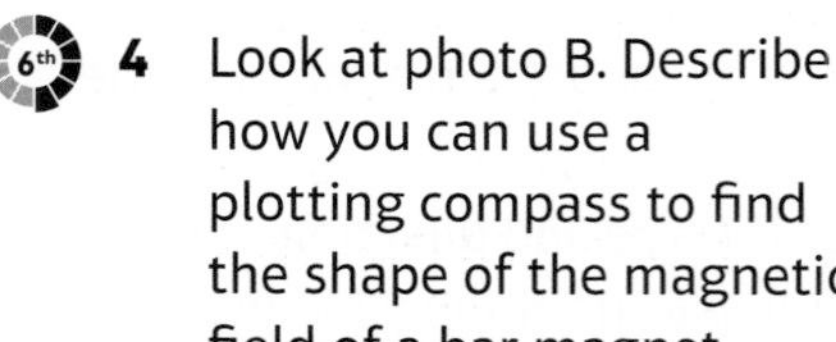

6th **4** Look at photo B. Describe how you can use a plotting compass to find the shape of the magnetic field of a bar magnet.

B You can draw lines to show the shape of a magnetic field using a plotting compass

Diagram C shows the shape of the magnetic field around a bar magnet. Diagram D shows how two magnets together can form a uniform magnetic field. This has the same strength and direction everywhere.

5 Describe two differences between a uniform magnetic field and the field around a bar magnet.

Earth's magnetic field

The needle of a plotting compass is a very small magnet. Compasses can be used to help people to find their way, as the needle always points to a position near the Earth's North Pole. A magnet suspended on a string will tilt relative to the horizontal by different amounts in different places. Compass needles are weighted at one end to keep them level.

This behaviour of compasses is evidence that the Earth has a magnetic field, which is similar in shape to the magnetic field of a bar magnet. The Earth's magnetic field is thought to be caused by electric currents in the molten outer **core**, which is made from a mixture of iron and nickel.

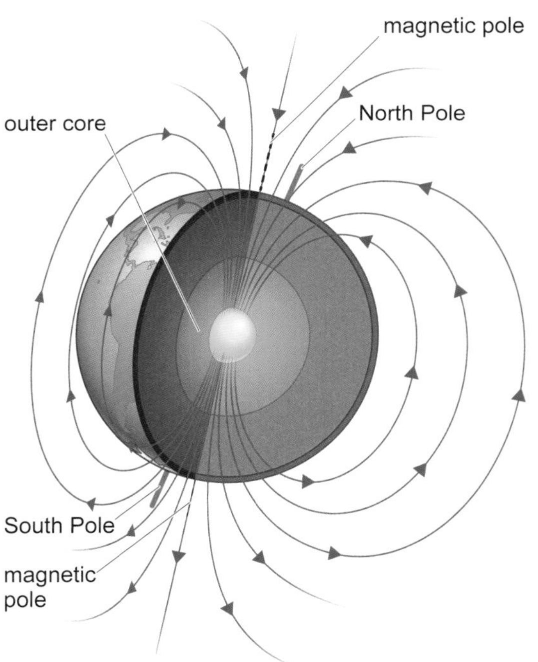

E The Earth's magnetic field has a similar shape to the magnetic field of a bar magnet.

Did you know?

In geography and everyday language, the Earth's magnetic pole in the northern hemisphere is called the 'north magnetic pole'. However, in physics a magnetic north pole is defined in terms of the direction of magnetic field lines, and so the magnetic pole in the northern hemisphere is actually a south magnetic pole. That is the reason that the north pole of a compass needle points north: it is attracted to the south magnetic pole of the Earth!

6 Look at diagram D. Explain where the Earth's magnetic field is:

a strongest **b** weakest.

Exam-style question

Describe the difference between a permanent magnet and an induced magnet. *(2 marks)*

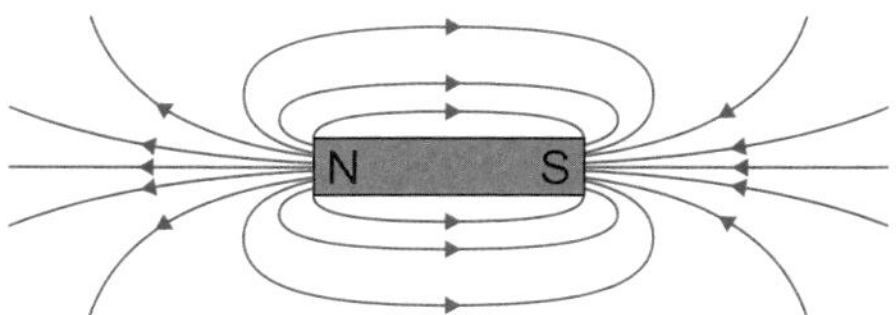

C The magnetic field occurs all around a bar magnet but we use a 2-dimensional diagram like this as a model.

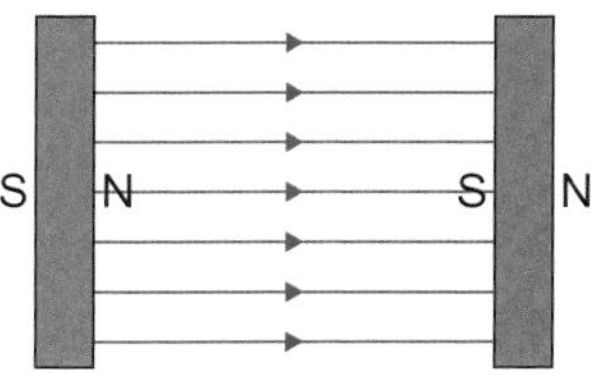

Two flat magnets produce a uniform magnetic field between them.

D These two magnets are producing a uniform magnetic field.

Checkpoint

How confidently can you answer the Progression questions?

Strengthen

S1 Explain why compasses point north. Use the words attract, repel, pole and magnetic field in your answer.

Extend

E1 Describe how a magnet suspended on a thread can be used to find the shape of the Earth's magnetic field. Explain why this works.

SP12b Electromagnetism

Specification reference: P12.7; P12.8; P12.9

Progression questions

- How is the magnetic field around a wire related to the current?
- What factors affect the strength of the magnetic field around a wire?
- How does the magnetic field around a wire change when the wire is made into a coil?

A Earphones and loudspeakers work because of the magnetic effect of an electric current.

A current flowing through a wire causes a magnetic field. Electric motors and many other devices depend on the magnetic effect of electric currents.

Photo B shows a wire passing through a piece of card. When a current flows through the wire, iron filings on the card make circular patterns. This shows that the current is causing a magnetic field because the iron filings are lining up with the direction of the magnetic field.

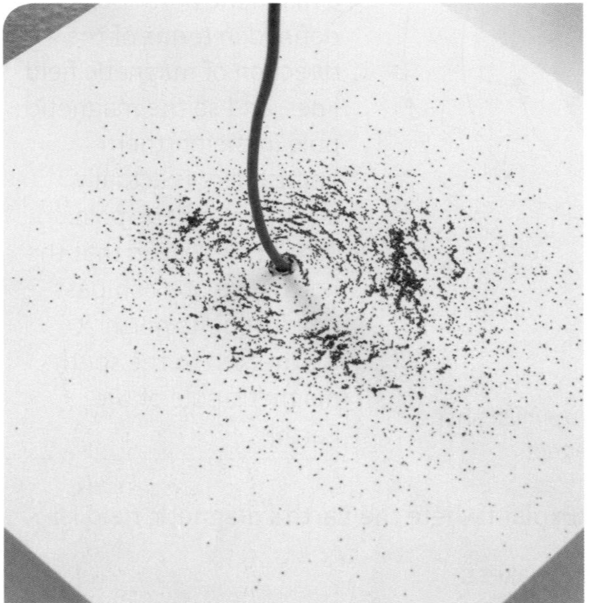

B The iron filings show the shape of the magnetic field around a wire carrying a current.

Did you know?

Hans Christian Ørsted (1777–1851) had been trying to show that electric currents had a magnetic effect for several years. He did not find any evidence until he was giving a lecture and noticed a compass needle move when current was switched on in a nearby wire. His previous experiments had been designed to find a magnetic field running in the same direction as the wire.

You can use plotting compasses to find the direction of the magnetic field. The direction of the magnetic field depends on the direction of the current. If the current changes direction, so does the direction of the magnetic field.

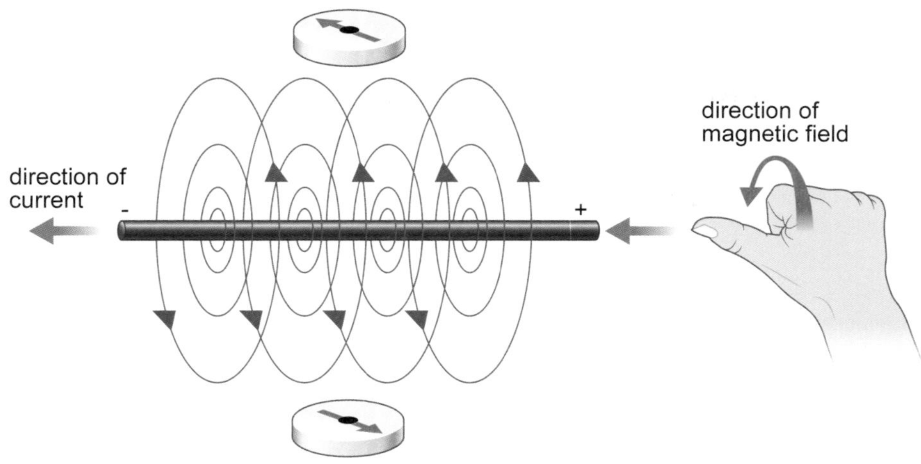

C If you point your right thumb in the direction of the current (from + to –), the magnetic field goes in the direction your fingers are pointing.

1 In photo B, the current is flowing down the wire. Draw a sketch to show the shape and direction of the magnetic field around the wire.

2 Look at diagram C. How does the diagram show that the magnetic field is strongest close to the wire?

The strength of the magnetic field depends on the size of the current – the higher the current the stronger the field. The magnetic field is strongest closer to the wire and gets weaker with increasing distance.

You can think of the magnetic field as forming a series of cylinders around the wire. If the wire is made into a coil (called a **solenoid**), the magnetic fields of all the different parts of the wire form an overall magnetic field like the one shown in part b of diagram D. The fields from individual coils add together to form a very strong field inside the solenoid. Outside the solenoid the fields from one side of the coil tend to cancel out the fields from the other side to give a weaker field outside. This is shown in part a of diagram D.

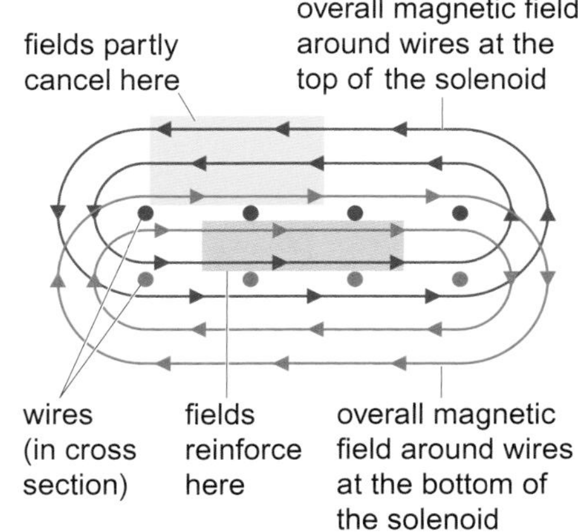

a cross section of a small coil of wire

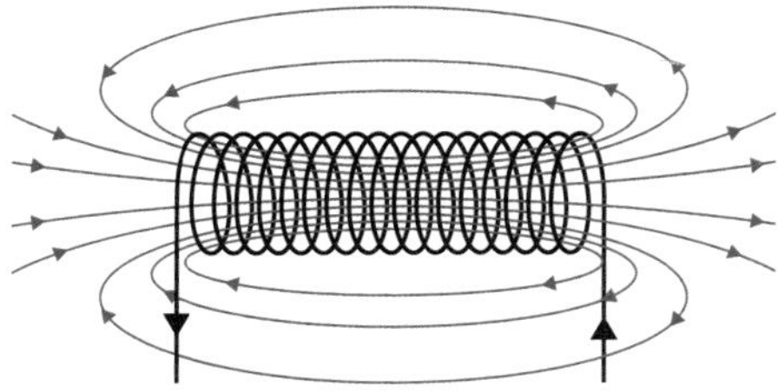

b The magnetic field inside the solenoid is almost uniform near the centre of the coil.

D

A coil of wire with a current flowing through it is an **electromagnet**. The magnetic field of an electromagnet can be made stronger by putting a piece of iron (an iron core) inside the coil. This iron becomes a **temporary magnet** – it is only magnetic while the field from the electromagnet is affecting it.

3 Where is the magnetic field around an electromagnet strongest? Explain your answer.

4 **a** Suggest why iron, rather than a metal like copper, is needed to make the magnetic field of an electromagnet stronger.

b Suggest one other way of making the field stronger.

c Suggest how you can reverse the direction of the magnetic field in an electromagnet.

Exam-style question

Describe two ways in which the magnetic field around a wire can be changed. *(2 marks)*

Checkpoint

How confidently can you answer the Progression questions?

Strengthen

S1 Describe two different shaped magnetic fields that can be made using current flowing through a wire.

Extend

E1 Describe how you could use a coil of wire and a plotting compass as a simple ammeter and explain how your idea will work.

SP12c Magnetic forces

Specification reference: H P12.10; H P12.11; H P12.12; H P12.13; H P12.14P

Progression questions

- H How can electricity and magnetism combine to produce forces?
- H How is the force on a wire in a magnetic field used to make an electric motor turn?
- H How can we calculate the size of the force produced by a current in a magnetic field?

H

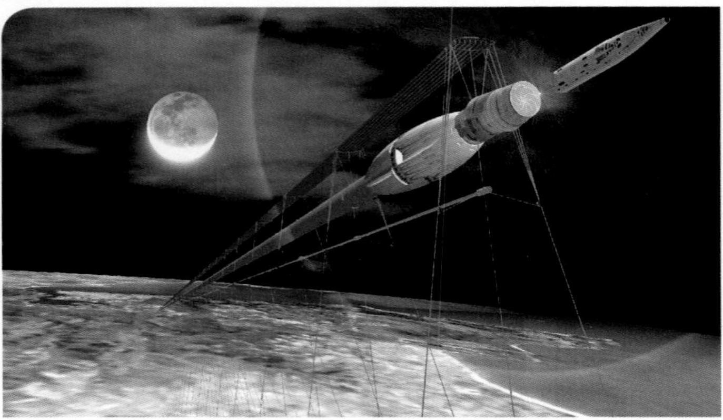

A artist's impression of a rail gun launcher

Photo A shows an idea for a 'rail gun' launcher, which uses a current flowing through two rails to produce a force. The force produced by the rail gun is an example of the **motor effect**.

A wire carrying a current experiences a force when it is placed between two magnets. The current in the wire creates a magnetic field around the wire and this interacts with the magnetic field between the magnets. The force is greatest when the wire is at right angles to the magnetic field produced by the magnets, and is zero when the wire is in the same direction as this magnetic field. There is an equal and opposite force on the magnets.

1 What is the motor effect?

The direction of the force depends on the directions of the magnetic field and the current. **Fleming's left-hand rule** shows how the directions are related, as shown in diagram B.

2 Look at diagram B. What will happen if the connections to the power supply are swapped over?

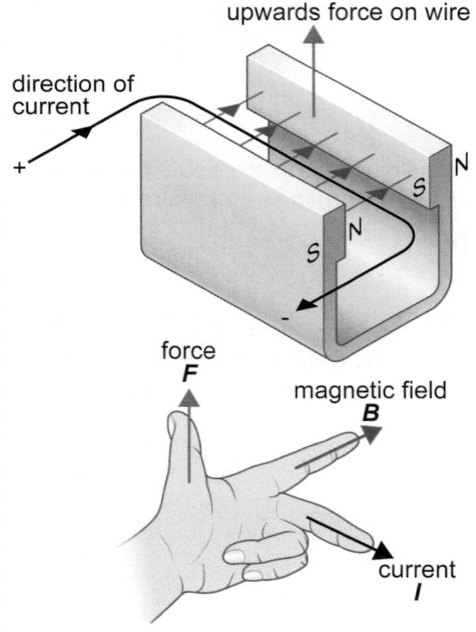

B Fleming's left-hand rule. The direction of the current is from + to -.

The size of the force depends on the magnetic field strength, the current and the length of the wire in the field. The strength of a magnetic field (the **magnetic flux density**) is measured in units of newtons per amp metre (N/A m) (also called **tesla**, **T**) and is given the symbol *B*.

force on conductor carrying current at right angles to magnetic field (N) = magnetic field strength (N/A m or T) × current (A) × length (m)

This can be shown in symbols as $F = B \times I \times l$.

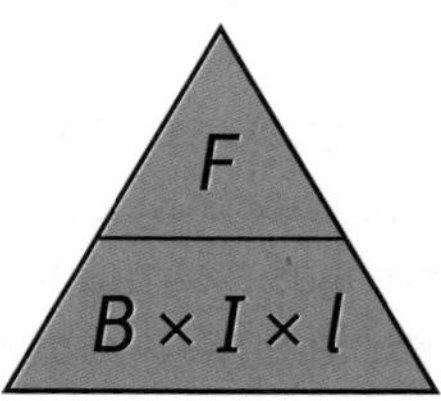

C

Worked example

A 200 m long wire carries a current of 3 A at right angles to the Earth's magnetic field. There is a force of 0.024 N on the wire. Calculate the strength of the magnetic field.

$$B = \frac{F}{I \times l} = \frac{0.024\text{ N}}{3\text{ A} \times 200\text{ m}} = 0.000\,04\text{ N/A m (or } 4 \times 10^{-5}\text{ N/A m)}$$

H Electric motors

The force on a conductor in a magnetic field is used to cause rotation in electric motors. Diagram D shows a simplified motor (a real motor would have many more turns of wire on the coil). There is a force on each part of the wire carrying a current in the magnetic field and so using a coil with many turns of wire increases the total force turning the coil.

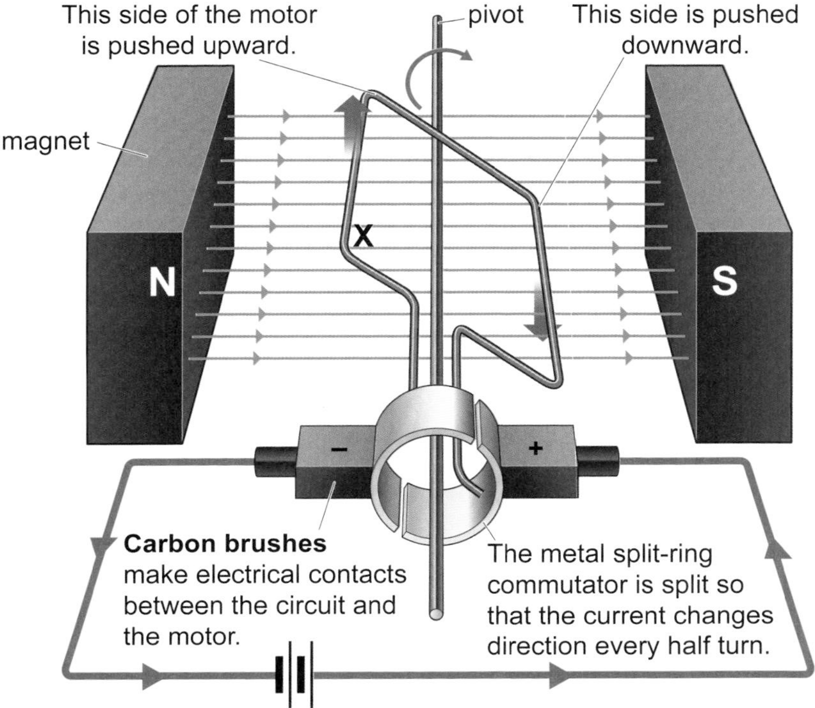

D an electric motor

The **split-ring commutator** ensures that the force on the coil always turns it in the same direction. In diagram D, the part of the coil labelled X is moving upwards. If the current in this part of the wire was still moving in the same direction (towards us on the diagram) half a turn later, the force on it would still be upwards, and this would have the effect of trying to turn the coil in the opposite direction. The commutator ensures that the current is always flowing in the correct direction to make the coil continue to spin.

5 Explain how the following changes will affect the motor.

a The current is doubled.

b The magnets are moved further apart (weakening the field strength).

c Twice as many turns of wire are put on the coil.

Exam-style question

Describe three factors that affect the force experienced by a current-carrying conductor in a magnetic field. State the effect of changing each factor. *(3 marks)*

3 The apparatus shown in diagram B is placed on a balance. The wire is held so that it cannot move. Explain how the reading on the balance changes when the current is switched on.

4 A magnetic field has a strength of 0.5 N/A m. A 10 cm wire is running at right angles to it.

a Calculate the force on the wire when the current is 0.3 A.

b Calculate the current needed to produce a force of 0.002 N.

Checkpoint

How confidently can you answer the Progression questions?

Strengthen

S1 a A 2 m long wire runs at right angles across a magnetic field with a strength of 0.2 N/A m. Calculate the force on the wire when the current is 0.5 A.

b Describe how an electric motor works.

Extend

E1 Two wires are held in clamp stands so that they run parallel to each other a few centimetres apart. When current flows through the wires they repel each other. Explain why this happens.

SP13a Electromagnetic induction

Specification reference: H P13.1P; H P13.2; H P13.3P; H P13.4P

Progression questions

- H How can you produce an electric current using a magnet and conductor?
- H What are the factors that affect the size and direction of an induced potential difference?
- H How is induction used in generators and microphones?

H

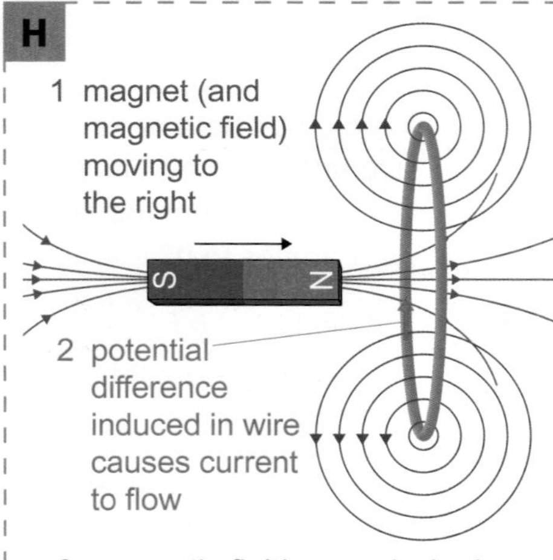

A a magnet being moved through a loop of wire

A changing magnetic field can **induce** a **voltage** or **potential difference** (**p.d.**) in a wire and this causes a current to flow. A p.d. can also be induced if a wire is moved in a magnetic field. The wire is often made into a coil so that there is more wire in the changing magnetic field.

The size of the induced potential difference depends on the number of turns in a coil of wire, the strength of the magnetic field, and on how fast the magnetic field changes or moves past the coil. Reversing the direction of change reverses the direction of the induced potential difference.

If the potential difference causes a current to flow, the magnetic field of the current opposes the original change.

1 Look at diagram A. The magnet is moved to the left.

a What will be the direction of the magnetic field induced around the wire?

b Explain how you could have worked out your answer without having the diagram to refer to.

2 Look at diagram A. Describe two ways of increasing the size of the current in the wire.

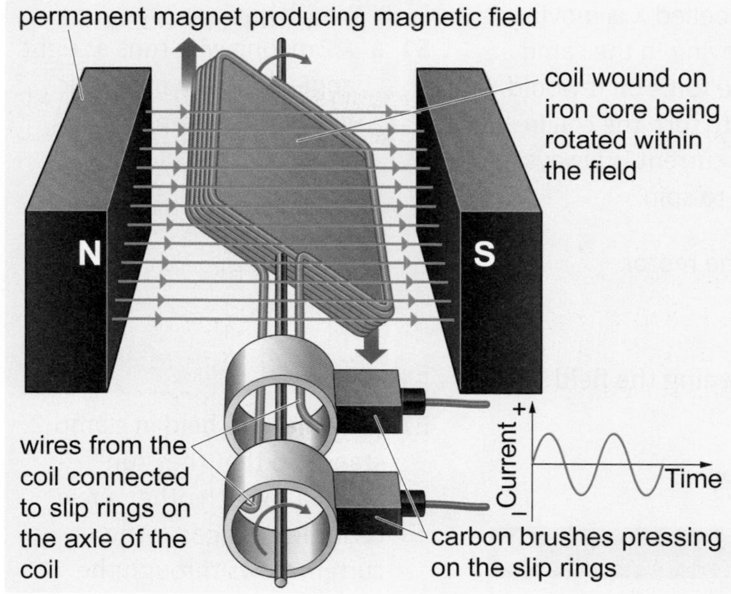

B An alternator produces a current that changes direction many times each second.

Generators

A **generator** consists of a coil of wire that is rotated inside a magnetic field. As the coil turns, a voltage is induced in the wire. The ends of the coil are connected to **slip rings**. Electrical contact with an external circuit is made using **carbon brushes**, which press on the slip rings**.** A generator like this produces **alternating current**, and is often called an **alternator**.

3 Suggest three ways of increasing the current produced by a generator.

Generators used in power stations and the alternators in cars work on similar principles but have a rotating electromagnet surrounded by coils of wire.

H

Many electrical appliances need **direct current**. A **commutator** switches over the connections every half-turn of the coil, and so produces a form of direct current. A generator with a commutator is often called a **dynamo**.

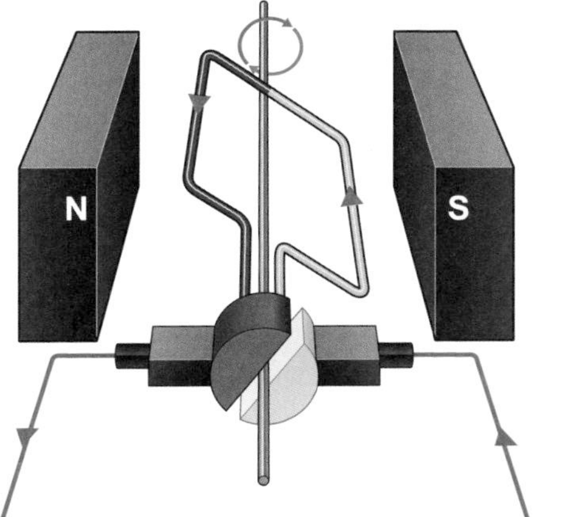
N
S

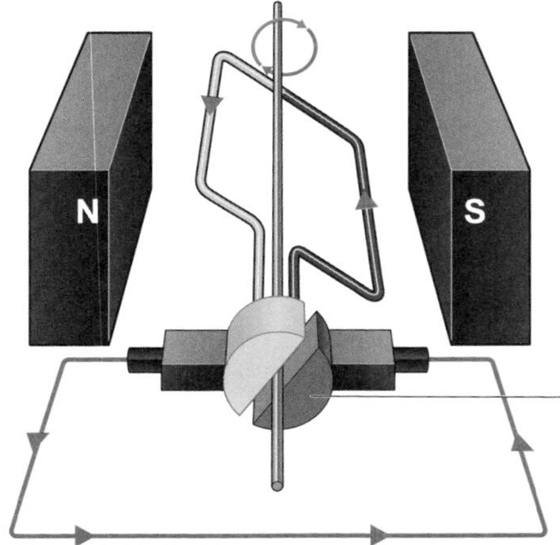
N
S

C The commutator allows the generator to produce direct current.

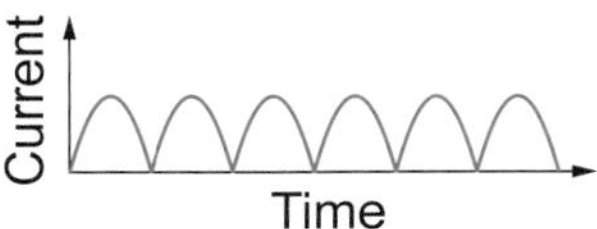
Current
Time

The commutator swaps the connections every half-turn so the current in the external circuit always flows in the same direction.

8th **4** State two similarities and one difference between slip rings and a commutator.

Microphones and loudspeakers

Microphones convert the pressure variations in sound waves into variations in current in electrical circuits.

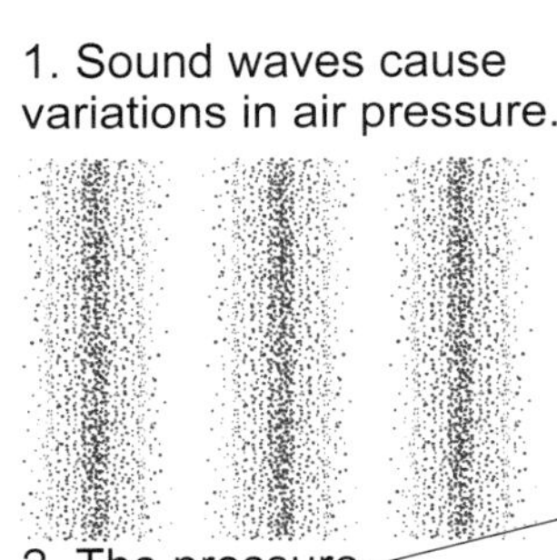

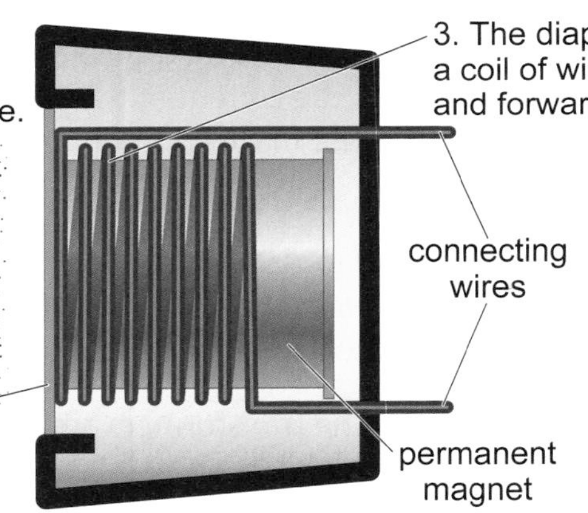
1. Sound waves cause variations in air pressure.
2. The pressure variations make a **diaphragm** vibrate.
3. The diaphragm moves a coil of wire backwards and forwards.
connecting wires
permanent magnet

D Some microphones use electromagnetic induction to produce a varying current.

8th **5** Look at diagram D. Describe how the microphone works.

Loudspeakers convert variations in an electrical current into sound waves. Loudspeakers have similar components to the microphone shown in diagram D. The varying current flows through a coil that is in a magnetic field. This causes a force on the coil, which moves backwards and forwards as the current varies. The coil is connected to a diaphragm, and the movements of the diaphragm produce sound waves.

8th **6** Compare and contrast a microphone and a loudspeaker.

Exam-style question

Compare and contrast an alternator and a dynamo. *(4 marks)*

Checkpoint

How confidently can you answer the Progression questions?

Strengthen

S1 Write a paragraph to describe how electricity is used to make microphones and loudspeakers work.

Extend

E1 Explain why the force needed to turn the coil in a generator becomes larger if the coil is spun faster.

SP13b The national grid

Specification reference: H P13.5; H P13.6; H P13.7P; P13.8; P13.9; H P13.11

Progression questions

- Why are transformers used to help transmit electricity around the country?
- H How does a transformer work?
- H How can you calculate the size of the voltage produced by a transformer?

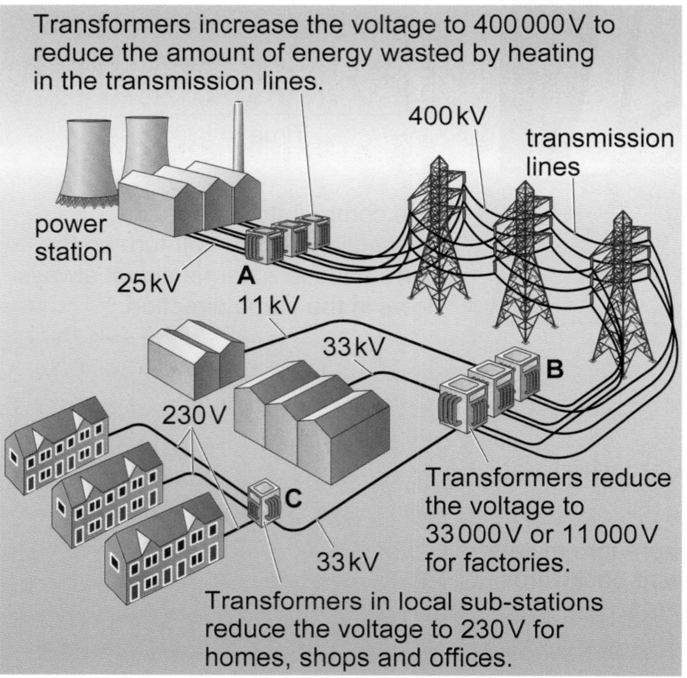

Transformers increase the voltage to 400 000 V to reduce the amount of energy wasted by heating in the transmission lines.
400 kV
transmission lines
power station
25 kV
A
11 kV
33 kV
B
230 V
C
33 kV
Transformers reduce the voltage to 33 000 V or 11 000 V for factories.
Transformers in local sub-stations reduce the voltage to 230 V for homes, shops and offices.

A the national grid

Electricity is sent from power stations to homes, schools and factories by a system of wires and cables called the **national grid**. When electricity flows through a wire, the wire gets warm. The amount of energy wasted by heating in wires inside buildings is quite small, but the amount is significant for the **transmission lines** in the national grid.

If the voltage (potential difference) of the electricity passing through a wire is increased, the current is decreased. When the current is smaller, less energy is transferred by heating and the efficiency is improved. Power stations produce electricity at 25 kV. This is changed to 400 kV by **transformers** before the electricity is sent around the country. The voltage is reduced before the electricity is sent to factories and other buildings.

A **step-up transformer** increases the voltage and decreases the current at the same time. A **step-down transformer** makes the voltage lower and the current higher.

Did you know?

There are over 25 000 km of transmission lines in the national grid.

1 Write these national grid voltages in the order that they occur in diagram A, starting with the voltage at the power station: 11 kV, 25 000 V, 33 kV, 0.23 kV, 400 kV.

2 Look at diagram A. The transformers are labelled A, B and C. For each transformer, say whether it is a step-up or step-down transformer and explain why it is used.

B Electricity sub-stations contain step-down transformers.

How transformers work

A transformer is made using two coils of insulated wire wound onto an iron core. There is no electrical connection between the two coils of wire.

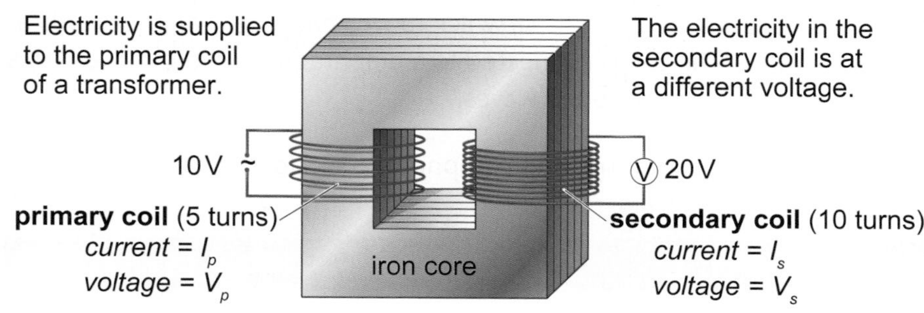

C the structure of a transformer

H

Transformers only work with alternating current, when the direction of the potential difference (and so also the current) changes many times each second. The alternating current in the primary coil creates a continuously changing magnetic field, and the iron core of the transformer carries this magnetic field to the secondary coil.

The changing magnetic field induces a changing potential difference in the secondary coil. The potential difference is greater in the secondary coil if it has more turns than the primary coil.

The potential differences across the coils of a transformer can be worked out using this equation:

$$\frac{\text{potential difference across primary coil}}{\text{potential difference across secondary coil}} = \frac{\text{number of turns in primary coil}}{\text{number of turns in secondary coil}}$$

This can be written in symbols as:

$\frac{V_p}{V_s} = \frac{N_p}{N_s}$ where N represents the number of turns in the coil.

 **3** Explain why the core of a transformer is made from iron.

4 The UK mains supply changes direction 100 times each second. How will changing this to 200 times per second affect the magnetic field induced in the core of a transformer?

Worked example

A radio runs off the 230 V mains supply but only needs 23 V. Its transformer has 100 turns of wire in the primary coil. How many turns are needed in the secondary coil?

$$\frac{V_p}{V_s} = \frac{N_p}{N_s}$$

$$\frac{230\text{ V}}{23\text{ V}} = \frac{100}{N_p}$$

$$10 = \frac{100}{N_p}$$

$$10 \times N_p = 100$$

$$N_p = \frac{100}{10} = 10$$

So the secondary coil must have 10 turns.

 **5** A transformer has 1500 turns on the primary coil and 30 000 turns on the secondary coil. What is the potential difference across the secondary coil for a primary coil potential difference of 20 000 V?

Exam-style question

State two ways in which transformers are used in the national grid.

(2 marks)

Checkpoint

How confidently can you answer the Progression questions?

Strengthen

S1 Explain why different voltages are used in the national grid. Present your answer as a paragraph.

Extend

E1 A student has recorded the following information for a step-down transformer: 500 V, 11 kV, 10 000 turns in primary coil. Identify the primary voltage and calculate the missing value.

SP13c Transformers and energy

Specification reference: P13.10; H P13.11P

Progression questions

- How is the law of conservation of energy applied to transformers?
- How can you calculate the current and voltage produced by transformers?
- H How can you carry out calculations to show the advantage of transmitting electricity at high voltages?

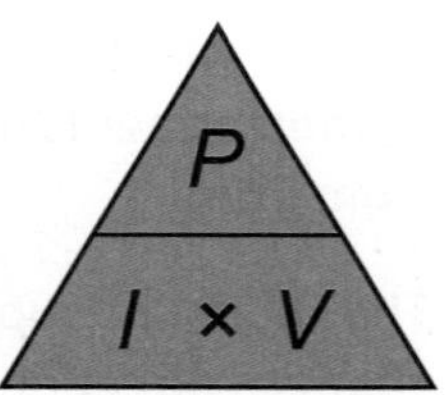

A equation triangle for calculating electrical power (*I* represents current, *V* represents potential difference)

The potential difference (voltage) is a measure of the energy transferred by each **coulomb** of charge that flows through a wire. The current measures the number of coulombs per second. So the energy transferred each second by an electric current (the **power**), is calculated using this equation.

electrical power	=	current	×	potential difference
(W)		(A)		(V)

1 A 20 W light bulb uses electricity from the mains supply (230 V). Calculate the size of the current.

B The core of this transformer is made of thin slices of iron glued together, which helps to make the transformer more efficient.

Energy cannot be created or destroyed, so the power supplied to a transformer in the primary coil must be equal to the power transferred away from the transformer in the secondary coil. If the transformer is 100% efficient (no energy is wasted by heating), then the power in the secondary coil equals the power in the primary coil.

potential difference across primary coil (V)	×	current in primary coil (A)	=	potential difference across secondary coil (V)	×	current in secondary coil (A)

This can also be written as:

$V_p \times I_p = V_s \times I_s$

Worked example W1

The primary coil of a transformer has a current of 0.5 A with a potential difference of 100 V. The current in the secondary coil is 25 A. What is the potential difference across the secondary coil? Use $V_p \times I_p = V_s \times I_s$

$100\text{ V} \times 0.5\text{ A} = V_s \times 25\text{ A}$

$50 = V_s \times 25$

$V_s = \frac{50}{25} = 2\text{ V}$

2 A transformer receives 5000 W of energy by electricity.

a How much power is transferred by the electricity coming out of the transformer?

b Explain how the power output will be different if the transformer is not 100% efficient.

3 Calculate the missing values for these transformers.

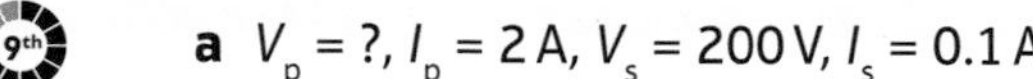

a $V_p = ?, I_p = 2\text{ A}, V_s = 200\text{ V}, I_s = 0.1\text{ A}$

b $V_p = 33\text{ kV}, I_p = 2\text{ A}, V_s = 230\text{ V}, I_s = ?\text{ A}$

H

You can carry out calculations to help you to explain why transmitting power at high voltages is more efficient. The box shows the equations you need.

power (W) = $\frac{\text{energy transferred (J)}}{\text{time taken (s)}}$, $P = \frac{E}{t}$

electrical power (W) = current (A) × potential difference (V), $P = I \times V$

electrical power (W) = current squared (A)^2 × resistance (Ω), $P = I^2 \times R$

Worked example W2

An electricity substation supplies 2 MW of power to a small housing estate. Electricity is sent to the substation along cables with a resistance of 0.08 Ω. The supply is at 230 V. Calculate the energy wasted every hour.

Current required: $I = \frac{P}{V}$

$= \frac{2 \times 10^6\,\text{W}}{230\,\text{V}}$

$= 8.7 \times 10^3\,\text{A}$

Power transferred by heating in the wires to the substation: $P = I^2 \times R$

$= (8.7 \times 10^3\,\text{A})^2 \times 0.08\,\Omega$

$= 6.05 \times 10^6\,\text{W}$

Energy transferred per hour: $E = P \times t$

$= 6.05 \times 10^6\,\text{W} \times 3600\,\text{s}$

Energy wasted = 2.18×10^{10} J

4 Look at worked example W2. Calculate how much energy is wasted every hour if the supply is at 33 kV.

5 A 2 MW substation converts the electricity supply to 230 V. The primary coil has 10 050 turns and the secondary coil has 70 turns.

a Calculate the voltage of the electricity arriving at the transformer.

b Explain why electricity is transmitted at this voltage.

Exam-style question

A step-up transformer is used to increase the potential difference of an electricity supply. Explain why the current decreases at the same time.

(3 marks)

Did you know?

The hoverboard in the photo works using rotating magnets that induce a current inside the metal ramp. A magnetic field caused by this current repels the magnets in the board.

C

Checkpoint

How confidently can you answer the Progression questions?

Strengthen

S1 Calculate the missing value.
$V_p = 500\,\text{V}$, $I_p = ?$, $V_s = 5\,\text{V}$, $I_s = 2\,\text{A}$

Extend

E1 Some types of halogen light bulbs need a 12 V supply. A transformer is used to convert the mains supply to run six 50 W bulbs. Use the transformer equation to calculate the current drawn from the mains supply.

E2 **H** A 100 km transmission line has a resistance of 2 Ω. Calculate the amount of energy wasted per year when it is supplying 1200 MW of power at 400 kV.

Magnetic fields

An electric current in a wire produces a magnetic field around the wire. A straight wire can be coiled to form a solenoid.

Describe the shapes of the magnetic fields around a straight wire and a solenoid, and the factors that affect the direction and strength of the field. **(6 marks)**

Student answer

The magnetic field around a straight wire is circular. The direction of the field depends on the direction of the current [1]. *It* [2] *gets stronger if the current gets bigger, and is strongest close to the wire and weaker further away* [3]. *When the wire is wound up to make a solenoid, all the fields from all the coils join up and make a field that looks like a bar magnet. The field is the same all the way inside the coil* [4].

[1] This is a good description of the shape of the field around a straight wire, and describes what affects its direction.

[2] 'It' can sometimes be misunderstood. This sentence would be better if it started with 'The magnetic field gets stronger if…'

[3] This is a good description of the factors affecting the strength of the field, including *how* each factor affects its strength.

[4] The student understands that the fields from the individual coils combine to form the field, and describes the shape. However the final sentence is incorrect – the field is strongest inside the coil, but it is not uniform.

Verdict

This is an acceptable answer. The student has described the shapes of the magnetic fields around a straight wire and around a solenoid, and has clearly described the effects of two factors that affect the strength of the field around a wire. The answer is well organised and includes links between scientific ideas, for example the field around a solenoid and the field around a bar magnet.

The answer could be improved by describing where the magnetic field of a solenoid is strongest and the factors that affect its strength.

Exam tip

The word 'describe' is often used in exam questions. You might be asked to describe a process, how something works or the effect of one thing on another. In a 'describe' question, you don't need to give reasons for *why* things happen. Instead, you need to write an account of *what* happens.

Paper 2

SP14 Particle Model / SP15 Forces and Matter

'Dry ice' is often used in theatres or concerts to produce fog that creeps along the floor. Dry ice is actually frozen carbon dioxide. When this becomes warmer than −78 °C it starts to sublimate (change directly from a solid to a gas). To make fog, the dry ice is put into warm water where it sublimates and makes bubbles of carbon dioxide gas. These bubbles are very cold, and when they escape from the water they cause water vapour in the air to condense and form very tiny droplets of water. These drops of liquid water form the fog that you can see.

In this unit you will learn how the particle model explains the properties of matter and what happens when energy is transferred to or from a substance. You will also learn about springs and the energy transfers in stretching them.

The learning journey

Previously you will have learnt at KS3:

- that mass is conserved during changes of state
- about the properties of solids, liquids and gases
- how particles are arranged in solids, liquids and gases, and how this is affected by temperature.

Previously you will have learnt in *SP2 Motion and Forces*:

- some of the effects that forces have on objects.

In this unit you will learn:

- how to explain different densities of substances and how to calculate density
- about specific heat capacity and specific latent heat
- how changing the temperature and volume of a gas affects its pressure and how to calculate temperatures and pressures
- about elastic and inelastic distortion
- about the relationship between force and extension, and how to calculate the extension and spring constant
- how to calculate the work done when stretching a spring
- how pressure in fluids depends on density and depth.

SP14a Particles and density

Specification reference: P14.1; P14.2; P14.4; P14.5

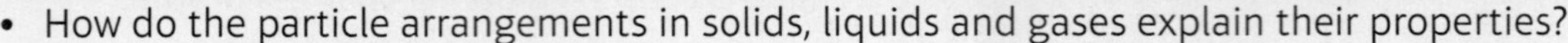

Progression questions

- How do the particle arrangements in solids, liquids and gases explain their properties?
- What happens to particles when a substance changes state?
- How can you calculate the density of a substance?

The ice in glaciers is added to by winter snowfall. Ice is lost from the top of a glacier by **sublimation**, when a solid turns straight into a gas without becoming a liquid first.

Ice, water and water vapour are three different **states of matter** with very different properties.

A a glacier in Greenland

1 Describe two properties of:

 **a** solids

 **b** liquids

 **c** gases.

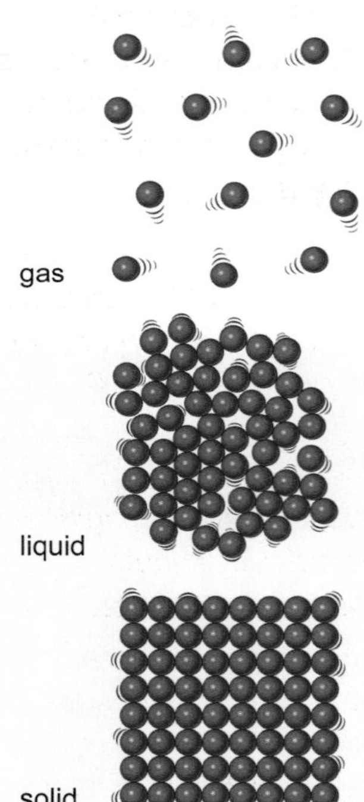

B Particles (e.g. **atoms**, **molecules**) are arranged differently in the three states of matter.

Kinetic theory

The **kinetic theory** states that everything is made of tiny particles.

In solids, forces of attraction hold particles closely together. The particles can vibrate but they cannot move around. This explains why solids keep their shape and usually cannot be **compressed**.

In liquids, the particles are moving faster and so the forces of attraction between the particles are not strong enough to hold them in fixed positions. The particles can move past each other so liquids flow and take the shape of their container. The particles are still very close together, so liquids usually can't be compressed.

In a gas, the particles are far apart and moving around quickly. Gases are compressible and expand to fill their container.

When a substance undergoes a **change of state** the particles end up in a different arrangement. There are the same number of particles so the mass stays the same (mass is **conserved**). This is a **physical change**, because no new substances are formed and the substance recovers its original properties if the change is reversed. Mass is also conserved in **chemical changes**, but the change in the substances often cannot be reversed.

2 Explain why:

 **a** liquids and gases can flow but solids cannot

 **b** gases can be compressed but solids and liquids cannot.

Density

The **density** of a substance is the mass of a certain volume of the substance. Almost all substances are most dense when they are solids and least dense when they are gases. The arrangement of particles can explain the differences in density between different states of matter. A solid is usually denser than the same substance as a liquid, because the particles in solids are closer together.

C Ice is less dense than liquid water. This is unlike most substances, which become more dense when they turn from liquid to solid.

Density can be calculated using the equation below. The units for density are usually kilograms per cubic metre (kg/m^3).

$$\text{density} = \frac{\text{mass}}{\text{volume}} \qquad \rho = \frac{m}{V}$$

Worked example

Calculate the density of a 7074 kg iron girder with a volume of $0.9\ m^3$.

$$\text{density} = \frac{\text{mass}}{\text{volume}}$$

$$= \frac{7074\ kg}{0.9\ m^3}$$

$$= 7860\ kg/m^3$$

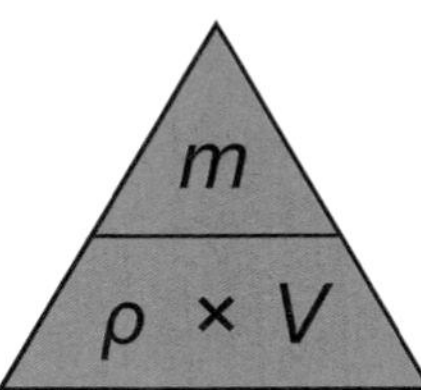

D You can rearrange the equation for density using this triangle. *m* represents mass and *V* represents volume. ρ is the Greek letter rho and represents density.

 **5** A 500 kg block of aluminium has a volume of $0.185\ m^3$. Calculate its density.

 **6** The density of solid copper is $8960\ kg/m^3$. Calculate the volume of a 5 tonne (5000 kg) delivery of copper.

Exam-style question

Use the kinetic theory model to explain *two* differences in the properties of solids and gases. *(4 marks)*

Did you know?

Water is an unusual substance because when it freezes the particles form a spaced-out arrangement, meaning that ice is less dense than liquid water. This is why icebergs and ice cubes float in water. Most other substances become denser when they change from a liquid to a solid.

 **3** Explain which beaker in photo C shows how the density of most substances changes when they freeze.

 **4** Look at diagram B. Explain why a substance becomes less dense when it changes from a liquid to a gas.

Checkpoint

How confidently can you answer the Progression questions?

Strengthen

S1 Explain how the arrangement of particles in a liquid explains its properties.

S2 A swimming pool contains $2500\ m^3$ of water. The water has a density of $1000\ kg/m^3$. Calculate the mass of water in the pool.

Extend

E1 The density of solid copper is $8960\ kg/m^3$ and the density of liquid copper is $8020\ kg/m^3$. Explain why the densities are different.

E2 Calculate the change in volume when 2000 kg of copper is melted.

SP14a Core practical – Investigating densities

Specification reference: P14.3

Aim

Investigate the densities of solids and liquids.

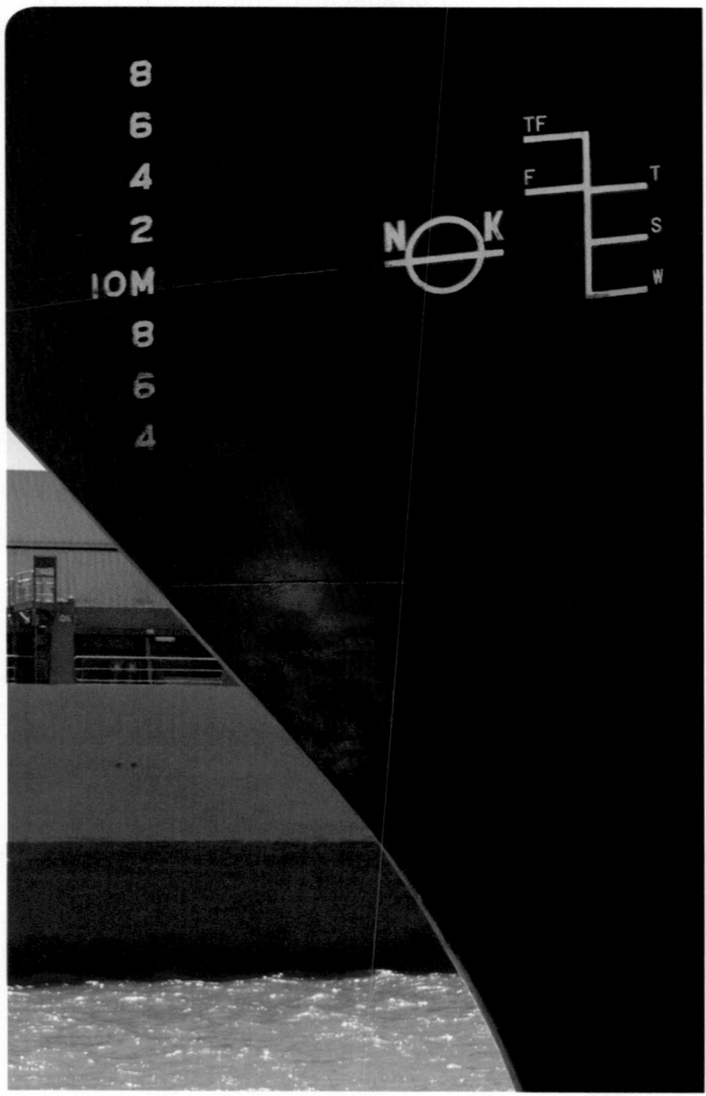

A Plimsoll line markings on the hull of a ship. TF stands for 'tropical fresh', F stands for 'fresh', T, S and W stand for Tropical, Summer and Winter seawater.

As ships are loaded they sink further down into the water. Ships have a 'Plimsoll line' marked on them to show how far into the water they can sink without becoming unsafe. The safe level depends on the density of the sea water. The density of sea water depends on its temperature and saltiness, so there are extra lines to show the safe level for different parts of the world and different times of the year.

Your task

To work out where to put the Plimsoll line markings on hulls, ship builders need to know the densities of different types of water. You are going to measure the densities of some different solids and liquids.

Method

Liquids

A Put an empty beaker on a balance, and set the balance to zero.

B Use a measuring cylinder to measure 50 cm^3 of a liquid and then pour it into the beaker. Write down the reading on the balance. This is the mass of 50 cm^3 of the liquid.

Solids

C Find the mass of the solid and write it down.

Diagram B shows how to find the volume of an irregular shape:

D Stand a displacement can on the bench with its spout over a bowl. Fill it with water until the water just starts to come out of the spout.

E Hold a measuring cylinder under the spout and carefully drop your object into the can. If your object floats, carefully push it down until all of it is under the water. Your finger should not be in the water.

F Stand the measuring cylinder on the bench and read the volume of water you have collected. This is the same as the volume of your object. Write it down.

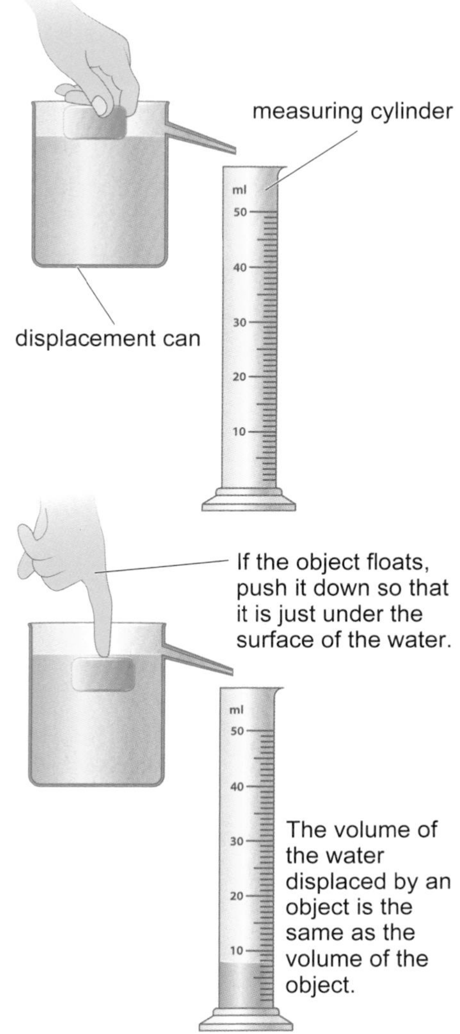

measuring cylinder
displacement can
ml
50
40
30
20
10
If the object floats, push it down so that it is just under the surface of the water.
The volume of the water displaced by an object is the same as the volume of the object.

B how to use a displacement can

Exam-style questions

1 Solids and liquids are both made up of tiny particles. Compare solids and liquids in terms of:

  a how the particles move *(2 marks)*

  b the spacing between the particles. *(1 mark)*

2 **a** Write down the equation for calculating the density of a substance. *(1 mark)*

  b Give suitable units for each of the quantities in the equation. *(1 mark)*

3 You need to find the differences in density between different concentrations of salty water.

  a List the apparatus you would need to carry out this investigation. *(3 marks)*

  b Write a method for your investigation. *(3 marks)*

  c State how you would make sure your investigation was a fair test. *(1 mark)*

4 A student found that the mass of 50 cm^3 (0.000 05 m^3) of cooking oil was 46 g. Calculate the density of the cooking oil. Give your answer in kg/m^3. *(3 marks)*

5 A large piece of wood is 2 m long, 50 cm wide and 2 cm thick. It has a mass of 12 kg. Calculate its density. *(3 marks)*

6 A student uses the method in steps E and F, and works out that the density of pure water is 980 kg/m^3. A textbook gives a value of 1000 kg/m^3.

  a Give a possible reason for the error in the student's result. *(1 mark)*

  b Describe a way of making the measurement of the density of fluids more accurate. *(1 mark)*

7 The values for the densities of substances given in reference books often state a temperature at which that density is correct. Explain why the density of a substance depends on its temperature. *(2 marks)*

SP14b Energy and changes of state

Specification reference: P14.6; P14.7; P14.10

Progression questions

- What effect does heating a substance have on the substance?
- How can we reduce unwanted energy transfers?
- What do specific heat capacity and specific latent heat mean?

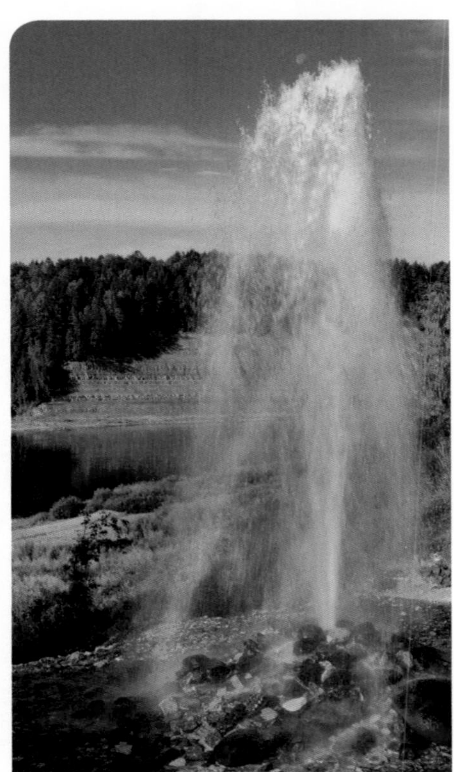

A A geyser shoots hot water into the air. Energy from hot rocks is transferred to the water, which gets hot enough to turn to steam and the water above this steam explodes out of the vent.

Energy transferred to a system by heating will be stored. The energy is stored in the movement of the particles that make up the substances in the system. Energy stored in this way is sometimes called **thermal energy**.

Temperature changes

When a solid stores more thermal energy, the vibrations of its particles increase. The speeds of the particles in liquids and gases increase when the liquid or gas is storing more energy. The **temperature** of a substance is a measure of the movement of the particles.

Temperature is not the same as thermal energy. For example, a kettle full of boiling water stores more energy than a cup full of water at the same temperature.

Did you know?

The temperature can affect the properties of a material. Flowers that are put into liquid nitrogen (about −196 °C) become very brittle and shatter if they are dropped.

B

To maintain a store of thermal energy, the amount of energy that is transferred to the surroundings by heating needs to be reduced. This can be done by surrounding the warm object with insulating materials such as wool, foam or bubble wrap. Insulation is also used in things like fridges, to stop energy from the surroundings being transferred to the inside of the fridge by heating.

The amount of thermal energy stored in something depends on:

- its temperature
- its mass
- the material it is made from.

The **specific heat capacity** of a material is the amount of energy it takes to increase the temperature of 1 kilogram of the substance by 1 °C.

1 What are the units used for measuring:

 **a** temperature

 **b** energy?

 **2** Explain why a saucepan full of soup stores more energy than a spoonful of soup at the same temperature.

Changes of state

When enough energy is transferred to a solid it reaches its melting point. If energy continues to be transferred, the temperature stops rising because the extra energy is used to overcome the forces between the particles and turn the solid into a liquid.

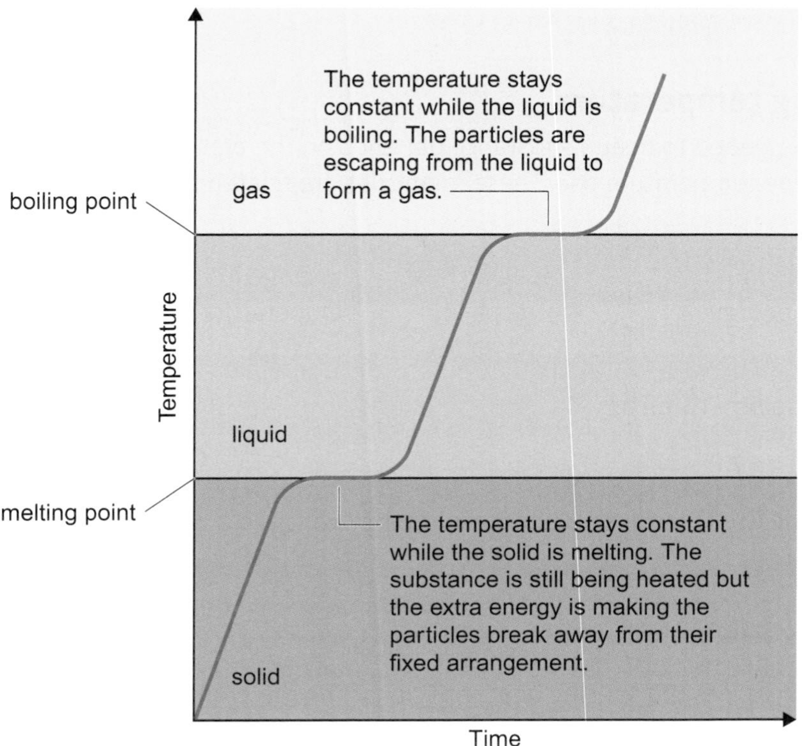

C This heating curve shows how the temperature of a substance changes as it absorbs energy.

The amount of energy it takes to make 1 kg of a substance change state is called the **specific latent heat**. It takes more energy to evaporate 1 kg of a substance than it does to melt 1 kg of the same substance, so these quantities are sometimes called the specific latent heats of melting and of evaporation. This energy is given out again when a substance freezes or condenses.

5 Sketch a labelled graph similar to graph C to show how the temperature of a substance changes as it cools from a gas into a liquid then into a solid.

6 Explain why the specific latent heat of melting for butanol is less than its specific latent heat of evaporation.

Exam-style question

a Sketch a graph to show how the temperature changes when a block of ice is heated from –5 °C until the melted water is at 10 °C. *(2 marks)*

b Explain the shape of your graph. *(3 marks)*

3 A cook heats 1 kg of water and 1 kg of cooking oil. The cooking oil reaches 50 °C before the water.

a Explain which substance has the higher specific heat capacity.

b What assumption did you make in your answer to part **a**?

4 Look at graph C. Explain why the temperature stops rising when the liquid is boiling.

Checkpoint

How confidently can you answer the Progression questions?

Strengthen

S1 A bag of food is put into a freezer. Explain the factors that affect how much energy is transferred from the food to the air in the freezer as its temperature falls.

S2 Water put into a freezer cools down and then starts to turn to ice. Explain why the temperature stops falling while the water is freezing.

Extend

E1 Explain this energy-saving advice: 'When you are boiling potatoes, turn the heat down as soon as the water starts to boil. Leaving the heat turned up high will not make your potatoes cook any faster!'

SP14c Energy calculations

Specification reference: P14.8; P14.9

Progression questions

- How is a change in thermal energy related to the mass, specific heat capacity and temperature difference?
- How can we calculate the energy needed to make a substance melt or evaporate?
- How can we calculate the energy released when a substance condenses or freezes?

A A masonry heater can have a mass of around 800 kg. It is designed to store energy from a fire and continue radiating energy into the house long after the fire has stopped burning.

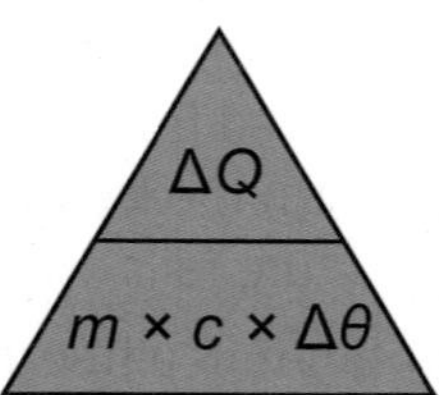

B This triangle can help you to rearrange the equation.

Changing temperature

The energy needed to heat a substance depends on the type of material, its mass and the temperature rise. These quantities are related by the following equation:

change in thermal energy (J) = mass (kg) × specific heat capacity (J/kg °C) × change in temperature (°C)

This can also be written as:

$$\Delta Q = m \times c \times \Delta\theta$$

where Δ (the Greek letter delta) represents the change in a quantity

Q represents energy

m represents mass

c represents the specific heat capacity of the material

θ (the Greek letter theta) represents the temperature.

Worked example W1

The specific heat capacity of water is 4182 J/kg °C. Calculate the energy needed to heat 2 kg of water from 10 °C to 60 °C.

$\Delta\theta = 60\,°C - 10\,°C$

$= 50\,°C$

$\Delta Q = 2\,kg \times 4182\,J/kg\,°C \times 50\,°C$

$= 418\,200\,J$

 **1** **a** Explain why the heater in photo A has a large mass.

 **b** Brick has a specific heat capacity of 840 J/kg °C. Calculate how much energy the 800 kg heater in photo A stores when the bricks are 40 °C above the air temperature in the room.

 **2** Calculate the temperature change when 25 000 J of energy is transferred to a 1 kg brick.

Changing state

Energy is needed to make a substance melt or evaporate. The amount of energy depends on the mass of the substance and on its specific latent heat. These quantities are related by the following equation:

$$\text{thermal energy needed for a change of state (J)} = \text{mass (kg)} \times \text{specific latent heat (J/kg)}$$

This can also be written as:

$Q = m \times L$

where Q represents energy

m represents mass

L represents the specific latent heat

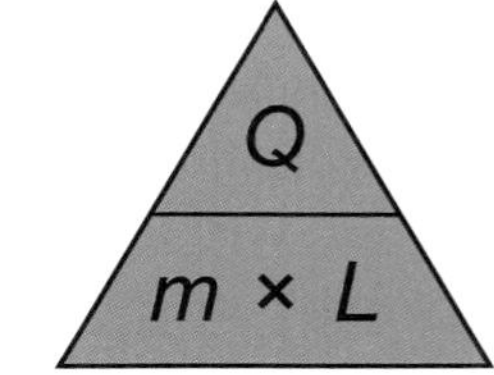

D This triangle can help you to rearrange the equation.

Worked example W2

The specific latent heat of evaporation for water is 2257 kJ/kg. How much energy does it take to evaporate 5 kg of water at 100 °C?

2257 kJ/kg = 2 257 000 J/kg

energy for change of state = mass × specific latent heat

energy = 5 kg × 2 257 000 J/kg

= 11 285 000 J

Did you know?

Getting a scald from steam hurts more than spilling the same mass of boiling water on your skin. This is because the steam releases the latent heat of evaporation when it condenses.

3 Look at Worked example W2. Explain why the question gives the temperature of the water.

4 The specific latent heat of melting of water is 334 kJ/kg. How much energy does it take to melt 10 kg of ice at 0 °C?

5 In an experiment, students transfer 100 000 J of energy to water at 100 °C. Calculate the mass of water that evaporates.

Exam-style question

A kettle is filled with 0.5 litres of water, which has a mass of 0.5 kg. The temperature of the water is 10 °C. Calculate how much energy is needed to bring the water to boiling point and then to evaporate all the water in the kettle. *(4 marks)*

C Energy is transferred from the stag's body to evaporate water from its lungs. This energy is transferred to the surroundings when the water vapour condenses again.

Checkpoint

How confidently can you answer the Progression questions?

Strengthen

S1 Calculate the temperature change when 8000 J of energy is transferred to 0.2 kg of water.

S2 10 000 J of energy is transferred to boiling water. Calculate the mass of water that evaporates.

Extend

E1 1 g of steam at 100 °C condenses on your hand and the water cools to your skin temperature (30 °C). Calculate how much energy this releases compared to spilling 1 g of boiling water onto your skin. (*Hint*: calculate the heat released during condensation and then as the water on your skin cools.)

SP14c Core practical – Investigating water

Specification reference: P14.11

Aim

Investigate the properties of water by determining the specific heat capacity of water and obtaining a temperature–time graph for melting ice.

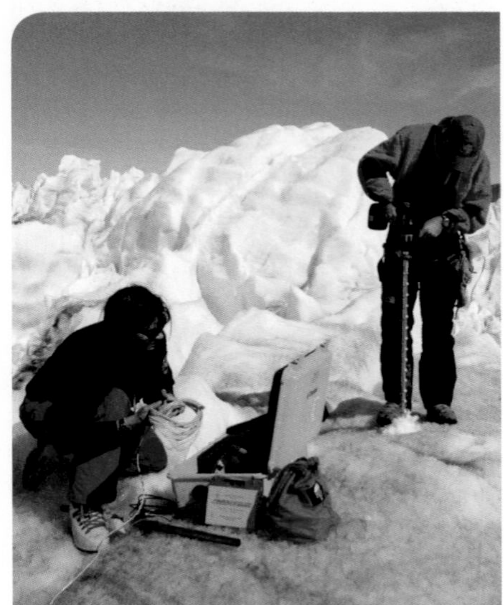

A glaciologists drilling for an ice core sample in a glacier

The world is getting gradually warmer due to the effects of climate change. One effect of this is that glaciers in various parts of the world are melting and getting smaller. Glaciers can also get smaller due to sublimation. Scientists monitoring the changes in glaciers need to understand the properties of water in its solid (ice), liquid and gas forms.

Your task

You will find out what happens to the temperature of ice as it melts, and how much energy is needed to increase the temperature of a certain mass of water by 1 °C.

Method

Melting ice

Wear eye protection.

A Put a boiling tube full of crushed ice into a Pyrex [or heatproof] beaker. Put a thermometer in the ice and note the temperature.

B Put the beaker onto a tripod and gauze. Pour hot water from a kettle into the beaker, and keep it warm using a Bunsen burner.

C Measure the temperature of the ice every minute and record your results in a table. Stop taking readings three minutes after all the ice has melted.

D Note the times at which the ice starts to melt and when it appears to be completely melted.

Specific heat capacity

E Put a polystyrene cup in a beaker onto a battery-powered balance and zero the balance. Then fill the cup almost to the top with water and write down the mass of the water. Carefully remove the cup from the balance.

F Put a thermometer in the water and support it as shown in photo B. Put a 12 V electric immersion heater into the water, making sure the heating element is completely below the water level. Connect the immersion heater to a joulemeter.

G Record the temperature of the water, and then switch the immersion heater on. Stir the water in the cup gently using the thermometer.

H After five minutes record the temperature of the water again and also write down the reading on the joulemeter.

B

Exam-style questions

1 Describe how the particles are arranged and held together in:

- **a** ice *(2 marks)*
- **b** liquid water. *(2 marks)*

2 Table C shows a set of results from the melting ice investigation. Plot a graph to present these results. Draw a line through the points. *(5 marks)*

3 Using ideas about kinetic theory:

- **a** explain why the graph has a level section in the middle *(2 marks)*
- **b** state why the temperature of the ice increases at a different rate to the temperature of the water. *(1 mark)*

4 What does the specific heat capacity of a substance tell us about the substance? *(1 mark)*

5 Look at photo B.

- **a** Describe the purpose of the glass beaker and the tripod. *(1 mark)*
- **b** Suggest why the water to be heated was put into a polystyrene cup instead of being put directly into the beaker. *(3 marks)*

6 Sam heated 250 g of water in a polystyrene cup. The joulemeter reading was 11 kJ and the temperature change was 10 °C.

- **a** Calculate the specific heat capacity of water using the following equation:

  specific heat capacity (J/kg °C) =

  $$\frac{\text{change in thermal energy (J)}}{\text{mass (kg)} \times \text{change in temperature (°C)}}$$

  (3 marks)

- **b** A textbook gives the specific heat capacity of water as 4181 J/kg °C. Suggest why you would expect Sam's result to be higher than this. *(3 marks)*
- **c** Suggest how the method described above could be improved to reduce these errors. *(1 mark)*

Time (min)	Temperature (°C)
0	−12
1	−8
2	−4
3	0
4	0
5	0
6	2
7	4
8	6

C

SP14d Gas temperature and pressure

Specification reference: P14.12; P14.13; P14.14; P14.15

Progression questions

- What causes gas pressure?
- How does the temperature of a gas affect its pressure?
- What is the difference between the kelvin and Celsius temperature scales?

The particles in a gas are far apart from each other and move around quickly. The temperature of a gas is a measure of the average **kinetic energy** of the particles in the gas. The faster the average speed of the particles, the higher the temperature. Heating a gas increases the kinetic energy of the particles, so they move faster and the temperature rises.

Did you know?

The Sun never shines on parts of these craters near the south pole of the Moon. The places in permanent shadow are the coldest recorded places in the Solar System and temperatures can get below −240 °C.

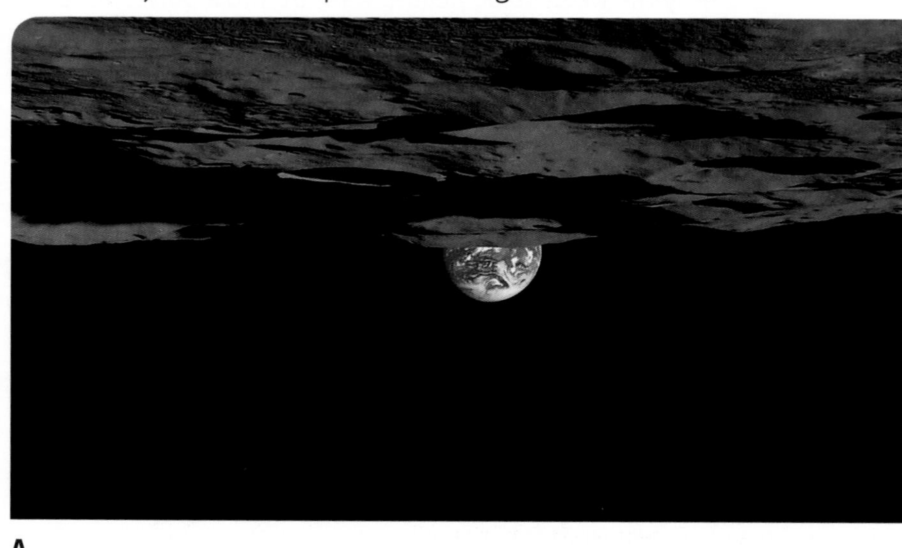

A

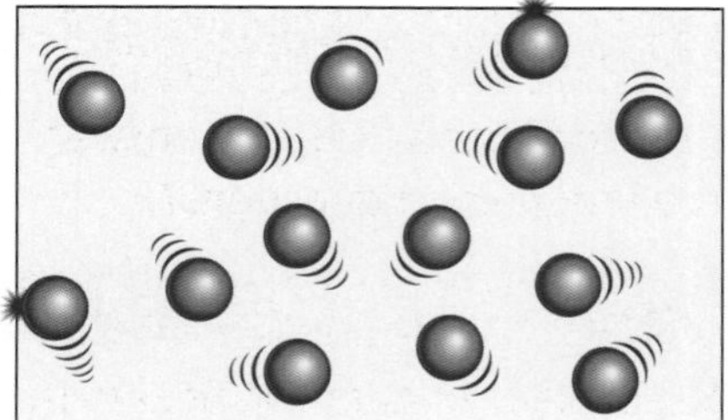

lower temperature

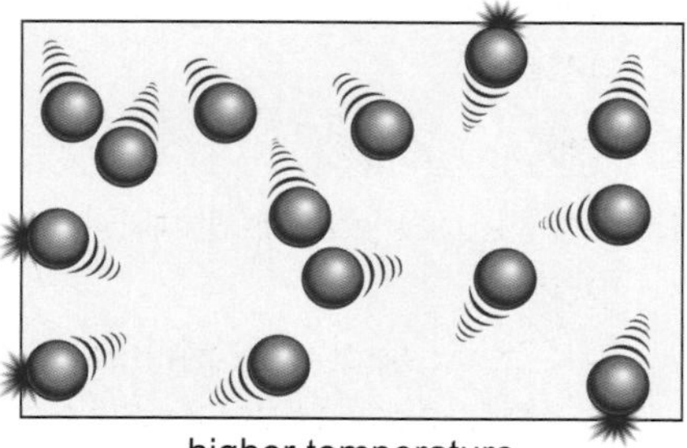

higher temperature

B The faster the average speed of the particles in a gas, the higher the temperature of the gas. The higher the temperature, the higher the pressure. This is because faster particles hit with more force and more often.

Particles and pressure

The **pressure** of a gas is due to forces on the walls of a container caused by the moving particles hitting the walls. The faster the particles are moving, the more frequent the collisions will be and the more force they will exert when they collide. Increasing the temperature of a gas increases the speed of the particles, so it also increases the pressure of the gas. For a fixed mass of gas in a fixed volume, the pressure increases when the temperature increases. The units for pressure are **pascals** (**Pa**), where 1 Pa = 1 N/m^2.

1 What causes pressure in a gas?

2 Why does increasing the temperature of a fixed volume of gas increase its pressure? Give two reasons.

Absolute zero

Graph B shows how the pressure of a fixed volume of gas changes with temperature. The measurements cannot continue below the boiling point of the substance, as the gas will condense to form a liquid. However, the same graph is obtained for all gases, and if the lines are extended to colder temperatures they meet the horizontal axis at −273 °C. The temperature of −273 °C is called **absolute zero**. If a gas could be made this cold its pressure would be zero and the particles would not be moving.

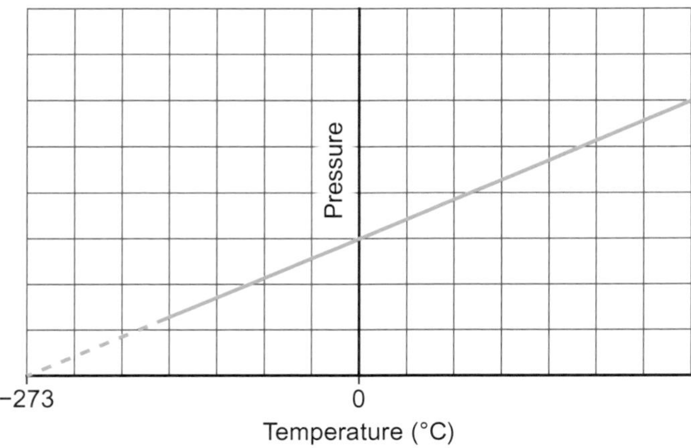

C the relationship between the pressure of a fixed volume of gas and the temperature of the gas (in degrees Celsius)

The **Kelvin temperature scale** measures temperatures relative to absolute zero. The units are **kelvin** (**K**), and 1 K is the same temperature interval as 1 °C. Absolute zero is 0 K on this scale.

To convert from kelvin to degrees Celsius, subtract 273.

To convert from degrees Celsius to kelvin, add 273.

The average kinetic energy of the particles in a gas is directly proportional to the kelvin temperature of the gas.

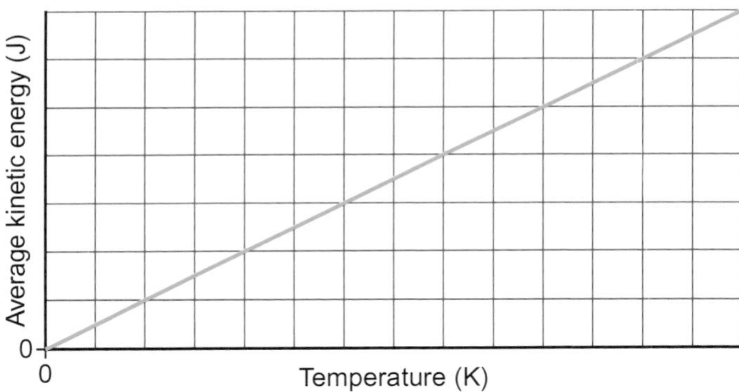

D the relationship between the average kinetic energy of gas particles and the kelvin temperature of the gas

Worked example

What is the boiling point of water in kelvin?

boiling point = 100 °C + 273 = 373 K

3 Convert the following temperatures to the Celsius scale.

a 500 K **b** 100 K

4 Convert the following temperatures to the kelvin scale.

a 500 °C **b** −100 °C

5 The temperature of a gas is increased from 200 K to 400 K. Explain what happens to the average kinetic energy of the particles.

Exam-style question

Look at graph D. Explain the significance of the point where the line crosses the horizontal axis. *(2 marks)*

Checkpoint

How confidently can you answer the Progression questions?

Strengthen

S1 Explain what absolute zero is. Use the terms kinetic energy and absolute zero in your answer.

Extend

E1 Look at the graphs on this page.

a Describe the relationship between the pressure of a gas and its temperature.

b Explain what the relationship is between the pressure of a gas and the average kinetic energy of its particles.

SP14e Gas pressure and volume

Specification reference: P14.16P; P14.17P; P14.18P; P14.19P; **H** P14.20P

Progression questions

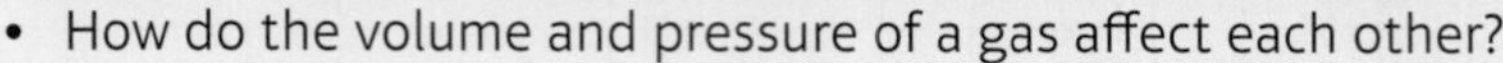

- How do the volume and pressure of a gas affect each other?
- How can you calculate the volume or pressure of a gas at a fixed temperature?
- **H** Why does doing work on a gas increase its temperature?

A This cylinder contains compressed carbon dioxide. It is being used to inflate the tyre instead of a bicycle pump.

When you pump up a bicycle tyre, pushing the handle on the pump increases the pressure on the air inside it and forces it into a smaller volume. This increases the pressure of the air inside the pump and the tyre. The higher the pressure of air inside the tyre, the harder and less squashy it feels.

The effect of gas particles hitting a surface causes a net (overall) force on the surface. The force acts at right angles to the surface and we detect this as the **gas pressure**.

Inside a container, the gas pressure acts at right angles on all its walls. If the same number of particles are forced into a smaller volume at the same temperature, they hit the walls more often and so the force on the walls increases.

1 **a** What will happen to the pressure of a gas if you put more particles into the same volume?

b Explain your answer.

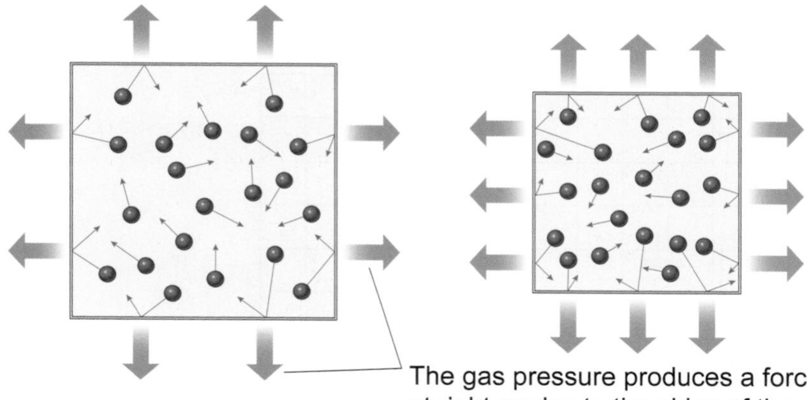

The gas pressure produces a force at right angles to the sides of the container.

B Decreasing the volume of a gas at a fixed temperature increases the pressure.

2 **Explain why decreasing the volume of a gas increases its pressure.**

3 **Explain why keeping the temperature the same is important if you are comparing the pressure of different volumes of gas.**

The volume and pressure of a fixed mass of gas at a constant temperature are related by this equation.

$$P_1 \times V_1 = P_2 \times V_2$$

In words, this means that the product of the starting pressure (P_1) and the starting volume (V_1) is equal to the product of the final pressure (P_2) and the final volume (V_2). The standard units for pressure and volume are pascals (Pa) and cubic metres (m^3), but you can use any units in this equation as long as you use the same one for both values of pressure and the same one for both values of volume.

Did you know?

Paintball guns and air guns work using compressed air. When the gun is fired, the high pressure air in a cylinder is released all at once, and drives the paintball or pellet out of the gun.

C

Worked example

A cylinder of oxygen contains 0.3 m^3 of gas at a pressure of 14 000 000 Pa. All the oxygen is released from the cylinder without changing the temperature. Calculate the volume of the gas at atmospheric pressure (approximately 100 000 Pa).

P_1 = 14 000 000 Pa, P_2 = 100 000 Pa, V_1 = 0.3 m^3

You need to use this version of the equation.

$$V_2 = \frac{P_1 \times V_1}{P_2}$$

$$= \frac{0.3\ m^3 \times 14\,000\,000\ Pa}{100\,000\ Pa}$$

$$= 42\ m^3$$

4 At a pressure of 230 000 000 Pa a gas has a volume of 0.15 m^3. What would its volume be at atmospheric pressure (100 000 Pa)?

H

When you use a bicycle pump it gets warm. Each time you push the pump handle, the force is transferring energy to the gas inside the pump. The energy transferred by a force is called the **work done**.

As the piston of the bicycle pump moves, the speed of any particles inside the pump will increase when they bounce off it. The average speed of the particles inside the pump will increase, and we detect this as an increase in temperature.

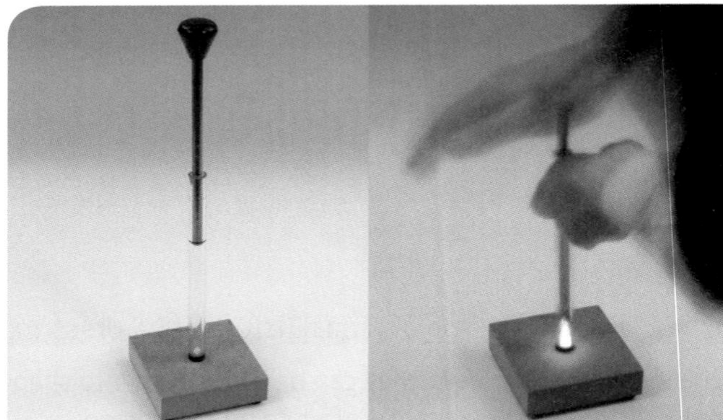

D A fire piston is an ancient device used to start fires. Hitting the top compresses the air inside it very quickly, and the air is hot enough to make a piece of tinder inside it start to burn.

5 Explain why a cylinder of compressed air will feel warm when it has just been filled.

Exam-style question

A fixed mass of gas at constant temperature is reduced in pressure. What happens to its volume? *(1 mark)*

Checkpoint

How confidently can you answer the Progression questions?

Strengthen

S1 Draw a balloon. Add arrows to the balloon to show the force on the balloon skin from the air inside it.

S2 A bicycle pump compresses 0.000 025 m^3 of air at atmospheric pressure (100 000 Pa) to a volume of 0.000 010 m^3. What is the pressure of the compressed gas? (Use the formula $P_2 = \frac{P_1 \times V_1}{V_2}$)

Extend

E1 Hospitals use a lot of oxygen and other gases. Explain why these gases are stored above atmospheric pressure and how engineers can calculate the volume of gas at atmospheric pressure that can be stored in a cylinder.

E2 **H** When you fire a paintball gun, the temperature of the compressed air left in the gun goes down. Explain why this happens in terms of work done and of particle movement.

SP15a Bending and stretching

Specification reference: P15.1; P15.2; P15.5

Progression questions

- How do forces cause objects to change shape?
- What is the difference between elastic and inelastic distortion?
- What is the relationship between force and extension when an object is deformed?

A This pole bends when forces are applied to both ends of it.

Forces can deform or change the shape of an object. It requires more than one force to stretch, bend or compress an object. For example, the weight of the pole vaulter in photo A is making the pole bend. But this only happens because there are also forces on the other end of the pole holding that end still.

The pole in photo A is **elastic**. This means that it will return to its original shape when the forces are removed. Some materials or objects are **inelastic**, which means that they will keep their new shape after the forces are removed.

B Metals are described as malleable, which means they can be hammered into shape. This hot iron is inelastic because it keeps its new shape after being hammered.

Did you know?

The tendons in our bodies that connect muscles to bones are springy. Tendons in our legs help us to save energy when we are running. The tendons store elastic potential energy as they stretch and this energy is transferred later in the stride to help to push us forwards.

Some objects are elastic when the forces are small but behave inelastically if the forces are too big. Metals can be made into springs that behave elastically, but if the forces used to stretch them are too big they become permanently deformed.

1 Suggest two materials that always deform inelastically.

2 A diving board usually bends when a diver stands on the end.

a Is the deformation elastic or inelastic? Explain your answer.

b Describe the two forces that cause the board to bend.

Force and extension

The **extension** of a spring (or other object) is the change in length when forces are applied.

For a metal spring, there is usually a **linear relationship** between the force and the length. This means a graph of force against length will be a straight line. If force is plotted against *extension* the line passes through the origin, so the extension is **directly proportional** to the force. This means that the extension doubles if the force doubles. However, the relationship becomes **non-linear** if the spring is stretched too far. Other objects, such as rubber bands, have non-linear relationships between force and extension.

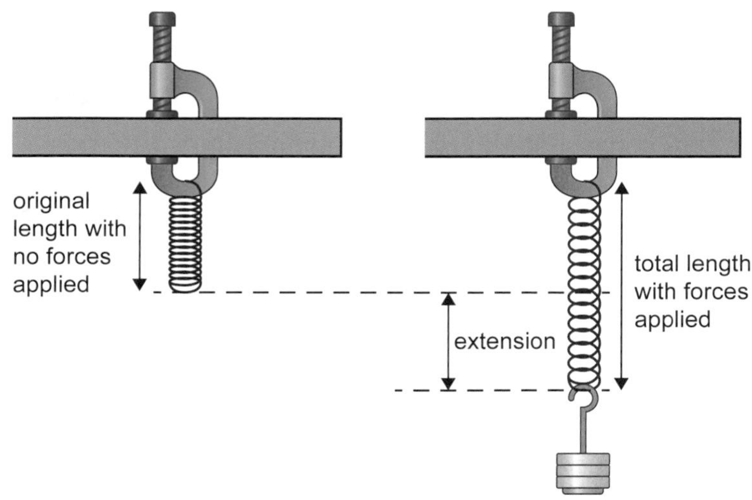

C The *extension* of a spring is not the same as its length.

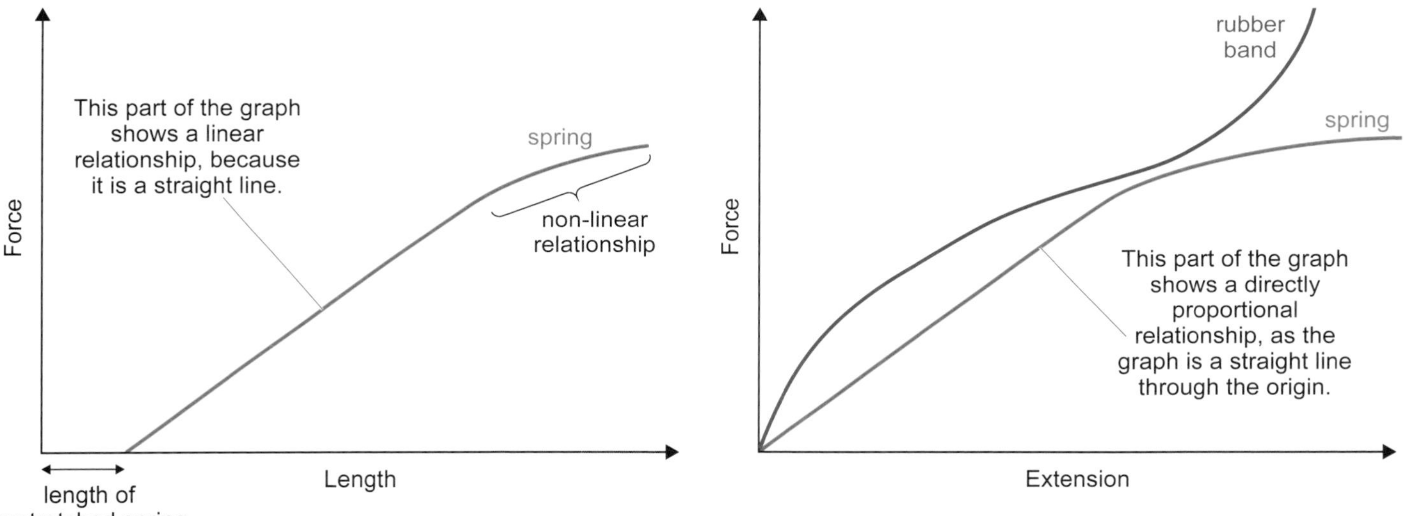

D force and extension relationships for springs and rubber bands

3 Look at diagram D. A spring has an extension of 4 cm when a 2 N weight is hung from it.

a What will the extension be when the weight is 6 N?

b Explain how you worked out your answer, including any assumptions you have made.

4 Look at diagram D. Describe how the force needed to produce greater and greater extensions changes for:

a a spring **b** a rubber band.

Exam-style question

Springs and rubber bands both stretch when forces are applied to them. Compare and contrast the way these items stretch. *(3 marks)*

Checkpoint

How confidently can you answer the Progression questions?

Strengthen

S1 Write glossary entries for all the key terms on these pages.

Extend

E1 A piece of material is 10 cm long. As five weights of 1 N each are added, its length becomes: 10.5, 11.5, 13.0, 14.5, 15.5 cm. Explain how this data shows that the material is probably a rubber band.

SP15b Extension and energy transfers

Specification reference: P15.3; P15.4

Progression questions

- What is the spring constant of a spring?
- What is the equation that relates the force and extension of a spring?
- How do we calculate the work done in stretching a spring?

Did you know?

Very small springs are used in many things, including medical devices such as hearing aids and in miniature cameras. Some of the smallest springs have a diameter of 45 µm (1 µm = 1×10^{-6} m or 0.000 001 m or 0.001 mm).

Graph A shows the force and extension for two different springs. Spring X needs a bigger force than spring Y to produce the same extension. The **spring constant** for a spring is the force needed to produce an extension of 1 metre. Spring X has a higher spring constant than spring Y.

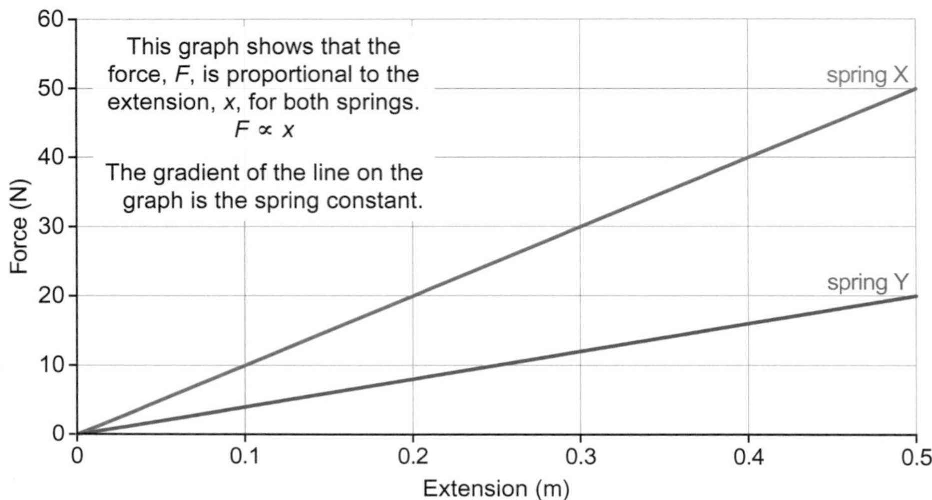

A Force-extension graph for two springs. The ∝ symbol means 'directly proportional to'.

B The spring used here needs a large spring constant.

The spring used in a pogo stick (shown in photo B) needs to be much stiffer than a spring used in a force meter to measure very small forces. It needs a larger spring constant.

The force, extension and spring constant are related by this equation:

force (N) = spring constant (N/m) × extension (m)

This can be written as: $F = k \times x$

where F represents force
k represents the spring constant
x represents extension.

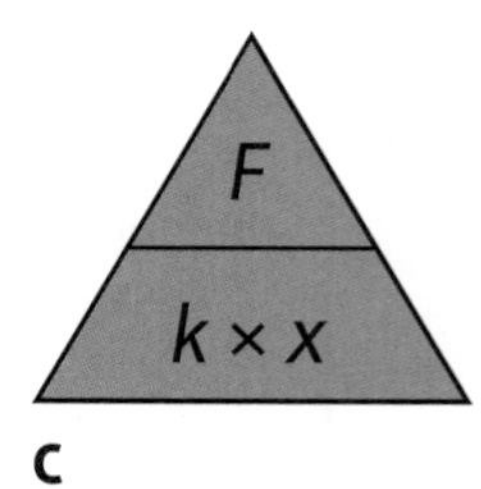

C

Worked example W1

Calculate the spring constant for spring X in graph A.

$k = \frac{F}{x}$

$= \frac{50\,\text{N}}{0.5\,\text{m}}$

$= 100\,\text{N/m}$

You can choose any point on the graph to read off a force and extension.

1 A spring has a spring constant of 400 N/m. Calculate the force needed to make it extend by:

a 0.8 m **b** 10 cm.

2 Show that the spring constant of spring Y in graph A is 40 N/m.

Energy transferred in stretching

Energy is needed to stretch a spring. The energy transferred when a force moves through a distance is calculated from the force multiplied by the distance. This is called the **work done** (see *SP8a Work and power*). However, when you stretch a spring the force needed increases as the extension increases, so the equation is a little more complicated.

The energy transferred in stretching a spring is calculated using this equation:

$$\underset{(\text{J})}{\text{energy transferred in stretching}} = \frac{1}{2} \times \underset{(\text{N/m})}{\text{spring constant}} \times \underset{(\text{m})^2}{(\text{extension})^2}$$

This can be written as:

$E = \frac{1}{2} \times k \times x^2$

where E represents energy transferred
k represents the spring constant
x represents extension.

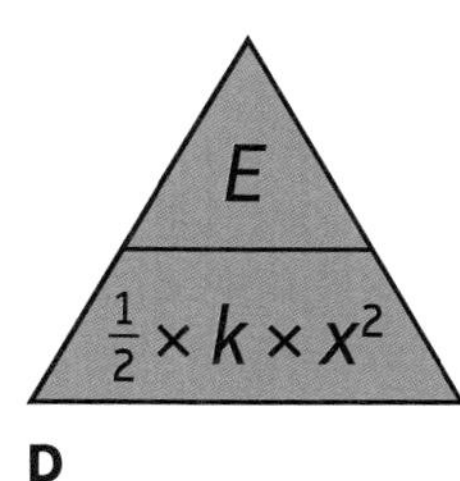

D

Worked example W2

Calculate the energy transferred when a spring with a spring constant of 100 N/m is stretched by 0.2 m.

$E = \frac{1}{2} \times k \times x^2$

$= \frac{1}{2} \times 100\text{ N/m} \times (0.2\text{ m})^2$

$= 2\text{ J}$

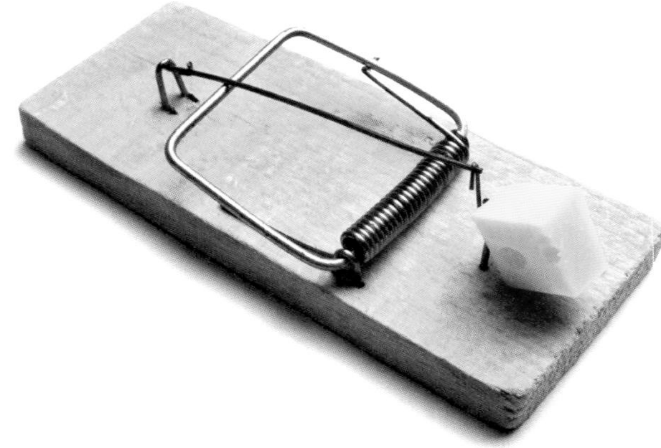

E Springs can also store energy when they are twisted. In this mousetrap, the spring is twisted to set the trap. When a mouse touches the pin the spring is released and the stored elastic potential energy is transferred to kinetic energy stored in the moving bar.

 **3** Springs X and Y in graph A are stretched by the same amount. Explain which one needs the greater transfer of energy.

 **4** Calculate the energy transferred when a spring with a spring constant of 200 N/m is stretched by 50 cm.

 **5** It takes 5 J of energy to stretch a spring by 10 cm. Calculate the spring constant.

Exam-style question

Explain how a designer would use the spring constant to choose a type of spring to use in a force meter. *(2 marks)*

Checkpoint

How confidently can you answer the Progression questions?

Strengthen

S1 A spring has a spring constant of 400 N/m. A force of 20 N is applied to it. Calculate the extension.

S2 The same spring has an extension of 5 cm. Calculate the energy transferred to produce this extension.

Extend

E1 A spring stretches by 10 cm when a 50 N force is applied to it.

- **a** Calculate the spring constant of the spring.
- **b** Calculate how far the spring must be stretched to transfer 10 J of energy.

SP15b Core practical – Investigating springs

Specification reference: P15.6

Aim

Investigate the extension and work done when applying forces to a spring.

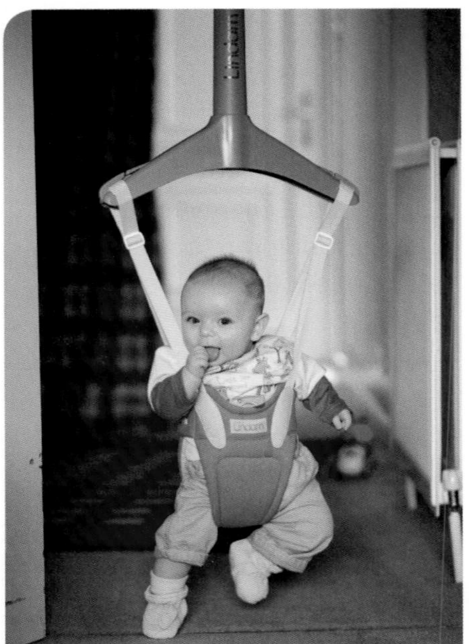

A A spring in the top of this door bouncer lets the baby bob up and down.

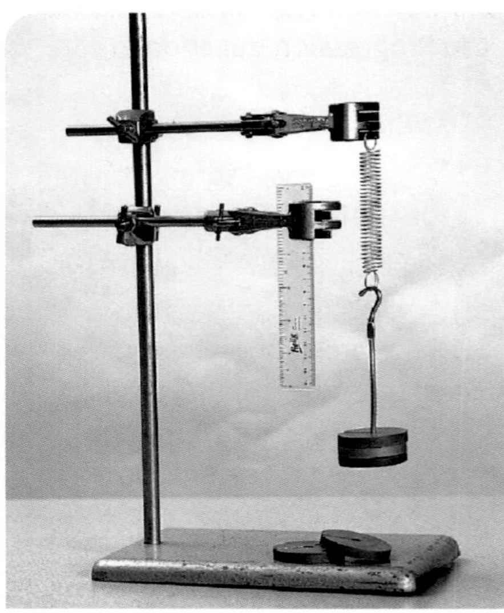

B

Photo A shows a baby in a door bouncer. The spring used in the door bouncer must be stretchy enough to allow the baby to bounce up and down, but not so stretchy that the weight of the baby stretches it too far and the baby ends up on the floor. Designers need to know the characteristics of springs so they can choose the type that best suits their purpose.

Your task

Investigate the force needed to stretch springs and calculate how much work is done when a spring is stretched.

The work done to stretch a spring is calculated using the following equation:

$$\underset{(\text{J})}{\text{energy transferred in stretching}} = \frac{1}{2} \times \underset{(\text{N/m})}{\text{spring constant}} \times \underset{(\text{m})^2}{(\text{extension})^2}$$

Method

Wear eye protection while you carry out this investigation.

A Set up the apparatus as shown in photo B. The zero on the ruler should be level with the bottom of the unstretched spring.

B Measure the length of the spring with no weights hanging on it and write it down.

C Hang a 1 newton weight on the spring. Record the extension of the spring (the length shown on the ruler).

D Repeat step C until you have found the extension of the spring with 10 different masses.

E Repeat steps A to D for a different spring.

F Use your results to calculate the spring constant for each spring.

Exam-style questions

1 **a** State what is meant by the spring constant of a spring. *(1 mark)*

b State the equation that can be used to find the spring constant of a spring. *(1 mark)*

c Give the units that should be used in the equation. *(1 mark)*

2 **a** When a force moves an object, the work done in moving the object can be calculated using the equation: work done = force × distance. Explain why this equation cannot be used to calculate the work done in stretching a spring. *(2 marks)*

b A spring has an extension of 0.5 m when there is a force of 20 N pulling on it. Calculate the spring constant, and then calculate the energy transferred in stretching this spring. *(4 marks)*

3 Look at photo A.

a Describe what happens to the spring in the door bouncer when the baby is first put into it. *(2 marks)*

b The baby's father pulls the baby down by 20 cm, and then releases her. Describe what happens to the energy transferred by the father pulling the baby downwards. *(2 marks)*

4 Write a list of the apparatus you need to carry out an investigation like the one described in the method. *(1 mark)*

5 Table C shows the results of one group's investigation.

a Draw a table of the results that shows the force and the *extension*, in metres. *(2 marks)*

b Plot a graph of force against extension to show their results. Draw the lines for both springs on the same axes. *(5 marks)*

c The readings for springs X and Y were taken by two different students. Use your graph to suggest which student has taken the more accurate readings for the length of the spring. *(1 mark)*

d Read a pair of values for force and extension from your graph for spring X and use these values to find the spring constant. *(3 marks)*

e Give a reason why you should use values read from the graph to find the spring constant, rather than taking data points from the table. *(2 marks)*

6 **a** Compare the two springs in terms of original length and how easily they stretch. *(2 marks)*

b Explain which spring stores more energy when it has an extension of 20 cm. *(2 marks)*

7 A student carried out the investigation described, adding 10 N to the spring between each measurement. The spring stretched by only 1 mm with 10 N hanging on it.

a Describe how this might affect the accuracy of the results. *(2 marks)*

b Explain how the method could be modified to improve the accuracy of the student's results. *(2 marks)*

Weight (N)	Length (cm)	
	spring X	spring Y
0	6.0	4.0
1	9.4	8.8
2	13.0	14.2
3	16.2	19.2
4	19.1	23.1
5	22.9	28.8
6	26.5	34.3
7	29.5	39.5
8	33.0	45.5
9	36.0	48.8
10	39.5	55.0

C results for a spring investigation

SP15c Pressure in fluids

Specification reference P15.7P; P15.8P; P15.9P; P15.10P; P15.11P; P15.12P

Progression questions

- How is pressure related to force and area?
- How can you calculate force, pressure or area?
- How does the pressure in a liquid change with depth and density?

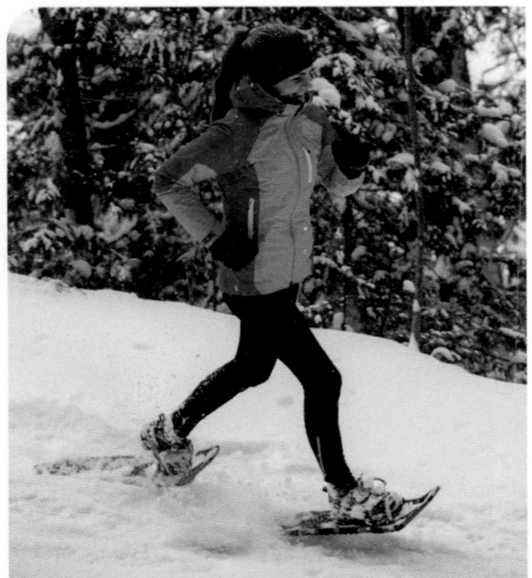

A Snowshoes reduce the pressure on the snow.

The person in photo A is wearing snowshoes to stop them sinking into the snow. The snowshoes spread their weight over a larger area so the **pressure** on the snow is less. Pressure is measured in **pascals** (**Pa**), 1 Pa = 1 N/m^2.

Pressure is a measure of the force on a unit of surface area, where the force is **normal** (at right angles) to the surface. Pressure, force and area are related by this equation.

$$\text{pressure (Pa)} = \frac{\text{force normal to the surface (N)}}{\text{area of surface (m}^2\text{)}}$$

Worked example

A woman's feet have a total area of 0.03 m^2 and she exerts a pressure of 20 000 Pa on the ground. Calculate her weight.

$$\text{pressure} = \frac{\text{force}}{\text{area}}$$

$$\text{force} = \text{pressure} \times \text{area}$$

$$= 20\,000 \text{ Pa} \times 0.03 \text{ m}^2$$

$$= 600 \text{ N}$$

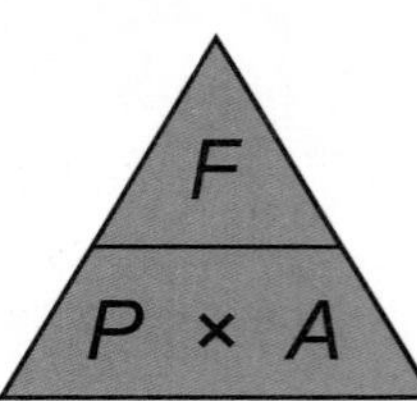

B Use this triangle to help you to rearrange the pressure equation.

1 A pair of snowshoes has a total area of 0.16 m^2. Calculate the pressure under a man with a weight of 900 N when he is wearing these snowshoes.

Did you know?

The deepest part of the ocean is the Challenger Deep. The bottom is over 10 900 m below the surface, where the pressure is about 1.1×10^8 Pa. Film director James Cameron went there in 2012, in the submersible *Deepsea Challenger*.

Pressure can also be exerted by **fluids** (liquids and gases). There is pressure on you all the time from the air around you. At sea level, this pressure from the air, or **atmospheric pressure**, is about 100 000 Pa. Pressure caused by fluids always acts normal to any surface.

The pressure exerted by a fluid depends on the depth of the fluid. The deeper you are, the more weight of fluid there is above you to exert pressure. When you are at sea level, you are at the bottom of the atmosphere and atmospheric pressure is at its maximum. If you go up a mountain, the air pressure decreases because there is less air above you.

2 Explain what happens to the air pressure if you go down a deep mine.

3 The area of your hand is about 0.01 m^2.

a Calculate the force acting downwards on your hand due to atmospheric pressure.

b Explain why you don't notice this force.

4 Look at photo C. Explain why the bag of crisps looks larger than usual.

Pressure also depends on the **density** of the fluid. Atmospheric pressure is due to the whole depth of the atmosphere above you. Water is over 800 times denser than air at sea level. If you dive 10 metres underwater, you will feel double the pressure that you felt at the surface. The total pressure on a diver is the pressure due to both the water and the air above.

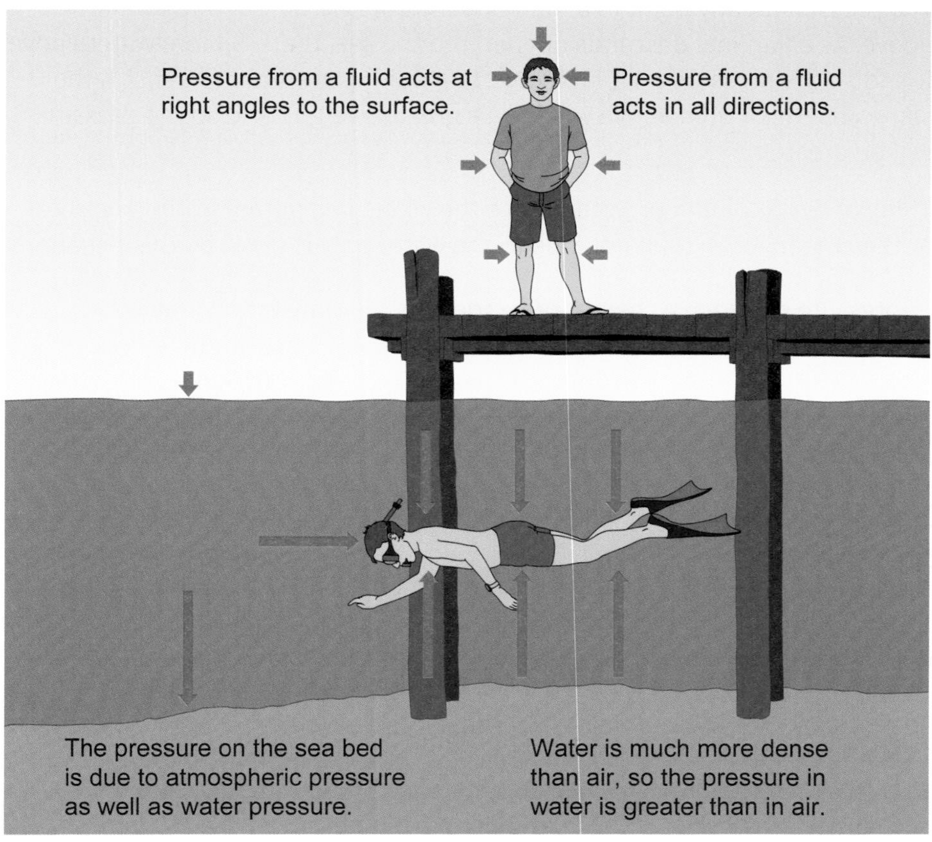

Pressure from a fluid acts at right angles to the surface.
Pressure from a fluid acts in all directions.
The pressure on the sea bed is due to atmospheric pressure as well as water pressure.
Water is much more dense than air, so the pressure in water is greater than in air.

D pressure due to air and water

5 A diver holds a ball below the surface of the sea.

a State the direction(s) in which pressure acts on the ball.

b State two things that cause pressure on the ball.

6 A diver is 1 m below the surface, where the pressure from the water is 10 000 Pa. Explain what the total pressure on the diver will be.

7 Sea water has a greater density than fresh water. Explain how this would affect the pressure on a fish at 50 m below the surface.

Exam-style question

Compare and contrast the pressure caused by the atmosphere and by the sea. *(3 marks)*

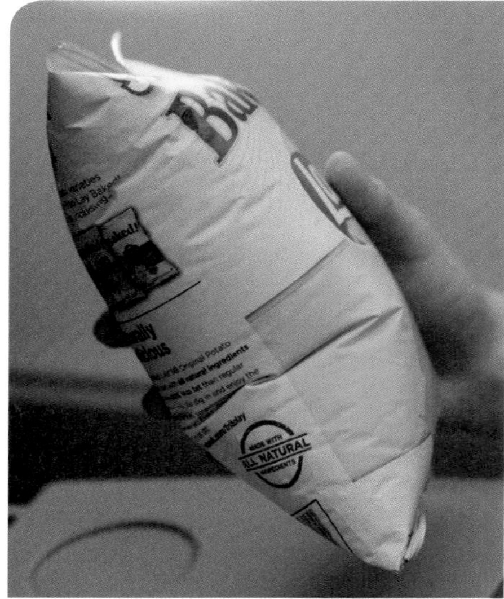

C This bag of crisps was sealed at sea level.

Checkpoint

How confidently can you answer the Progression questions?

Strengthen

S1 Mountaineers who need to walk on ice use crampons strapped to their boots. There are usually 10 tiny points on each crampon. Explain how wearing crampons helps their feet to grip the ice.

S2 Explain how atmospheric pressure changes as a mountaineer goes up a mountain.

Extend

E1 The surface of Lake Titicaca in South America is nearly 4000 m above sea level. A diver is 1 metre below the surface of the water. State what is causing pressure on her body, and explain how this pressure would be different if she were diving in the sea.

SP15d Pressure and upthrust

Specification reference: H P15.13P; H P15.14P; H P15.15P; H P15.16P; H P15.17P

Progression questions

- H Why does pressure in a liquid depend on density and depth?
- H How can you calculate the pressure in a liquid?
- H Why do some things float and some things sink?

H

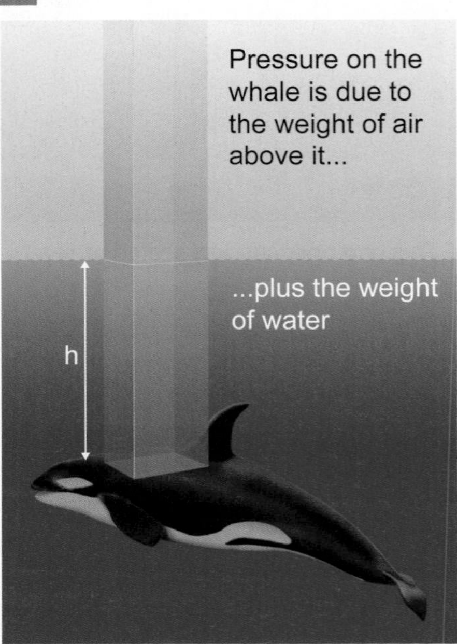
Pressure on the whale is due to the weight of air above it...
...plus the weight of water
h

A The pressure in water is due to the weight of fluid (water and air) above.

The pressure at any point in a fluid depends on the weight of the fluid above. As an animal descends deeper into the sea, there is more water above it and so the pressure on it increases. In a fluid with a greater density, the same volume of fluid above will have a greater weight, and so the pressure will be greater.

Think about a column of water above the whale in diagram A. The pressure at the bottom of this column due to the water only is given by this equation:

$$\underset{(\text{Pa})}{\text{pressure due to a column of liquid}} = \underset{(\text{m})}{\text{height of column}} \times \underset{(\text{kg/m}^3)}{\text{density of liquid}} \times \underset{(\text{N/kg})}{\text{gravitational field strength}}$$

or

$$P = h \times \rho \times g$$

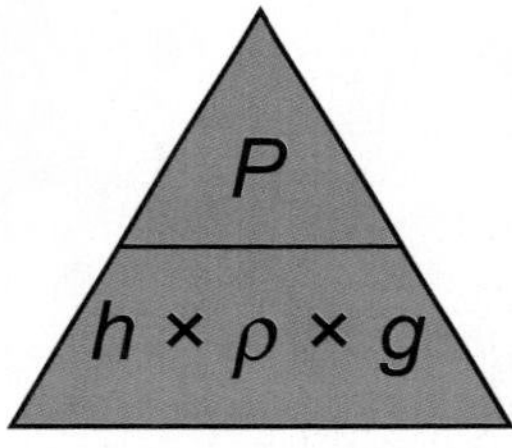
P
h × ρ × g

B ρ is the Greek letter rho, and represents density.

atmospheric pressure	100 000 Pa
density of sea water	1030 kg/m³
density of fresh water	1000 kg/m³
gravitational field strength	10 N/kg

C standard values of pressure, density and gravitational field strength

1 Calculate the pressure due to water on a whale at a depth of 2992 metres in the sea.

2 Calculate the depth of the whale if the water pressure on it is 500 000 Pa.

Did you know?

The deepest recorded dive for a mammal was by a Cuvier's beaked whale. A satellite tag on the whale showed it had reached a depth of 2992 metres.

Upthrust and floating

Objects in a fluid have a force called **upthrust** acting on them. This force is due to the difference in pressure above and below the object.

H

Worked example

Look at photo D. There is an average of 0.75 m depth between the top and bottom surfaces of the shark.

a Calculate the difference in pressure between the top and bottom surfaces.

pressure difference = depth difference × ρ × g

= 0.75 m × 1030 kg/m^3 × 10 N/kg

= 7725 N/m^2

b This pressure difference will produce a net upthrust. Calculate the size of this force. The horizontal area of the bottom of the shark is 8 m^2.

force = pressure difference × area = 7725 N/m^2 × 8 m^2 = 61 800 N

D The pressure on the top of this great white shark is less than the pressure below it.

The upthrust calculated in the worked example is also the weight of the water **displaced** (pushed out of the way) by the whale.

Diagram E shows blocks of different types of wood in water. For three of the blocks, the upthrust is equal to the weight of the block and so they are floating. The heavier blocks are floating deeper in the water because they need a greater pressure beneath them to balance their weight. The upthrust is equal to the weight of water displaced.

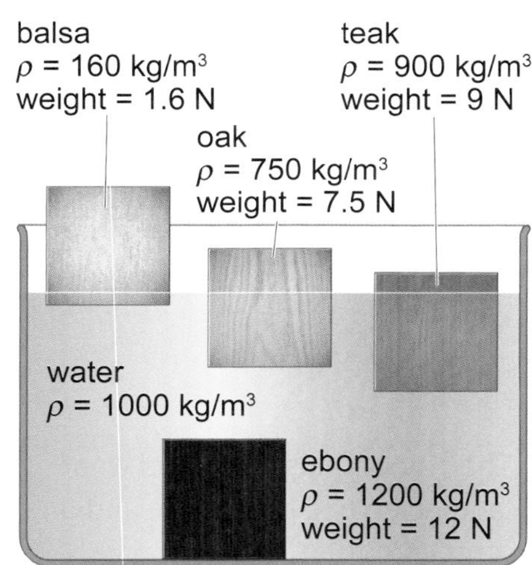

E Each block is a cube with sides 0.1 m long, and a volume of 0.001 m^3.

 **3** What is the upthrust on the oak block in diagram E? Explain your answer.

 **4** **a** Calculate the upwards force on the ebony block in diagram E using the method in the worked example. Use the density of fresh water in your calculation.

 **b** Explain why the ebony block does not float.

 **5** An object will float if its density is less than the density of the fluid. Explain why this is so.

Exam-style question

Explain why a submarine designed to dive to 5000 m has to be stronger than a submarine that operates at 500 m. *(2 marks)*

Checkpoint

How confidently can you answer the Progression questions?

Strengthen

S1 Explain why the pressure on a fish 5 m below the surface of a fresh-water lake is less than the pressure on a fish 30 m below the surface of the sea.

S2 Explain why a hot air balloon can float in the air.

Extend

E1 Dry cleaning fluid has a density of 1622 kg/m^3. Explain how diagram E would be different if the container held dry cleaning fluid instead of water. Include calculations to illustrate your answer.

Electric storage heaters

Electric storage heaters use cheap night-time electricity to raise the temperature of a storage material. During the day, the stored energy is transferred to the surroundings to keep homes warm.

Water has a specific heat capacity of 4182 J/kg °C. Brick has a specific heat capacity of 900 J/kg °C, and is more dense than water.

Assess the suitability of these materials to store energy in a storage heater. **(6 marks)**

Student answer

Specific heat capacity is the amount of energy it takes to raise the temperature of 1 kg of a substance by 1 °C [1]. *It will take more energy to heat a storage heater full of water than one with the same mass of bricks, but this also means that the water will store more energy. So there will be more energy released when the water cools down again than if there is brick inside* [2]. *But if you damage a heater with water in it, the water will spill on to the floor* [3].

[1] This shows that the student understands what specific heat capacity means.

[2] This is an advantage of using water inside the heater.

[3] This is a disadvantage of using water inside the heater.

Verdict

This is an acceptable answer. The student has explained what specific heat capacity means, and given one advantage and one disadvantage of using water compared to using brick.

The answer could be improved by linking some scientific ideas. For example, stating that the higher density of brick means that the same volume storage heater can hold a higher mass of brick, and how this will affect the energy stored. The command word for the question is 'assess', which means that the answer should include a conclusion about which material is best and which facts are important.

Exam tip

When a question asks you to assess something, you need to consider all the facts and explain which facts are the most important. Your answer also needs to make a judgement or reach a conclusion – in this case, which material is best.

Equations

Equations in the left hand column are ones you may be asked to *recall and apply* in your exam.

You do not need to recall the equations in the right hand column, but you should be able to *select and apply* them in an exam.

Equations for Higher tier only are marked with the Higher icon.

Recall and apply	Select and apply
Unit SP1 Motion distance travelled = average speed × time $x = s \times t$ acceleration = change in velocity ÷ time taken $a = \frac{(v - u)}{t}$	(final velocity)^2 – (initial velocity)^2 = 2 × acceleration × distance $v^2 - u^2 = 2 \times a \times x$
Unit SP2 Motion and Forces force = mass × acceleration $F = m \times a$ weigh = mass × gravitational field strength $W = m \times g$ H momentum = mass × velocity $p = m \times v$ work done = force × distance moved in the direction of the force $E = F \times d$ kinetic energy = $\frac{1}{2}$ × mass × (speed)^2 $KE = \frac{1}{2} \times m \times v^2$	H force = change in momentum ÷ time $F = \frac{(mv - mu)}{t}$
Unit SP3 Conservation of Energy change in gravitational potential energy = mass × gravitational field strength × change in vertical height $GPE = m \times g \times h$ efficiency = $\frac{\text{(useful energy transferred by the device)}}{\text{(total energy supplied to the device)}}$	

Recall and apply	Select and apply
Unit SP4 Waves wave speed = frequency × wavelength $v = f \times \lambda$ wave speed = distance ÷ time $v = \frac{x}{t}$	
Unit SP8 Energy – Forces Doing Work power = work done ÷ time taken $P = \frac{E}{t}$	
Unit SP9 Forces and their Effects moment of a force = force × distance normal to the direction of the force	
Unit SP10 Electricity and Circuits charge = current × time $Q = I \times t$ energy transferred = charge moved × potential difference $E = Q \times V$ potential difference = current × resistance $V = I \times R$ power = energy transferred ÷ time taken $P = \frac{E}{t}$ electrical power = current × potential difference $P = I \times V$ electrical power = current squared × resistance $P = I^2 \times R$	energy transferred = current × potential difference × time $E = I \times V \times t$
Unit SP12 Magnetism and the Motor Effect	 force on a conductor at right angles to a magnetic field carrying a current = magnetic flux density × current × length $F = B \times I \times l$

Recall and apply	Select and apply
Unit SP13 Electromagnetic Induction	**H** potential difference across primary coil / potential difference across secondary coil = number of turns in primary coil / number of turns in secondary coil $\frac{V_p}{V_s} = \frac{N_p}{N_s}$ For transformers with 100% efficiency, potential difference across primary coil × current in primary coil = potential difference across secondary coil × current in secondary coil $V_p \times I_p = V_s \times I_s$
Unit SP14 Particle Model density = mass ÷ volume $\rho = \frac{m}{V}$	change in thermal energy = mass × specific heat capacity × change in temperature $\Delta Q = m \times c \times \Delta\theta$ thermal energy for a change of state = mass × specific latent heat $Q = M \times L$ to calculate pressure or volume for gases of fixed mass at constant temperature $P_1 \times V_1 = P_2 \times V_2$
Unit SP15 Forces and Matter force exerted on a spring = spring constant × extension $F = k \times x$ pressure = force normal to surface ÷ area of surface $P = \frac{F}{A}$	energy transferred in stretching = 0.5 × spring constant × (extension)^2 $E = \frac{1}{2} \times k \times x^2$ **H** pressure due to a column of liquid = height of column × density of liquid × gravitational field strength $P = h \times \rho \times g$

Glossary

absolute zero The temperature at which the pressure of a gas drops to zero. It is −273 °C or 0 K.

absorb To soak up or take in – for waves, it is when the wave disappears as the energy it is carrying is transferred to a material.

absorption spectrum A spectrum of light (or electromagnetic radiation) that includes black lines. These are caused by some wavelengths being absorbed by the materials that the light (or radiation) passes through.

acceleration A measure of how quickly the velocity of something is changing. It can be positive if the object is speeding up or negative if it is slowing down. Acceleration is a vector quantity.

action–reaction forces Pairs of forces on interacting objects. Action–reaction forces are always the same size, in opposite directions, and acting on different objects. They are not the same as balanced forces (which act on a single object).

activity The number of emissions of ionising radiation from a sample in a given time. This is usually given in becquerels (bq).

alpha (α) particle A particle made of two protons and two neutrons, emitted as ionising radiation from some radioactive isotopes.

alternating current (a.c.) Current which changes direction many times each second.

ammeter An instrument for measuring the size of a current. It is put into a circuit in series with other components.

ampere (A) The unit of electric current. One ampere is a flow of 1 coulomb of charge per second.

amplitude The size of vibrations or the maximum distance a particle moves away from its resting position when a wave passes.

amplify To make bigger.

angle of incidence The angle between an incoming light ray and the normal.

angle of reflection The angle between the normal and a ray of light that has been reflected.

artificial satellite A satellite made by humans.

asteroid A small lump of rock orbiting the Sun.

atmospheric pressure The pressure exerted by the weight of the air around us.

atom The smallest neutral part of an element that can take part in chemical reactions.

atomic energy A term used to describe energy when it is stored inside atoms. It is another name for 'nuclear energy'.

atomic number The number of protons in the nucleus of an atom (symbol *Z*). It is also known as the proton number.

auditory nerve The nerve that carries impulses from an ear to the brain.

average speed The speed worked out from the total distance travelled divided by the total time taken for a journey. Speed = distance travelled/time.

---

background radiation Ionising radiation that is around us all the time from a number of sources. Some background radiation is naturally occurring, but some comes from human activities.

balanced forces When the forces in opposite directions on an object are the same size so that there is a zero resultant force.

battery A number of electrical cells in series.

becquerel (Bq) The unit for the activity of a radioactive object. One becquerel is one radioactive decay per second.

beta (χ) particle A particle of radiation emitted from the nucleus of a radioactive atom when it decays. It is an electron.

Big Bang theory The theory that says that the universe began from a tiny point with huge energy, and has been expanding ever since.

bio-fuel A fuel made from plants or animal wastes.

black hole Core of a red supergiant that has collapsed. Black holes are formed if the remaining core has a mass more than three or four times the mass of the Sun.

braking distance The distance travelled by a vehicle while the brakes are working to bring it to a halt.

---

carbon neutral E.g. a fuel which releases the same amount of carbon dioxide when burnt as it took from the atmosphere when it was formed.

cell A device for producing electricity, usually from a chemical reaction.

centripetal force A force that causes objects to follow a circular path. The force acts towards the centre of the circle.

chain reaction The sequence of reactions produced when a nuclear fission reaction triggers one or more further fissions.

change of state The changing of matter from one state to another e.g. from solid to liquid.

charge A conserved property of some particles (e.g. electron, proton) which causes them to exert forces on each other.

chemical change A change that results in the formation of new substances.

chemical energy A term used to describe energy when it is stored in chemical substances. Food, fuel and batteries all store chemical energy.

circuit breaker An electrical component that interrupts the current in a circuit if there is a fault and the current rises to dangerous levels.

climate change Changes that will happen to the weather as a result of global warming, which is caused by the increase in the amount of carbon dioxide in the atmosphere.

cochlea The part of the ear that changes vibrations into electrical impulses.

comet A small lump of dirty ice orbiting the Sun.

component (forces) When a single force is resolved into two forces at right angles to each other, these two forces are referred to as the component forces.

component A part of something e.g. a lamp might be a component of an electrical circuit.

compress To squash something together to make it shorter or smaller.

conduction The way energy is transferred through solids by heating. Vibrations are passed on from particle to particle.

conservation of momentum The total momentum of moving objects before a collision is the same as the total momentum afterwards as long as no external forces are acting.

conserved A quantity that is kept the same throughout e.g. momentum.

contact force A force that occurs when two objects are touching (e.g. friction or upthrust).

contamination An unwanted addition that makes something unsuitable or impure e.g. pure gold may become contaminated with another metal, or a person may be contaminated with a radioactive substance.

convection The movement of particles in a fluid (gas or liquid) depending on their temperature. Hotter, less dense regions rise, and cooler, denser regions sink.

converging lens A lens that brings light rays together (converges).

control rod A rod that can be lowered into the core of a nuclear reactor, to absorb neutrons and slow down the nuclear chain reaction.

core The innermost part of something.

cosmic microwave background (CMB) radiation Microwave radiation received from all over the sky, originating at the Big Bang.

cosmic rays Charged particles with a high energy that come from stars, neutron stars, black holes and supernovae.

coulomb (C) The unit of electric charge. One coulomb is the charge that passes a point in a circuit when there is a current of 1 ampere for 1 second.

count rate The number of alpha or beta particles or gamma rays detected by a Geiger-Müller tube in a certain time.

critical angle The angle of incidence above which total internal reflection occurs inside a material such as glass or water.

crumple zone A vehicle safety device in which part of the vehicle is designed to crumple in a crash, reducing the force of the impact.

---

daughter nucleus A nucleus produced when the nucleus of an unstable atom splits into two during fission or when a radioactive nucleus decays by emitting an alpha or beta particle.

decay When a radioactive isotope emits ionising radiation.

deceleration When an object is slowing down. A negative acceleration.

decommission Dismantle safely.

density The mass of a substance per unit volume. It has units such as g/cm^3.

diffuse reflection Reflection from a rough surface, where the reflected light is scattered in all directions.

diode A component that lets electric current flow through it in one direction only.

direct current (d.c.) A current that flows in one direction only, such as the current produced by a battery.

directly proportional A relationship between two variables where one variable doubles when the other doubles. The graph is a straight line through (0,0). We say that one variable is directly proportional to the other.

discharge To remove an electric charge by conduction.

displacement The distance travelled in a particular direction. Displacement is a vector, distance is not.

dissipated Spread out.

distance How far something has travelled. Distance is a scalar and has no direction.

distance/time graph A graph of the distance travelled against time for a moving object. The gradient of a line on a distance/time graph gives the speed.

diverging lens A lens that spreads out (diverges) light rays.

Doppler effect The change in the pitch of a sound heard when the source of sound is moving relative to the observer.

DNA Deoxyribonucleic acid. A polymer made of deoxyribose sugar molecules and phosphate groups joined to bases.

dose The total amount of something received e.g. medicine, ionising radiation.

dwarf planet A rocky body orbiting the sun that is not quite big enough to be called a planet (e.g. Pluto).

---

ear canal The tube in the head that leads to the eardrum.

eardrum A thin membrane inside the ear that vibrates when sound reaches it.

earth wire A low-resistance path for electric current to flow through for safety if there is a fault in an appliance.

earthed Connected to earth so that any electrostatic charges can flow away.

efficiency The proportion of input energy that is transferred to a useful form. A more efficient machine wastes less energy.

effort The force exerted on a lever, for example.

effort distance The distance the effort on a lever moves through. This distance is typically larger than the distance the load moves through, which makes it easier to move the load.

elastic An elastic material changes shape when there is a force on it, but returns to its original shape when the force is removed.

elastic potential energy	A name used to describe energy when it is stored in stretched or squashed things that can change back to their original shape. Another name for strain energy.
electric field	The space around an object with an electric charge where it can affect other objects.
electrical power	Power transferred by electricity.
electromagnet	A magnet made using a coil of wire with electricity flowing through it.
electromagnetic induction	A process that creates a current in a wire when the wire is moved relative to a magnetic field, or when the magnetic field around it changes.
electromagnetic radiation	A form of energy transfer, including radio waves, microwaves, infrared, visible light, ultraviolet, X-rays and gamma rays.
electromagnetic spectrum	The entire frequency range of electromagnetic waves.
electromagnetic waves	A group of waves that all travel at the same speed in a vacuum, and are all transverse.
electron	A tiny particle with a negative charge and very little mass.
electron shell	Areas around a nucleus that can be occupied by electrons and are usually drawn as circles. Also called an electron energy level or an 'orbit'.
electronic configuration	The arrangement of electrons in shells around the nucleus of an atom.
electrostatic	Relating to non-moving electric fields (and not to those produced by electric currents).
electrostatic repulsion	A force between two electrical charges that have the same sign that pushes them apart.
electrostatic field	The space around an object with a charge of static electricity where it can affect other objects.
element	A simple substance made up of only one type of atom.
elliptical	A shape like a squashed circle.
emission spectrum	A set of wavelengths of light or electromagnetic radiation showing which wavelengths have been given out (emitted) by a substance.
emit	To give out.
energy	Something that is needed to make things happen or change.
equilibrium	When a situation is not changing because all the things affecting it balance out.
extension	The amount by which a spring or other stretchy material has stretched. It is worked out from the stretched length minus the original length.
external radiotherapy	Treatment of cancer by sending radiation into the body from outside.
filter	Something that only lets certain colours through and absorbs the rest.
field lines	Lines which show where a force is stronger or weaker.
Fleming's left-hand rule	A way of remembering the direction of the force when a current flows in a magnetic field.
fluid	A liquid or a gas.
fluorescence	Absorbing radiation of one wavelength and re-emitting the energy at a different wavelength (usually so that it becomes visible).
focal length	The distance from a lens to the focal point.
focal point	The point at which parallel light rays converge after passing through a converging lens, or appear to spread out from after passing through a diverging lens.
force	At the simplest level a force is a push, pull or twist. Forces acting on an object can cause it to accelerate. Force is a vector quantity.
force field	The space around something where a non-contact force affects things. Examples include magnetic fields and gravitational fields.
force meter	A meter, often containing a spring, which measures forces in newtons.
fossil fuel	A fuel formed from the dead remains of organisms over millions of years (e.g. coal, oil or natural gas).
free body force diagram	A diagram of an object showing all the forces acting upon it and the size and direction of those forces.
frequency	The number of vibrations (or the number of waves) per second, measured in hertz
friction	A force between two surfaces that resists motion.
fuel rod	A rod containing the nuclear fuel for a nuclear reactor.
fuse	A small safety device containing a length of wire that is designed to melt if the current in a circuit gets too hot.
gamma camera	A camera that detects gamma rays.
gamma ray (γ)	A high-frequency electromagnetic wave emitted from the nucleus of a radioactive atom. Gamma rays have the highest frequencies in the electromagnetic spectrum.
gas pressure	The force on a surface caused by the collisions of gas particles with the surface. Gas pressure acts at right angles to a surface.
gears	A system of toothed wheels. The teeth interlock so that turning one wheel turns the one in contact with it. If gears of different sizes are used, the speed of rotation or the force transmitted can be changed.
Geiger-Müller (GM) tube	A device that can detect ionising radiation and is used to measure the activity of a radioactive source.
geocentric	Earth-centred.
gradient	A measurement describing the steepness of a line on a graph. It is calculated by taking the vertical distance between two points and dividing by the horizontal distance between the same two points. Also called the slope.
gravitational field	The space around any object with mass where its gravity attracts other masses.
gravitational field strength	A measure of how strong the force of gravity is somewhere. The units are newtons per kilogram (N/kg).
gravitational potential energy	A term used to describe energy when it is stored in objects that can fall.
greenhouse effect	The warming effect on the Earth's surface caused by greenhouse gases absorbing energy emitted from the warm surface of the Earth and re-emitting it back to the surface.
greenhouse gas	A gas, such as carbon dioxide, water vapour or methane, in the Earth's atmosphere, which absorbs energy emitted from the Earth's surface and then emits it back to the surface.
half-life	The average time taken for half of the radioactive nuclei in a sample of radioactive material to have decayed.
heliocentric	Sun-centred.
hertz (Hz)	The unit for frequency, 1 hertz is 1 wave per second.
hydroelectricity	Electricity generated by moving water, usually falling from a reservoir, to turn turbines and generators.
impulse	An electrical signal that travels in the nervous system.
in equilibrium	When things are balanced and not changing they are 'in equilibrium'.
incident ray	A ray of light going towards an interface or object.
induce	To create. For example, a wire in a changing magnetic field has a current in it.
induced magnet	A piece of material that becomes a magnet because it is in the magnetic field of another magnet.
(charging by) induction	When an object is charged by bringing another charged object near to it.
inelastic	A material that changes shape when there is a force on it but does not return to its original shape when the force is removed.
inertial mass	The mass of an object found from the ratio of force divided by acceleration. The value is the same as the mass calculated from the weight of an object and gravitational field strength.
infrared radiation	Electromagnetic radiation that we can feel as heat.
infrasound	Sound waves with a frequency below 20 Hz, which is too low for the human ear to detect.
instantaneous speed	The speed at one particular moment in a journey.
insulation	The method of reducing energy transfer (often using insulating materials).
insulator	A thermal insulator acts as a barrier to the transfer of energy by heating; an electrical insulator does not conduct electricity.
interface	The boundary between two materials.
internal radiotherapy	Treatment of cancer by putting a radioactive source inside the body.
ion	An atom or group of atoms with an electrical charge due to the gain or loss of electrons.
ionising radiation	Radiation that can cause charged particles (ions) to be formed. It can cause tissue damage and DNA mutations.
irradiated	Something has been irradiated if it has been exposed to ionising radiation e.g. to sterilise food or medical equipment with gamma rays.
isotope	Atoms of an element with the same number of protons (atomic number) but different mass numbers due to different numbers of neutrons.
joule (J)	A unit for measuring energy.
kelvin (K)	The unit in the kelvin temperature scale. One kelvin is the same temperature interval as 1 °C.
kelvin temperature scale	A temperature scale that measures temperatures relative to absolute zero.
kinetic energy	A term used to describe energy when it is stored in moving things.
kinetic theory	The model that explains the properties of different states of matter in terms of the movement of particles.
law of conservation of energy	The idea that energy can never be created or destroyed, only transferred from one store to another.
law of reflection	The law that says the angle of incidence and the angle of reflection are equal.
lever	A simple machine that consists of a long bar and a pivot. It can increase the size of a force or increase the distance the force moves.
light gate	A piece of apparatus containing an infrared beam that is transmitted from a source onto a detector. If the beam is cut, the light gate measures how long it is cut for, giving you a reading for time.
light-dependent resistor (LDR)	A resistor whose resistance gets lower when light shines on it.
linear relationship	A relationship between two variables shown by a straight line on a graph.
live wire	The wire that carries the oscillating voltage of an a.c. supply.
load	The force exerted on on one of a lever.

load distance	The distance the load on a lever moves through. This is typically less than the distance the effort moves through, meaning the load can be moved with less effort than if the force had been applied directly.
longitudinal wave	A wave where the vibrations are parallel to the direction in which the wave is travelling, e.g. in a sound wave.
lubrication	To reduce friction by putting a substance (usually a liquid) between two surfaces.
luminous	Giving off light. The Sun and light bulbs are luminous objects.
magnet	An object that has its own magnetic field around it.
magnetic field	The area around a magnet where it can affect magnetic materials or induce a current.
magnetic flux density	A way of describing the strength of a magnetic field. It is measured in tesla (T).
magnetic materials	Materials that are attracted to magnets, e.g. iron.
magnetism	The force caused by magnets on magnetic materials.
magnitude	The size of something, such as the size of a force or the measurement of a distance.
main sequence star	A star during the main part of its life cycle, when it is using hydrogen fuel.
mass	A measure of the amount of material there is in an object. The units are kilograms (kg). Mass is a scalar quantity.
mass number	The total number of protons and neutrons in the nucleus of an atom (symbol *A*). It is also known as the nucleon number.
medium	Material through which electromagnetic waves travel.
microwave	A type of electromagnetic wave, towards the lower frequency end of the spectrum.
moderator	A substance in a nuclear reactor that slows down neutrons, so that they can be absorbed by the nuclear fuel more easily.
moment	The turning effect of a force. It is calculated by multiplying the force by the distance between the force and the pivot, measured at right angles to the force (this is called the normal distance).
momentum	The mass of an object multiplied by its velocity. Momentum is a vector quantity measured in kilogram metres per second (kg m/s).
moon	A natural satellite of a planet.
mutation	A change to a gene caused by a mistake in copying the DNA base pairs during cell division, or by the effects of radiation or certain chemicals.
national grid	The system of wires and transformers that distributes electricity around the country.
natural satellite	Anything that orbits a planet that has not been made by humans.
nebula	(plural: nebulae). A cloud of gas in space. Some objects that look like nebulae are actually clusters of stars or other galaxies.
net force	Another term for resultant force.
neutral wire	A neutral wire is held at or near earth potential while the voltage in the live wire cycles between positive and negative in an a.c. supply.
neutron	A particle found in the nucleus of an atom having zero charge and mass of 1 (relative to a proton).
neutron star	The core of a red supergiant that has collapsed.
newton metre (N m)	The unit for the moment of a force.
non-contact force	Forces that can occur without objects touching. Non-contact forces are magnetism, gravity and static electricity.
non-linear	A relationship between two variables that does not produce a straight line on a graph.
non-renewable	Any energy resource that will run out because it cannot be renewed, e.g. oil.
normal	An imaginary line drawn at right angles to the surface of a mirror or lens where a ray of light hits it.
normal contact force	A force that acts at right angles to a surface as a reaction to a force on that surface.
nuclear energy	A name used to describe energy when it is stored inside atoms. Another term for atomic energy.
nuclear equation	An equation representing a change in an atomic nucleus due to radioactive decay. The atomic numbers and mass numbers must balance.
nuclear fission	The reaction in which the nucleus of a large atom, such as uranium, splits into two smaller nuclei.
nuclear fuel	A radioactive metal such as uranium. Nuclear fuels are used in nuclear power stations to generate electricity.
nuclear fusion	The reaction in which nuclei of light atoms, such as hydrogen, combine to make the nucleus of a heavier atom.
nucleon	A particle found in the nucleus (neutron or proton).
nucleon number	Another term for mass number.
nucleus	The central part of an atom or ion.
object	The thing looked at through a lens or other optical instrument.
ohm (Ω)	The unit for measuring electrical resistance.
orbit	The path taken by a planet around a star, or a satellite around a planet. Sometimes also used to describe electron shells.
oscillation	Movement backwards and forwards.
P waves	Longitudinal seismic waves that travel through the Earth.
parallel	An arrangement of an electrical circuit that allows the current to take different routes.
particle theory	A model for understanding matter that states that matter is made up of particles (atoms).
pascal (Pa)	A unit of pressure. 1 Pa = 1 N/m^2.
penetrate	Go through.
period	The time taken for one complete wave to pass a point. It is measured in seconds.
permanent magnet	A magnet that is always magnetic such as a bar magnet.
PET scanner	A medical scanning technique that detects gamma rays caused by the interaction of a positron from a radioactive source with an electron.
physical change	A change in which no new substances are formed, such as changes of state.
pitch	Whether a sound is low or high.
pivot	The axis around which a lever, for example, can rotate.
planet	A large body in space that orbits a star. The Earth is a planet.
point charge	A charge with a very small volume; a uniform sphere whose charge acts as if it is concentrated at the centre.
plotting compass	A small compass used to find the shape of a magnetic field.
positive ion	An atom that has lost electrons and so has an overall positive charge.
positron	The anti-particle of an electron, having the same mass but opposite charge. Positron emission is a type of beta decay.
potential difference (p.d.)	The difference in the energy carried by electrons before and after they have flowed through a component. Another term for voltage.
power (energy transfers)	The amount of energy (in joules, J) transferred every second. It is measured in watts (W).
power (lenses)	A measure of how much the lens bends light rays passing through it. A more powerful lens bends rays more and has a shorter focal length.
power rating	The energy transferred per second by an appliance.
pressure	The force on a certain area. It is measured in pascals or N/m^2.
primary coil	The coil on a transformer to which the electricity supply is connected.
probability	The likelihood of an event happening. It can be shown as a fraction from 0 to 1, a decimal from 0 to 1 or as a percentage from 0% to 100%.
proton	A particle found in the nucleus of an atom, having a positive charge and the same mass as a neutron.
proton number	The number of protons in an atomic nucleus. Another term for atomic number.
protostar	A cloud of gas drawn together by gravity that has not yet started to produce its own energy.
radiation	A way of transferring energy. Often used to signify the transfer of energy by heating, which is better referred to as infrared radiation.
radio wave	A low-frequency part of the electromagnetic spectrum.
radiotherapy	The use of ionising radiation to treat diseases, such as to kill cancer cells.
random	Any process that cannot be predicted and can happen at any time is said to be random.
rate	How quickly something happens.
ray diagram	A diagram that represents the path of light using arrows.
real image	An image through which light rays pass, so that it can be seen on a screen placed at that point.
reaction time	The time taken to respond to a stimulus, which is affected by the speed of activity in the brain and nervous system.
red giant	A star that has used up all the hydrogen in its core and is now using helium as a fuel. It is bigger than a normal star.
red supergiant	A star that has used up all the hydrogen in its core and has a mass much higher than the Sun.
red-shift	Waves emitted by something moving away from an observer have their wavelength increased and frequency decreased compared to waves from a stationary object.
reflected	When a wave is bounced off a surface instead of passing through it or being absorbed.
reflected ray	A ray of light that has been reflected from a surface.
refracted	When a wave changes direction as it passes from one substance to another.
refracted ray	A ray of light that has just passed through the interface between two materials.
refraction	The change in direction when a wave goes from one medium to another.
relative mass	The mass of something compared to the mass of something else, which is often given the mass of 1.
renewable	An energy resource that will never run out, e.g. solar power.
resistance	A way of saying how difficult it is for electricity to flow through something.
resolving	Representing a single force as two forces at right angles to each other.
response	An action that occurs due to a stimulus.
resultant force	The total force that results from two or more forces acting upon a single object. It is found by adding together the forces, taking into account their directions. Another term for net force.

S waves Transverse seismic waves that travel through the Earth.

Sankey diagram A diagram showing energy transfers, where the width of each arrow is proportional to the amount of energy it represents.

scalar quantity A quantity that has a magnitude (size) but not a direction. Examples include mass, distance, energy and speed.

scale diagram A way of working out the resultant forces or component forces by drawing a diagram where the lengths of arrows represent the sizes of the forces.

secondary coil The coil on a transformer where the changed voltage is obtained.

seismic waves Waves produced by an explosion or earthquake and which travel through the Earth. They include P waves and S waves.

seismometer An instrument that detects seismic waves.

series An arrangement of an electrical circuit which gives the current only one route to flow around the circuit.

shadow zone A part of the Earth's surface that P waves or S waves from an earthquake do not reach because of the way they have been reflected or refracted within the Earth.

skin cancer A cancer or cancerous tumour on the skin.

solar cell A flat plate that uses energy transferred by the light to produce electricity.

solar energy Energy from the Sun.

solenoid A coil of wire with electricity flowing in it. Also called an electromagnet.

sonar A way of finding the distance to an underwater object (such as the sea floor) by timing how long it takes for a pulse of ultrasound to be reflected.

sound wave Vibrations in the particles of a solid, liquid or gas, which are detected by our ears and 'heard' as sounds. Sound waves are longitudinal waves.

specific heat capacity The energy needed to raise the temperature of 1 kg of a substance by 1 °C.

specific latent heat The energy taken in or released when 1 kg of a substance changes state.

specular reflection When light is reflected evenly, so that all reflected light goes off in the same direction. Mirrors produce specular reflection.

speed A measure of the distance an object travels in a given time. Often measured in metres per second (m/s), miles per hour (mph) or kilometres per hour (km/h). It is a scalar quantity.

spring constant A measure of how stiff a spring is. The spring constant is the force needed to stretch a spring by 1 m.

star A huge ball of gas that radiates energy.

states of matter There are three different forms that a substance can be in: solid, liquid or gas. These are the three states of matter.

static electricity Unbalanced electric charges on the surface or within a material.

Steady State theory The theory that the universe is expanding but new matter is continually being created, so the universe will always appear the same.

step-down transformer A transformer that reduces the voltage.

step-up transformer A transformer that increases the voltage.

sterilise To destroy organisms (e.g. bacteria, viruses, fungi) in or on an object. It can be carried out using radioactive sources.

stimulus (plural: stimuli). A change in a factor (inside or outside the body) that is detected by receptors such as sight or sound.

stopping distance The distance in which a car stops, which is the sum of the thinking and braking distances.

strain energy The name used to describe energy when it is stored in stretched or squashed things that can change back to their original shapes. Another term for elastic potential energy.

subatomic particle A particle that is smaller than an atom, such as a proton, neutron or electron.

sublimation When a solid changes directly to a gas without becoming a liquid first.

supernova (plural: supernovae). An explosion produced when the core of a red supergiant collapses.

system A set of things being studied. For example, an electric kettle and its surroundings form a simple system.

telescope An instrument that is used to gather light from distant objects and make them look bigger.

temperature A measure of how hot something is.

temporary magnet A magnet that is not always magnetic, such as an electromagnet or an induced magnet.

tesla (T) The unit for magnetic flux density, also given as newtons per ampere metre (N/A m).

thermal conductivity A measure of how easily energy can pass through a material by heating. A material with a low thermal conductivity is a good insulating material.

thermal conductor A material that allows energy to be transferred through it easily by heating.

thermal energy A term used to describe energy when it is stored in hot objects. The hotter something is, the more thermal energy it has. It is sometimes called 'heat energy'.

thermal insulator A material that does not allow energy to be transferred through it easily by heating.

thermistor A component whose resistance gets lower as it heats up.

thinking distance The distance travelled by a vehicle while the driver reacts.

tidal power Generating electricity using the movement of the tides.

total internal reflection The reflection of a ray of light inside a medium such as glass or water when it reaches an interface. Total internal reflection only happens when the angle of incidence inside the material is greater than the critical angle.

tracer A radioactive substance that is deliberately injected into the body or into moving water. It allows the movement of the substance to be followed by detecting the ionising radiation emitted.

transformer A device that can change the voltage of an electricity supply.

transmission lines The wires (overhead or underground) that take electricity from power stations to towns and cities.

transmit For waves, when the wave passes through something and is not absorbed or reflected.

transverse wave A wave where the vibrations are at right angles to the direction in which the wave is travelling.

tumour A lump formed of cancer cells.

ultrasound Sound waves with a frequency above 20 000 Hz, which is too high for the human ear to detect.

ultrasound scan A way of making an image of part of the body (usually a fetus) using ultrasound waves reflected from parts of the inside of the body.

ultraviolet (UV) Electromagnetic radiation that has a shorter wavelength than visible light but a longer wavelength than X-rays.

unbalanced forces When the forces in opposite directions on an object do not cancel out, so there is a non-zero resultant force.

uniform The same in all places.

Universe All the stars, galaxies and space itself.

unstable An unstable nucleus in an atom is one that will decay and give out ionising radiation.

upthrust The upward force a liquid or gas exerts on a body floating in it.

uranium A radioactive metal that can be used as a nuclear fuel.

vacuum A place where there is no matter at all.

vector A quantity that has both a size and a direction. Examples include force, velocity, displacement, momentum and acceleration.

vector quantity A quantity that has both a size and a direction.

velocity The speed of an object in a particular direction. Usually measured in metres per second (m/s). Velocity is a vector, speed is not.

velocity/time graph A graph of velocity against time for a moving object. The gradient of a line on the graph gives the acceleration and the area under the graph gives the distance travelled.

virtual image An image that light rays do not pass through; they only appear to come from the image.

visible light Electromagnetic waves that can be detected by the human eye.

visible spectrum The seven colours that make up white light.

volt (V) The unit for measuring potential difference (voltage).

voltage The difference in the energy carried by electrons before and after they have flowed through a component. Another term for potential difference.

voltmeter An instrument for measuring the potential difference across a component. Connected in parallel to a circuit.

watt (W) The unit for measuring power. 1 watt = 1 joule of energy transferred every second.

wave A way of transferring energy or information. Many waves travel when particles pass on vibrations.

wavelength The distance between a point on one wave and the same point on the next wave.

weight The force pulling an object downwards. It depends on the mass of the object and the gravitational field strength. The units are newtons (N). Weight is a vector.

white dwarf A very dense star that is not very bright. A red giant turns into a white dwarf.

white light Normal daylight, or the light from light bulbs, is white light.

wind turbine A kind of windmill that generates electricity using energy transferred by the wind.

work The energy transferred when a force moves an object. It is calculated using the size of the force and the distance moved by the force. The unit for work is the joule (J).

work done A measure of the energy transferred when a force acts through a distance.

X-ray Electromagnetic radiation that has a shorter wavelength than UV but a longer wavelength than gamma rays.

Index